Lecture Notes in Mathematics

Edited by A. Dold and B. Eckmann

544

Robert P. Langlands

On the Functional Equations Satisfied by Eisenstein Series

Springer-Verlag
Berlin · Heidelberg · New York 1976

Author
Robert P. Langlands
School of Mathematics
Institute for Advanced Study
Princeton, N.J. 08540/USA

Library of Congress Cataloging in Publication Data

Langlands, Robert P 1936-
On the functional equations satisfied by Eisenstein series.

(Lecture notes in mathematics ; 544)
Bibliography: p.
Includes index.
1. Eisenstein series. 2. Functional equations. 3. Lie groups. 4. Automorphic forms. I. Title. II. Series: Lecture notes in mathematics (Berlin) ; 544.
QA3.L28 no. 544 [QA404] 510'.8s [515'.7] 76-41389

AMS Subject Classifications (1970): 10C15, 10C99, 22E40, 30A58, 32N10, 42A16, 43A65

ISBN 3-540-07872-X Springer-Verlag Berlin · Heidelberg · New York
ISBN 0-387-07872-X Springer-Verlag New York · Heidelberg · Berlin

Printed in Germany
Printing and binding: Beltz Offsetdruck, Hemsbach/Bergstr.

PREFACE

In these days of dizzying scientific progress some apology is called for when offering to the mathematical public a work written twelve years ago. It certainly bears the stamp of a juvenile hand, and I had always hoped to revise it, but my inclination to a real effort grew ever slighter, and the manuscript was becoming an albatross about my neck. There were two possibilities: to forget about it completely, or to publish it as it stood; and I preferred the second.

There were, when it was first written, other reasons for delaying publication. The study of Eisenstein series is a preliminary to the development of a trace formula, and the trace formula has been a long time evolving. Not only does it present serious analytic difficulties, but also the uses to which it should be put have not been clear. A sustained attack on the analytic difficulties is now being carried out, by Arthur and others, and, thanks to a large extent to developments within the theory of Eisenstein series itself, we now have a clearer picture of the theorems that will flow from the trace formula. However a great deal remains to be done, and a complete treatment of Eisenstein series, even imperfect, may be useful to those wishing to try their hand at developing or using the trace formula.

Much of the material in §2-§6 is included in Harish-Chandra's notes (Lecture Notes 62). He, following an idea of Selberg with which I was not familiar, uses the Maass-Selberg relations. Since I was not aware of them when I wrote it, they do not figure in the present text; they would have simplified the exposition at places.

In §2-§6 the Eisenstein series associated to cusp forms are treated. However the central concern is with the spectral decomposition, and for this one needs all Eisenstein series. The strategy of these notes is, the preliminary discussion of §2-§6 completed, to carry out the spectral decomposition and the study of the general Eisenstein series simultaneously, by an inductive procedure; so §7 is the heart of the text.

It has proven almost impenetrable. In an attempt to alleviate the situation, I have added some appendices. The first is an old and elementary manuscript, dating from 1962. Its aim when written was to expose a technique, discovered by Godement and Selberg as well, for handling some Eisenstein series in several variables. The method, involving a form of Hartog's lemma, has not yet proved to be of much importance; but it should not be forgotten. In addition, and this is the reason for including it, it contains in nascent form the method of treating Eisenstein series associated to forms which are not cuspidal employed in §7.

The second appendix may be viewed as an introduction to §7. The principal theorems proved there are stated as clearly as I could manage. The language of adèles is employed, because it is simpler and because it is the adèlic form of the theorems which is most frequently applied. I caution the reader that he will not appreciate the relation between §7 and this appendix until he has an intimate understanding of §7. The appendix should be read first however.

It is also difficult to come to terms with §7 without a feeling for examples. Some were given in my lecture on Eisenstein series in Algebraic Groups and Discontinuous Subgroups. Others exhibiting the more complicated phenomena that can occur are given in the third appendix, whose first few pages should be glanced at before §7 is tackled.

The last appendix has nothing to do with §7. It is included at the suggestion of Serge Lang, and is an exposition of the Selberg method in the context in which it was originally discovered.

In the Introduction I thank those who encouraged me during my study of Eisenstein series. Here I would like to thank those, Godement and Harish-Chandra, who encouraged me after the notes were written. Harish-Chandra's encouragement was generous in the extreme and came at what was otherwise a difficult time. Its importance to me cannot be exaggerated.

It has been my great good fortune to have had these notes typed by Margaret (Peggy) Murray, whose skills as a mathematical typist are known to all visitors to the IAS. I thank her for another superb job.

TABLE OF CONTENTS

1. Introduction.

One problem in the theory of automorphic forms that has come to the fore recently is that of explicitly describing the decomposition, into irreducible representations, of the regular representation of certain topological groups on Hilbert spaces of the form $\mathcal{L}^2(\Gamma\backslash G)$ when Γ is a discrete subgroup of G. Usually Γ is such that the volume of $\Gamma\backslash G$ is finite. Except for some abelian groups, this problem is far from solved. However, Selberg has discovered that the gross features of the decomposition are determined by simple properties of the group Γ and this discovery has led to the development, mostly by Selberg himself, of the theory of Eisenstein series. Of course he has preferred to state the problems in terms of eigenfunction expansions for partial differential equations or integral operators. At present the theory is developed only for connected reductive Lie groups which, without real loss of generality, may be assumed to have compact centres. Even for these groups some difficulties remain. However, some of the problems mentioned in [19] are resolved in this paper, which is an exposition of that part of the theory which asserts that all Eisenstein series are meromorphic functions which satisfy functional equations and that the decomposition of $\mathcal{L}^2(\Gamma\backslash G)$ is determined by the representations occurring discretely in $\mathcal{L}^2(\Gamma\backslash G)$ and certain related Hilbert spaces. For precise statements the reader may refer to Section 7.

At present it is expected that the main assertions of this paper are true if the volume of $\Gamma\backslash G$ is finite. It is of course assumed that G is a connected reductive Lie group. Unfortunately not enough is known about the geometry of such discrete groups to allow one to work with this assumption alone. However, the property which is described in Section 2 and which I thereafter assume Γ possesses is possessed by all discrete groups known to me which have a fundamental domain with finite volume. Indeed it is abstracted from the results of Borel [2] on arithmetically defined groups. Section 2 is devoted to a discussion of the consequences of this property. In Section 3 the notion of a cusp form is introduced and some preliminary estimates are derived. In Section 4 we begin the discussion of Eisenstein series, while Section 5 contains some important technical results. In Section 6 the functional equations for Eisenstein series associated to cusp forms are proved. For series in one variable the argument is essentially the same as one sketched to me by Professor Selberg nearly two years ago, but for the series in several variables new arguments of a different nature are necessary. In Section 7 the functional equations for the remaining Eisenstein series are derived in the course of decomposing $\mathcal{L}^2(\Gamma \backslash G)$ into irreducible representations.

I have been helped and encouraged by many people while investigating the Eisenstein series but for now I would like to thank, as I hope I may without presumption, only Professors Bochner and Gunning for their kind and generous encouragement, three years ago, of the first results of this investigation.

2. The assumptions.

Let G be a Lie group with Lie algebra $\mathfrak{g}$. It will be supposed that G has only a finite number of connected components and that $\mathfrak{g}$ is the direct sum of an abelian subalgebra and a semi-simple Lie algebra $\mathfrak{g}^s$. It will also be supposed that the centre of G^s, the connected subgroup of G with Lie algebra $\mathfrak{g}^s$, is finite. Suppose $\mathfrak{a}$ is a maximal abelian subalgebra of $\mathfrak{g}^s$ whose image in ad $\mathfrak{g}$ is diagonalizable. Choose an order on the space of real linear functions on $\mathfrak{a}$ and let Q be the set of positive linear functions α on $\mathfrak{a}$ such that there is a non-zero element X in $\mathfrak{g}$ so that $[H, X] = \alpha(H)X$ for all H in $\mathfrak{a}$. Q is called the set of positive roots of $\mathfrak{a}$. Suppose $\mathfrak{a}'$ is another such subalgebra and Q' is the set of positive roots of $\mathfrak{a}'$ with respect to some order. It is known that there is some g in G^s so that Ad $g(\mathfrak{a}) = \mathfrak{a}'$ and so that if $\alpha' \in Q'$ then the linear function α defined by $\alpha(H) = \alpha'(\text{Ad } g(H))$ belongs to Q. Moreover any two elements of G^s with this property belong to the same right coset of the centralizer of $\mathfrak{a}$ in G^s. G itself possesses the first of these two properties and it will be assumed that it also possesses the second. Then the centralizer of $\mathfrak{a}$ meets each component of G.

For the purposes of this paper it is best to define a parabolic subgroup P of G to be the normalizer in G of a subalgebra $\mathfrak{p}$ of $\mathfrak{g}$ such that the complexification $\mathfrak{p}_c = \mathfrak{p} \otimes_{\mathbb{R}} \mathbb{C}$ of $\mathfrak{p}$ contains a Cartan subalgebra $\mathfrak{j}_c$ of $\mathfrak{g}_c$ together with the root vectors belonging to the roots of $\mathfrak{j}_c$ which are positive with respect to some order on $\mathfrak{j}_c$. It is readily verified that the Lie algebra of P is $\mathfrak{p}$ so that P is its own normalizer. Let $\mathfrak{n}$ be a maximal normal subalgebra of $\mathfrak{p}^s = \mathfrak{p} \cap \mathfrak{g}^s$ which consists entirely of elements whose adjoints are nilpotent and let $\mathfrak{m}'$ be a maximal subalgebra of $\mathfrak{p}$ whose image in ad $\mathfrak{g}$ is fully reducible. It follows from [16] that $\mathfrak{p} = \mathfrak{m}' + \mathfrak{n}$ and that $\mathfrak{m}'$ contains a Cartan subalgebra of $\mathfrak{g}$. Let $\mathfrak{a}$ be a subalgebra of the centre of $\mathfrak{m}' \cap \mathfrak{g}^s$ whose image in ad $\mathfrak{g}$ is diagonalizable. If $\mathfrak{m}$ is the orthogonal complement of $\mathfrak{a}$, with respect to the Killing form on $\mathfrak{g}$, in $\mathfrak{m}'$ then $\mathfrak{a} \cap \mathfrak{m} = \{0\}$. There is a set Q of real linear functions on $\mathfrak{a}$ such that $\mathfrak{n} = \Sigma_{\alpha \in Q} \mathfrak{n}_\alpha$ where

$$\mathfrak{n}_\alpha = \{X \in \mathfrak{n} \,|\, [H, X] = \alpha(H)X \text{ for all } H \text{ in } \mathfrak{a}\}$$

$\mathfrak{a}$ or A, the connected subgroup of P with the Lie algebra $\mathfrak{a}$, will be called a split component of P if the trace of the restriction of ad Y to $\mathfrak{n}_\alpha$ is zero for any Y in $\mathfrak{m}$ and any α in Q. There is a Cartan subalgebra $\mathfrak{j}$ of $\mathfrak{g}$ and an order on the real linear functions on $\mathfrak{j}_c$ so that $\mathfrak{a} \subseteq \mathfrak{j} \subseteq \mathfrak{m}'$ and so that Q consists of the restrictions of the positive roots to $\mathfrak{a}$ except perhaps for zero. Let Q'_α be the set of positive roots whose restriction to $\mathfrak{a}$ equals α; then

$$1/\dim \mathfrak{n}_\alpha \, \Sigma_{\alpha' \in Q'_\alpha} \alpha'$$

is zero on $\mathfrak{j} \cap \mathfrak{m}$ and equals α on $\mathfrak{a}$. Thus if $\Sigma_{\alpha \in Q} c_\alpha \alpha = 0$ and $c_\alpha \geq 0$ for all α then

$$\Sigma_{\alpha \in Q} \Sigma_{\alpha' \in Q'_\alpha} (\dim \mathfrak{n}_\alpha)^{-1} c_\alpha \alpha' = 0$$

which implies that $c_\alpha = 0$ for all α. In particular zero does not belong to Q so that $\mathfrak{m}'$ is the centralizer and normalizer of $\mathfrak{a}$ in $\mathfrak{g}$.

Since $\mathfrak{m}'$ contains a Cartan subalgebra it is its own normalizer. Let us show that if M' is the normalizer of $\mathfrak{m}'$ in P then the connected component of M' is of finite index in M'. M' is the inverse image in G of the intersection of an algebraic group with Ad G. Since Ad G contains the connected component, in the topological sense, of the group of automorphisms of $\mathfrak{g}$ which leave each element of the centre fixed the assertion follows from Theorem 4 of [23]. Since the Lie algebra of M' is $\mathfrak{m}'$ it follows from Lemma 3.1 of [16] that M' is the inverse image in G of a maximal fully reducible subgroup of the image of P in Ad G. Let N be the connected subgroup of G with the Lie algebra $\mathfrak{n}$. Since the image of N in Ad G is simply connected it follows readily from [16] that M' and N are closed, that $P = M' \cdot N$, and that $M' \cap N = \{1\}$.

We must also verify that M' is the centralizer of $\mathfrak{a}$ in G. M' certainly contains the centralizer of $\mathfrak{a}$ in G. Let $\mathfrak{b}$ be a maximal abelian subalgebra of $\mathfrak{g}^s$ which contains $\mathfrak{a}$ such that the image of $\mathfrak{b}$ in ad $\mathfrak{g}$ is diagonalizable. Certainly $\mathfrak{m}'$ contains $\mathfrak{b}$. Let $\mathfrak{b} = \mathfrak{b}_1 + \mathfrak{b}_2$ where $\mathfrak{b}_1$ is the intersection of $\mathfrak{b}$ with the centre of $\mathfrak{m}'$ and $\mathfrak{b}_2$ is the intersection of $\mathfrak{b}$ with the semi-simple part of $\mathfrak{m}'$. $\mathfrak{b}_2$ is a maximal abelian subalgebra of the semi-simple part of $\mathfrak{m}'$ whose image in ad $\mathfrak{m}'$ is diagonalizable. It may be supposed (cf. [11], p. 749) that the positive roots of $\mathfrak{b}$ are the roots whose root vectors either lie in $\mathfrak{n}_c$, the complexification of $\mathfrak{n}$, or lie in $\mathfrak{m}'_c$ and belong to positive roots of $\mathfrak{b}_2$. If m lies in M' then $\operatorname{Ad} m(\mathfrak{b}_1) = \mathfrak{b}_1$. Moreover replacing if necessary m by mm_0 where m_0 lies in the connected component of M' and hence in the centralizer of $\mathfrak{a}$ we may suppose that $\operatorname{Ad} m(\mathfrak{b}_2) = \mathfrak{b}_2$ and that Ad m takes positive roots of $\mathfrak{b}_2$ to positive roots of $\mathfrak{b}_2$. Thus $\operatorname{Ad} m(\mathfrak{b}) = \mathfrak{b}$ and Ad m leaves invariant the set of positive roots of $\mathfrak{b}$; consequently, by assumption, m lies in the centralizer of $\mathfrak{b}$ and hence of $\mathfrak{a}$. It should also be remarked that the centralizer of A meets each component of P and G and P meets each component of G.

If M is the group of all m in M' such that the restriction of Ad m to $\mathfrak{n}_\alpha$ has determinant ± 1 for all α then M is closed; since Q contains a basis for the space of linear functions on $\mathfrak{a}$ the intersection $A \cap M$ is $\{1\}$. Let $\alpha_1, \ldots, \alpha_p$ be such a basis. To see that $AM = M'$ introduce the group M_1 of all m in M' such that the restriction of Ad m to $\mathfrak{n}_{\alpha_i}$ has determinant ± 1 for $1 \leq i \leq p$. Certainly $AM_1 = M'$ so it has merely to be verified that M, which is contained in M_1, is equal to M_1. Since the Lie algebra of both M and M_1 is $\mathfrak{m}$ the group M contains the connected component of M_1. Since $A \cap M_1 = \{1\}$ the index $[M_1 : M]$ equals $[AM_1 : AM]$ which is finite. It follows readily that $M = M_1$. It is clear that M and $S = MN$ are uniquely determined by P and A. The pair (P, S) will be called a split parabolic subgroup with A as split component. Its rank is the dimension of A. Observe that A is not uniquely determined by the pair (P, S).

The next few lemmas serve to establish some simple properties of split parabolic subgroups which will be used repeatedly throughout the paper. If (P, S) and (P_1, S_1) are any two split parabolic subgroups then (P, S) is said to contain (P_1, S_1) if P contains P_1 and S contains S_1.

LEMMA 2.1. Suppose (P, S) contains (P_1, S_1). Let A be a split component of (P, S) and A_1 a split component of (P_1, S_1). There is an element p in the connected component of P so that pAp^{-1} is contained in A_1.

Since S is a normal subgroup of P, pAp^{-1} will be a split component of (P, S). According to Theorem 4.1 of [16] there is a p in the connected component of P so that $\mathfrak{n}_1 + \mathfrak{m}_1 \subseteq \mathrm{Ad}\, p(\mathfrak{n} + \mathfrak{m})$. Thus it suffices to show that if $\mathfrak{n}_1 + \mathfrak{m}_1$ is contained in $\mathfrak{n} + \mathfrak{m}$ then $\mathfrak{n}$ is contained in $\mathfrak{n}_1$. If $\mathfrak{n}_1 + \mathfrak{m}_1 \subseteq \mathfrak{n} + \mathfrak{m}$ then $\mathfrak{n}$ and $\mathfrak{n}_1$ commute so that $\mathfrak{n}$ is contained in $\mathfrak{n}_1 + \mathfrak{m}_1$; moreover $\mathfrak{m}$ contains $\mathfrak{m}_1$ because $\mathfrak{m} \cap \mathfrak{s}_1 = (\mathfrak{n} + \mathfrak{m}) \cap \mathfrak{s} \cap \mathfrak{s}_1 \supseteq (\mathfrak{n}_1 + \mathfrak{m}_1) \cap \mathfrak{s}_1 = \mathfrak{m}_1$. Consequently $\mathfrak{n}$ is orthogonal to $\mathfrak{m}_1$ with respect to the Killing form and hence is contained in $\mathfrak{n}_1$.

LEMMA 2.2. Suppose P is a parabolic subgroup and $\mathfrak{n}$ is a split component of P. Let $\{\alpha_{1,}, \ldots, \alpha_{p,}\}$ be a minimal subset of Q such that any α in Q can be written as a linear combination $\Sigma_{i=1}^{p} m_i \alpha_{i,}$ with non-negative integers m_i. Then the set $\{\alpha_{1,}, \ldots, \alpha_{p,}\}$ is linearly independent.

This lemma will be proved in the same manner as Lemma 1 of [13]. Let $\langle \lambda, \mu \rangle$ be the bilinear form on the space of linear functions on $\mathfrak{n}$ dual to the restriction of the Killing form to $\mathfrak{n}$. It is enough to show that if i and j are two distinct indices then $\alpha_{i,} - \alpha_{j,}$ neither equals zero nor belongs to Q and that if α and β belong to Q and neither $\alpha - \beta$ nor $\beta - \alpha$ belongs to Q or is zero then

$\langle \alpha, \beta \rangle \leq 0$. If this is so and $\Sigma_{i=1}^{p} a_i \alpha_{i,} = 0$ let $F = \{i \mid a_i \geq 0\}$ and $F' = \{i \mid a_i < 0\}$. Set

$$\lambda = \Sigma_{i\epsilon F} a_i \alpha_{i,} = -\Sigma_{i\epsilon F'} a_i \alpha_{i,}$$

then

$$0 \leq \langle \lambda, \lambda \rangle = -\Sigma_{i\epsilon F} \Sigma_{j\epsilon F'} a_i a_j \langle \alpha_{i,}, \alpha_{j,} \rangle \leq 0$$

which implies that $\lambda = 0$. As a consequence of a previous remark $a_i = 0$, $1 \leq i \leq p$. Certainly $\alpha_{i,} - \alpha_{j,}$ is not zero if $i \neq j$; suppose that $\alpha_{i,} - \alpha_{j,} = \alpha$ belongs to Q. Then

$$\alpha_{i,} - \alpha_{j,} = \Sigma_{k=1}^{p} m_k \alpha_{k,}$$

or

$$(m_i - 1)\alpha_{i,} + (m_j + 1)\alpha_{j,} + \Sigma_{k \neq i, j} m_k \alpha_{k,} = 0$$

so that $m_i - 1 < 0$. Hence $m_i = 0$ and

$$\alpha_{i,} = (m_j + 1)\alpha_{j,} + \Sigma_{k \neq i, j} m_k \alpha_{k,}$$

which is a contradiction. Suppose α and β belong to Q and neither $\alpha - \beta$ nor $\beta - \alpha$ belongs to Q or is zero. Choose the Cartan subalgebra $\mathfrak{j}$ as above and let (λ', μ') be the bilinear form on the space of linear functions on the complexification of $\mathfrak{j} \cap \mathfrak{g}^s$ dual to the Killing form. If μ is the restriction of μ' to $\mathfrak{a}$ then

$$\langle \alpha, \mu \rangle = 1/\dim \mathfrak{n}_\alpha \Sigma_{\alpha' \epsilon Q'_\alpha} (\alpha', \mu')$$

In particular if β' belongs to Q'_β then

$$\langle \alpha, \beta \rangle = 1/\dim \mathfrak{n}_\alpha \Sigma_{\alpha' \epsilon Q'_\alpha} (\alpha', \beta')$$

Because of the assumption on α and β, $\alpha' - \beta'$ is neither a root nor zero; thus (cf.

[15], Ch. IV) each term of the sum and hence the sum itself is non-positive. It is clear that the set $\{\alpha_{1}, \ldots, \alpha_{p}\}$ is unique and is a basis for the set of linear functions on $\mathfrak{a}$; it will be called the set of simple roots of $\mathfrak{a}$. It is also clear that if P_1 contains P and A_1 is a split component of P_1 contained in A then the set of simple roots of $\mathfrak{a}_1$ is contained in the set of linear functions on $\mathfrak{a}_1$ obtained by restricting the simple roots of $\mathfrak{a}$ to $\mathfrak{a}_1$.

LEMMA 2.3. *Suppose*

$$P = P_1 \subsetneqq P_2 \subsetneqq \cdots \subsetneqq P_k$$

is a sequence of parabolic subgroups with split components

$$A_1 \supset A_2 \supset \ldots \supset A_k$$

and

$$\dim A_{i+1} - \dim A_i = 1, \quad 1 \leq i < k$$

If $\{\alpha_{1}, \ldots, \alpha_{p}\}$ *is the set of simple roots of* $\mathfrak{a}$ *and* $\{\alpha_{j}, \ldots, \alpha_{p}\}$ *restricted to* $\mathfrak{a}_j$ *is the set of simple roots for* $\mathfrak{a}_j$, $1 \leq j \leq k$, *then*

$$\mathfrak{a}_j = \{H \in \mathfrak{a} \mid \alpha_{i}(H) = 0,\ i < j\}$$

and if

$$Q_j' = \{\alpha \in Q \mid \alpha(H) \neq 0 \text{ for some } H \text{ in } \mathfrak{a}_j\}$$

then

$$\mathfrak{n}_j = \Sigma_{\alpha \in Q_j'} \mathfrak{n}_\alpha, \quad 1 \leq j \leq k$$

Conversely if F *is a subset of* $\{1, \ldots, p\}$; *if*

$${}^*\mathfrak{a} = \{H \in \mathfrak{a} \mid \alpha_{i}(H) = 0 \text{ for all } i \in F\}$$

if

$$^{*}Q' = \{\alpha \in Q \mid \alpha(H) \neq 0 \text{ for some } H \in {}^{*}\mathfrak{a}\}$$

if

$$^{*}\mathfrak{n} = \Sigma_{\alpha \in {}^{*}Q'} \mathfrak{n}_{\alpha}$$

and if $^{*}\mathfrak{m}$ is the orthogonal complement of $^{*}\mathfrak{a}$ in the centralizer of $^{*}\mathfrak{a}$ in $\mathfrak{g}$ then $^{*}\mathfrak{p} = {}^{*}\mathfrak{a} + {}^{*}\mathfrak{m} + {}^{*}\mathfrak{n}$ is the Lie algebra of a parabolic subgroup $^{*}P$ of G which contains P and has $^{*}\mathfrak{a}$ as a split component.

In the discussions above various objects such as A, Q, $\mathfrak{n}$ have been associated to a parabolic subgroup P; the corresponding objects associated to another parabolic group, say P_1, will be denoted by the same symbols, for example A_1, Q_1, $\mathfrak{n}_1$, with the appropriate indices attached. It is enough to prove the direct part of the lemma for k = 2. Since P_2 properly contains P_1 and since, as is readily seen, P_j is the normalizer of $\mathfrak{n}_j$, j = 1, 2 the algebra $\mathfrak{n}_2$ must be properly contained in $\mathfrak{n}_1$. Consequently there is an $\alpha \in Q$ whose restriction to $\mathfrak{a}_2$ is zero and $\alpha = \Sigma_{i=1}^{p} m_i \alpha_i$, with non-negative integers m_i. Let $\bar{\alpha}_{i,}$ be the restriction of $\alpha_{i,}$ to $\mathfrak{a}_2$ and let $\bar{\alpha}_{1,} = \Sigma_{j=2}^{n} n_j \bar{\alpha}_{j,}$; then

$$0 = \Sigma_{j=2}^{p} (m_j + m_1 n_j) \bar{\alpha}_{j,}$$

so $m_j = 0$, $j \geq 2$ and $\alpha = m_1 \alpha_{1,}$. Since $\dim \mathfrak{a}_2 - \dim \mathfrak{a}_1 = 1$ the direct part of the lemma is proved. Proceeding to the converse we see that if $^{*}P$ is taken to be the normalizer of $^{*}\mathfrak{p}$ in G then $^{*}P$ is parabolic by definition. $^{*}P$ contains the connected component of P and the centralizer of A in G so it contains all of P. Moreover the image of $^{*}\mathfrak{a} + {}^{*}\mathfrak{m}$ in ad $\mathfrak{g}$ is fully reducible and $^{*}\mathfrak{n}$ is a normal subalgebra of $^{*}\mathfrak{p}$ so to prove that $^{*}\mathfrak{a}$ is a split component of $^{*}P$ it has to be shown that if α belongs to $^{*}Q'$ and

$$Q_{\alpha} = \{\beta \in Q \mid \alpha(H) = \beta(H) \text{ for all } H \text{ in } {}^{*}\mathfrak{a}\}$$

then the trace of the restriction of ad X to $\Sigma_{\beta \in Q_\alpha} \mathfrak{n}_\beta$ is zero for all X in ${}^*\mathfrak{m}$. It is enough to show this when X belongs to the centre of ${}^*\mathfrak{m}$. But then X commutes with $\mathfrak{a}$ and so lies in $\mathfrak{a} + \mathfrak{m}$; say $X = Y + Z$. If i belongs to F the trace of the restriction of ad X to $\mathfrak{n}_{\alpha_{i,}}$ is $\alpha_{i,}(Y) \dim \mathfrak{n}_{\alpha_{i,}}$; on the other hand it is zero because $\mathfrak{n}_{\alpha_{i,}}$ belongs to ${}^*\mathfrak{m}$. Thus $\alpha_{i,}(Y) = 0$ for all $i \in F$ so that Y belongs to ${}^*\mathfrak{a}$ and hence is zero. Since the assertion is certainly true for Z it is true for X.

There are some simple conventions which will be useful later. If $\mathfrak{j}_c$ and $\mathfrak{j}'_c$ are two Cartan subalgebras of $\mathfrak{g}_c$ and an order is given on the set of real linear functions on $\mathfrak{j}_c$ and $\mathfrak{j}'_c$ then there is exactly one map from $\mathfrak{j}_c$ to $\mathfrak{j}'_c$ which takes positive roots to positive roots and is induced by an element of the adjoint group of $\mathfrak{g}_c$. Thus one can introduce an abstract Lie algebra which is provided with a set of positive roots and a uniquely defined isomorphism of this Lie algebra with each Cartan subalgebra such that positive roots correspond to positive roots. Call this <u>the</u> Cartan subalgebra of $\mathfrak{g}_c$. Suppose (P, S) is a split parabolic subgroup with A and A' as split components. Let $\mathfrak{j}$ be a Cartan subalgebra containing $\mathfrak{a}$ and let $\mathfrak{j}'$ be a Cartan subalgebra containing $\mathfrak{a}'$. Choose orders on $\mathfrak{j}_c$ and $\mathfrak{j}'_c$ so that the root vectors belonging to positive roots lie in $\mathfrak{p}_c$. There is a p_1 in P so that $\mathrm{Ad}\, p_1(\mathfrak{a}) = \mathfrak{a}'$; since the centralizer of A meets each component of P there is a p in the connected component of P such that $\mathrm{Ad}\, p(H) = \mathrm{Ad}\, p_1(H)$ for all H in $\mathfrak{a}$. Let $\mathrm{Ad}\, p(\mathfrak{j}) = \mathfrak{j}''$. There is an element m in the adjoint group of $\mathfrak{m}_c$ so that $\mathrm{Ad}\, p\, \mathrm{Ad}\, m(\mathfrak{a}) = \mathfrak{a}'$, $\mathrm{Ad}\, p\, \mathrm{Ad}\, m(\mathfrak{j}) = \mathfrak{j}'$, and $\mathrm{Ad}\, p\, \mathrm{Ad}\, m'$ takes positive roots of $\mathfrak{j}'$ to positive roots of $\mathfrak{j}$. The maps $\mathfrak{a} \longrightarrow \mathfrak{j} \longrightarrow \mathfrak{j}_c$ and $\mathfrak{a}' \longrightarrow \mathfrak{j}' \longrightarrow \mathfrak{j}'_c$ determine maps of $\mathfrak{a}$ and $\mathfrak{a}'$ into <u>the</u> Cartan subalgebra of $\mathfrak{g}_c$ and if H belongs to $\mathfrak{a}$ then H and $\mathrm{Ad}\, p_1(H)$ have the same image. The image of $\mathfrak{a}$ will be called the split component of (P, S). Usually the context will indicate whether it is a split component or the split component which is being referred to. If F is a subset of the set of simple roots of the split component it determines a subset of the set of simple roots of any split component which, according to the previous lemma, determines another split parabolic subgroup. The latter depends only on F and will be called simply the split parabolic subgroup determined by F; such a subgroup will be said to belong to (P, S).

If (P, S) is a split parabolic subgroup with the split component $\mathfrak{a}$ let $\alpha_{,1}, \ldots, \alpha_{,p}$ be the linear functions on $\mathfrak{a}$ such that $\langle \alpha_{,i}, \alpha_{j,} \rangle = \delta_{ij}$, $1 \leq i, j \leq p$. Of course $\alpha_{1,}, \ldots, \alpha_{p,}$ are the simple roots of $\mathfrak{a}$. If $-\infty \leq c_1 < c_2 \leq \infty$ let

$$\mathfrak{a}^+(c_1, c_2) = \{H \in \mathfrak{a} \mid c_1 < \alpha_{i,}(H) < c_2,\ 1 \leq i \leq p\}$$

and let

$${}^+\mathfrak{a}(c_1, c_2) = \{H \in \mathfrak{a} \mid c_1 < \alpha_{,i}(H) < c_2,\ 1 \leq i \leq p\}$$

It will be convenient to set $\mathfrak{a}^+(0, \infty) = \mathfrak{a}^+$ and ${}^+\mathfrak{a}(0, \infty) = {}^+\mathfrak{a}$. If A is the simply-connected abstract Lie group with the Lie algebra $\mathfrak{a}$ then A will also be called the split component of (P, S). The map $H \longrightarrow \exp H$ is bijective; if λ is a linear function on $\mathfrak{a}$ set $\xi_\lambda(\exp H) = \exp(\lambda(H))$. If $0 \leq c_1 \leq c_2 \leq \infty$ we let

$$A^+(c_1, c_2) = \{a \in A \mid c_1 < \xi_{\alpha_{i,}}(a) < c_2,\ 1 \leq i \leq p\}$$

and

$${}^+A(c_1, c_2) = \{a \in A \mid c_1 < \xi_{\alpha_{,i}}(a) < c_2,\ 1 \leq i \leq p\}$$

We shall make frequent use of the two following geometrical lemmas.

LEMMA 2.4. *For each* $s < \infty$ *there is a* $t < \infty$ *so that* $\mathfrak{a}^+(s, \infty)$ *is contained in* ${}^+\mathfrak{a}(t, \infty)$. *In particular* $\mathfrak{a}^+$ *is contained in* ${}^+\mathfrak{a}$.

For each s there is an element H in $\mathfrak{a}$ so that $\mathfrak{a}^+(s, \infty)$ is contained in $H + \mathfrak{a}^+$; thus it is enough to show that $\mathfrak{a}^+$ is contained in ${}^+\mathfrak{a}$. Suppose we could show that $\langle \alpha^i, \alpha^j \rangle \geq 0$, $1 \leq i, j \leq p$. Then $\alpha_{,i} = \Sigma_{j=1}^{p} a^i_j \alpha_{j,}$ with $a^i_j \geq 0$ and it follows immediately that $\mathfrak{n}^+$ is contained in ${}^+\mathfrak{n}$. Since $\langle \alpha_{i,}, \alpha_{j,} \rangle \leq 0$ if $i \neq j$ this lemma is a consequence of the next.

LEMMA 2.5. Suppose V is a Euclidean space of dimension n and $\lambda_{1,}, \ldots, \lambda_{n,}$ is a basis for V such that $(\lambda_{i,}, \lambda_{j,}) \leq 0$ if $i \neq j$. If $\lambda_{,i}$, $1 \leq i \leq n$, are such that $(\lambda_{,i}, \lambda_{j,}) = \delta_{ij}$ then either there are two non-empty disjoint subsets F_1 and F_2 of $\{1, \ldots, n\}$ such that $F_1 \cup F_2 = \{1, \ldots, n\}$ and $(\lambda_{i,}, \lambda_{j,}) = 0$ if $i \in F_1$, $j \in F_2$ or $(\lambda_{,i}, \lambda_{,j}) > 0$ for all i and j.

The lemma is easily proved if $n \leq 2$ so suppose that $n > 2$ and that the lemma is true for $n-1$. Suppose that, for some i and j, $(\lambda_{,i}, \lambda_{,j}) \leq 0$. Choose k different from i and j and project $\{\lambda_{\ell,} \mid \ell \neq k\}$ on the orthogonal complement of $\lambda_{k,}$ to obtain $\{\mu_{\ell,} \mid \ell \neq k\}$. Certainly for $\ell \neq k$ the vector $\lambda_{,\ell}$ is orthogonal to $\lambda_{,k}$ and $(\lambda_{,\ell}, \mu_{m,}) = \mu_{\ell m}$. Moreover

$$(\mu_{\ell,}, \mu_{m,}) = (\lambda_{\ell,}, \lambda_{m,}) - (\lambda_{\ell,}, \lambda_{k,})(\lambda_{m,}, \lambda_{k,})/(\lambda_{k,}, \lambda_{k,}) \leq (\lambda_{\ell,}, \lambda_{m,})$$

with equality only if $\lambda_{\ell,}$ or $\lambda_{m,}$ is orthogonal to $\lambda_{k,}$. By the induction assumption there are two disjoint subsets F'_1 and F'_2 of $\{\ell \mid 1 \leq \ell \leq n, \ell \neq k\}$ such that $(\mu_{\ell,}, \mu_{m,}) = 0$ if $\ell \in F'_1$ and $m \in F'_2$. For such a pair $(\mu_{\ell,}, \mu_{m}) = (\lambda_{\ell,}, \lambda_{m,})$ so either $(\lambda_{\ell,}, \lambda_{k,}) = 0$ for all $\ell \in F'_1$ or $(\lambda_{m,}, \lambda_{k,}) = 0$ for all $m \in F'_2$. This proves the assertion.

Suppose that $\mathfrak{n}$ is just a split component of P and F is a subset of Q. Let $\mathfrak{t} = \{H \in \mathfrak{n} \mid \alpha(H) = 0 \text{ for all } \alpha \in F\}$; if F is a subset of the set of simple roots $\mathfrak{t}$ is called a distinguished subspace of $\mathfrak{n}$. Let $\mathfrak{g}_{\mathfrak{t}}$ be the orthogonal complement of $\mathfrak{t}$ in the centralizer of $\mathfrak{t}$ in $\mathfrak{g}$ and let $G_{\mathfrak{t}}$ be any subgroup of G with the Lie algebra $\mathfrak{g}_{\mathfrak{t}}$ which satisfies the same conditions as G. Then $\mathfrak{p} \cap \mathfrak{g}_{\mathfrak{t}}$ is the Lie algebra of a parabolic subgroup P' of $G_{\mathfrak{t}}$ and $\mathfrak{b}$, the orthogonal complement of $\mathfrak{t}$ in $\mathfrak{n}$, is a split component of P'. We regard the dual space of $\mathfrak{b}$ as the set of linear functions on $\mathfrak{a}$ which vanish on $\mathfrak{t}$. Let $\beta_{1,}, \dots, \beta_{q,}$ be the simple roots of $\mathfrak{b}$; $\{\beta_{1,}, \dots, \beta_{q,}\}$ is a subset of Q. There are two quadratic forms on the dual space of $\mathfrak{b}$, namely the one dual to the restriction of the Killing form on $\mathfrak{g}$ to $\mathfrak{b}$ and the one dual to the restriction of the Killing form on $\mathfrak{g}_{\iota}$ to $\mathfrak{b}$. Thus there are two possible definitions of $\beta_{,1}, \dots, \beta_{,q}$ and hence two possible definitions of $^+\mathfrak{b}$. In the proof of Theorem 7.7 it will be necessary to know that both definitions give the same $^+\mathfrak{b}$. A little thought convinces us that this is a consequence of the next lemma.

The split parabolic subgroup (P, S) will be called reducible if $\mathfrak{g}$ can be written as the direct sum of two ideals $\mathfrak{g}_1$ and $\mathfrak{g}_2$ in such a way that $\mathfrak{p} = \mathfrak{p}_1 + \mathfrak{p}_2$ with $\mathfrak{p}_i = \mathfrak{p} \cap \mathfrak{g}_i$ and $\mathfrak{s} = \mathfrak{s}_1 + \mathfrak{s}_2$ with $\mathfrak{s}_i = \mathfrak{s} \cap \mathfrak{g}_i$. Then $\mathfrak{n} = \mathfrak{n}_1 + \mathfrak{n}_2$ with $\mathfrak{n}_i = \mathfrak{n} \cap \mathfrak{g}_i$. If $\mathfrak{a}$ is a split component of (P, S) and $\mathfrak{m}'$ is the centralizer of $\mathfrak{a}$ in $\mathfrak{g}$ then $\mathfrak{m}' = \mathfrak{m}'_1 + \mathfrak{m}'_2$ with $\mathfrak{m}'_i = \mathfrak{m}' \cap \mathfrak{g}_i$. Since $\mathfrak{m} = \mathfrak{m}' \cap \mathfrak{s}$ it is also the direct sum of $\mathfrak{m}_1$ and $\mathfrak{m}_2$ and $\mathfrak{a}$, being the orthogonal complement of $\mathfrak{m}$ in $\mathfrak{m}'$, is the direct sum of $\mathfrak{a}_1$ and $\mathfrak{a}_2$. If (P, S) is not reducible it will be called irreducible.

LEMMA 2.6. *Suppose that the split parabolic subgroup* (P, S) *is irreducible and suppose that* π *is a representation of* $\mathfrak{g}$ *on the finite-dimensional vector space* V *such that if* α *is a linear function on* $\mathfrak{a}$ *and*

$$V_\alpha = \{v \in V \mid \pi(H)v = \alpha(H)v \text{ for all } H \text{ in } \mathfrak{a}\}$$

then the trace of the restriction of $\pi(X)$ *to* V_α *is zero for all* X *in* $\mathfrak{m}$. *Then there is a constant* c *so that* trace $\{\pi(H_1)\pi(H_2)\} = c\langle H_1, H_2\rangle$ *for all* H_1 *and* H_2 *in* $\mathfrak{a}$.

If $\mathfrak{g}$ contains a non-trivial centre then $\mathfrak{g}$ is abelian and $\mathfrak{a} = \{0\}$ so there is nothing to prove. We suppose then that $\mathfrak{g}$ is semi-simple so that the Killing form $\langle X, Y\rangle$ is non-degenerate. Consider the bilinear form trace $\pi(X)\pi(Y) = (X, Y)$ on $\mathfrak{g}$. It is readily verified that

$$([X, Y], Z) + (Y, [X, Z]) = 0$$

Let T be the linear transformation on $\mathfrak{g}$ such that $(X, Y) = \langle TX, Y \rangle$; then $\langle TX, Y \rangle = \langle X, TY \rangle$ and $T([X, Y]) = [X, TY]$. If H belongs to $\mathfrak{a}$ and X belongs to $\mathfrak{m}$ the assumption of the lemma implies that $(H, X) = 0$. Moreover choosing a basis for V with respect to which the transformations $\pi(H)$, $H \in \mathfrak{a}$, are in diagonal form we see that $(X, Y) = 0$ if X belongs to $\mathfrak{p}$ and Y belongs to $\mathfrak{n}$. If α belongs to Q let

$$\mathfrak{n}_\alpha^- = \{X \in \mathfrak{g} \mid [H, X] = -\alpha(H)X \text{ for all } H \text{ in } \mathfrak{g}\}$$

and let $\mathfrak{n}^- = \Sigma_{\alpha \in Q} \mathfrak{n}_\alpha^-$. If X belongs to $\mathfrak{a} + \mathfrak{m} + \mathfrak{n}^-$ and Y belongs to $\mathfrak{n}^-$ then $(X, Y) = 0$. Thus if H belongs to $\mathfrak{a}$ then $\langle TH, Y \rangle = 0$ for Y in $\mathfrak{m} + \mathfrak{n} + \mathfrak{n}^-$ so that TH lies in $\mathfrak{a}$. For the same reason $T\mathfrak{m} \subseteq \mathfrak{m}$ and $T\mathfrak{n} \subseteq \mathfrak{n}$. Let $\lambda_1, \ldots, \lambda_r$ be the eigenvalues of T and let $\mathfrak{g}_i = \{X \in \mathfrak{g} \mid (T-\lambda_i)^n X = 0$ for some $n\}$. $\mathfrak{g}_i$, $1 \le i \le r$, is an ideal of $\mathfrak{g}$ and $\mathfrak{g} = \oplus\, \mathfrak{g}_i$, $\mathfrak{p} = \oplus\, \mathfrak{p} \cap \mathfrak{g}_i$, and $\mathfrak{S} = \oplus\, \mathfrak{S} \cap \mathfrak{g}_i$. We conclude that $r = 1$. The restriction of T to $\mathfrak{a}$ is symmetric with respect to the restriction of the Killing form to $\mathfrak{a}$. Since the latter is positive definite the restriction of T to $\mathfrak{a}$ is a multiple of the identity. This certainly implies the assertion of the lemma.

Now suppose Γ is a discrete subgroup of G. When describing the conditions to be imposed on Γ we should be aware of the following fact.

LEMMA 2.7. *If* Γ *is a discrete subgroup of* G, *if* (P_1, S_1) *and* (P_2, S_2) *are two split parabolic subgroups, if* $\Gamma \cap P_i \subseteq S_i$, $i = 1, 2$, *if the volume of* $\Gamma \cap S_i \backslash S_i$ *is finite for* $i = 1, 2$, *and if* $P_1 \supseteq P_2$ *then* $S_1 \supseteq S_2$.

Let $S = S_1 \cap S_2$; then $\Gamma \cap P_2 \subseteq S$. S is a normal subgroup of S_2 and $S \backslash S_2$ is isomorphic to $S_1 \backslash S_1 S_2$ and is consequently abelian. It follows readily from the definition of a split parabolic subgroup that the Haar measure on S_x is left and right invariant. This is also true of the Haar measure on $S\backslash S_2$ and hence it is true of the Haar measure on S. Thus

$$\int_{\Gamma\cap S_2\backslash S_2} ds_2 = \int_{S\backslash S_2} ds_2 \int_{\Gamma\cap S_2\backslash S} ds = \mu(S\backslash S_2)\mu(\Gamma\cap S\backslash S)$$

Consequently $\mu(S\backslash S_2)$ is finite and $S\backslash S_2$ is compact. Since the natural mapping from $S\backslash S_2$ to $S_1\backslash S_1S_2$ is continuous $S_1\backslash S_1S_2$ is also compact. But $S_1\backslash S_1S_2$ is a subgroup of $S_1\backslash P_1$ which is isomorphic to A_1 and A_1 contains no non-trivial compact subgroups. We conclude that S_2 is contained in S_1.

If Γ is a discrete subgroup of G and (P, S) is a split parabolic subgroup then (P, S) will be called cuspidal if every split parabolic subgroup (P', S') belonging to (P, S) is such that $\Gamma \cap P' \subseteq S'$, $\Gamma \cap N'\backslash N'$ is compact, and $\Gamma \cap S'\backslash S'$ has finite volume. A cuspidal subgroup such that $\Gamma \cap S\backslash S$ is compact will be called percuspidal. Since the last lemma implies that S is uniquely determined by P and Γ we will speak of P as cuspidal or percuspidal. If P is a cuspidal subgroup the group $N\backslash S$ which is isomorphic to M satisfies the same conditions as G. It will usually be identified with M. The image Θ of $\Gamma \cap S$ in M is a discrete subgroup of M. If $({}^*P, {}^*S)$ is a split parabolic group belonging to (P, S) then $({}^\Psi P, {}^\Psi S) = ({}^*N\backslash P \cap {}^*S, {}^*N\backslash S)$ is a split parabolic subgroup of ${}^*M$. If (P, S) is a cuspidal subgroup of G then $({}^\Psi P, {}^\Psi S)$ is a cuspidal subgroup of ${}^*M$ with respect to the group ${}^*\Theta$.

Once we have defined the notion of a Siegel domain we shall state the condition to be imposed on Γ. Fix once and for all a maximal compact subgroup K of G which contains a maximal compact subgroup of G_0. If (P, S) is a split parabolic subgroup, if c is a positive number, and if ω is a compact subset of S then a Siegel domain $\mathfrak{S} = \mathfrak{S}(c, \omega)$ associated to (P, S) is

$$\{g = sak \mid s \in \omega,\ a \in A^+(c, \infty),\ k \in K\}$$

A is any split component of (P, S).

A set E of percuspidal subgroups will be said to be complete if when (P_1, S_1) and (P_2, S_2) belong to E there is a g in G so that $gP_1g^{-1} = P_2$ and $gS_1g^{-1} = S_2$ and when (P, S) belongs to E and γ belongs to Γ the pair $(\gamma P\gamma^{-1}, \gamma S\gamma^{-1})$ belongs to E.

ASSUMPTION. *There is a complete set* E *of percuspidal subgroups such that if* P *is any cuspidal subgroup belonging to an element of* E *there is a subset* $\{P_1, \ldots, P_r\}$ *of* E *such that* P *belongs to* P_i, $1 \leq i \leq r$, *and Siegel domains* $\mathfrak{S}_i$ *associated to* $(N\backslash P_i \cap S, N\backslash S_i)$ *such that* $M = \bigcup_{i=1}^{r} \Theta\, \mathfrak{S}_i$. *Moreover there is a finite subset* F *of* E *so that* $E = \bigcup_{\gamma\in\Gamma} \bigcup_{P\in F} \gamma P \gamma^{-1}$.

Henceforth a cuspidal subgroup will mean a cuspidal subgroup belonging to an element of E and a percuspidal subgroup will mean an element of E. It is apparent that the assumption has been so formulated that if $^*P$ is a cuspidal subgroup then it is still satisfied if the pair Γ, G is replaced by the pair $^*\Theta$, $^*M$. Let us verify that this is so if E is replaced by the set of subgroups $^*N\backslash P \cap {}^*S$ where P belongs to E and $^*P$ belongs to P. It is enough to verify that if $(^*P, {}^*S)$ is a split parabolic group belonging to (P_1, S_1) and to (P_2, S_2) and $gP_1g^{-1} = P_2$, $gS_1g^{-1} = S_2$ then g lies in $^*P$. Let $^*\mathfrak{a}$ be a split component of $(^*P, {}^*S)$; let $\mathfrak{a}_1$ be a split component of (P_1, S_1) containing $^*\mathfrak{a}$; and let $\mathfrak{h}$ be a maximal abelian subalgebra of $\mathfrak{g}^s$ containing $\mathfrak{a}_1$ whose image in ad $\mathfrak{g}$ is diagonalizable. Choose p in $^*P$ so that (pP_2p^{-1}, pS_2p^{-1}) has a split component $\mathfrak{a}_2$ which contains $^*\mathfrak{a}$ and is contained in $\mathfrak{h}$ and so that Ad $pg(\mathfrak{a}_1) = \mathfrak{a}_2$ and Ad $pg(\mathfrak{h}) = \mathfrak{h}$. Replacing g by pg if necessary we may suppose that $p = 1$. Choose an order on the real linear functions on $\mathfrak{h}$ so that any root whose restriction to $\mathfrak{a}_1$ lies in Q_1 is positive. If the restriction of the positive root α to $^*\mathfrak{a}$ lies in $^*Q$ then the restriction to $\mathfrak{a}_1$ of the root α' defined by

$$\alpha'(H) = \alpha(\mathrm{Ad}\ g(H))$$

lies in Q_1 and is thus positive. The roots whose restriction to ${}^*\mathfrak{n}$ are zero are determined by their restrictions to the intersection of $\mathfrak{b}$ with the semi-simple part of ${}^*\mathfrak{m}$. It is possible (cf. [11]) to choose an order on the linear functions on this intersection so that the positive roots are the restrictions of those roots α of $\mathfrak{b}$ such that α', with $\alpha'(H) = \alpha(\mathrm{Ad}\ g(H))$, is positive. It is also possible to choose an order so that the positive roots are the restrictions of the positive roots of $\mathfrak{b}$. Consequently there is an m in ${}^*M$ so that $\mathrm{Ad}\ mg(\mathfrak{b}) = \mathfrak{b}$ and $\mathrm{Ad}\ mg$ takes positive roots to positive roots. mg then belongs to the centralizer of $\mathfrak{b}$ and hence to ${}^*P$ so the assertion is proved.

Some consequences of the assumption which are necessary for the analysis of this paper will now be deduced. If (P, S) is a split parabolic subgroup of G the map $(p, k) \longrightarrow pk$ is an analytic map from $P \times K$ onto G. If A is a split component of (P, S) then every element p of P may be written uniquely as a product $p = as$ with a in A and s in S. Although in the decomposition $g = pk$ the factor p may not be uniquely determined by g the factor $a = a(g)$ of the product $p = as$ is. In fact the image of $a(g)$ in the split component of (P, S) is. Henceforth, for the sake of definiteness, $a(g)$ will denote this image. Every percuspidal subgroup has the same split component which we call $\mathfrak{f}$. Suppose the rank of $\mathfrak{f}$ is p and $\alpha_{,1}, \ldots, \alpha_{,p}$ are the linear functions on $\mathfrak{f}$ dual to the simple roots.

LEMMA 2.8. <u>If</u> P <u>is a percuspidal subgroup there is a constant</u> μ <u>so that</u> $\xi_{\alpha_{,i}}(a(\gamma)) \leq \mu$, $1 \leq i \leq p$, <u>for all</u> γ in Γ.

If C is a compact subset of G so is KC and $\{a(h) \mid h \in KC\}$ is compact; thus there are two positive numbers μ_1 and μ_2 with $\mu_1 < \mu_2$ so that it is contained in ${}^+A(\mu_1, \mu_2)$. If $g = ask$ with a in A, s in S, and k in K and h belongs to C then $a(gh) = a(kh)a(g)$ so that

$$\mu_1 \xi_{\alpha_{,i}}(a) < \xi_{\alpha_{,i}}(a(gh)) < \mu_2 \xi_{\alpha_{,i}}(a), \ 1 \le i \le p$$

In particular in proving the lemma we may replace Γ by a subgroup of finite index, which will still satisfy the basic assumption, and hence may suppose that G is connected. If P is a cuspidal subgroup let $\Delta = \Gamma \cap S$. If P is percuspidal there is a compact set ω in S so that every left coset $\Delta\gamma$ of Δ in Γ contains an element $\gamma' = sa(\gamma)k$ with s in ω. For the purposes of the lemma only those γ such that $\gamma' = \gamma$ need be considered. It is not difficult to see (cf. [22], App. II) that there is a finite number of elements $\delta_1, \ldots, \delta_n$ in $\Gamma \cap N$ so that the connected component of the centralizer of $\{\delta_1, \ldots, \delta_n\}$ in $\Gamma \cap N$ is N^c, the centre of N. A variant of Lemma 2 of [18] then shows that $\Gamma \cap N^c \backslash N^c$ is compact so that, in particular, there is an element $\delta \neq 1$ in $\Gamma \cap N^c$. If Q^c is the set of α in Q such that $\mathfrak{n}_\alpha \cap \mathfrak{n}^c \neq \{0\}$ there is a constant ν so that, for all γ in Γ, $\xi_\alpha(a(\gamma)) \le \nu$ for at least one α in Q^c. If not there would be a sequence $\{\gamma_\ell\}$ with $\{\xi_\alpha^{-1}(a(\gamma_\ell))\}$ converging to zero for all α in Q^c so that

$$\gamma^{-1}\delta\gamma = k_\ell^{-1}(a^{-1}(\gamma_\ell)(s_\ell^{-1}\delta s_\ell)a(\gamma_\ell))k_\ell$$

would converge to 1 which is impossible.

If $\mathfrak{g} = \oplus_{i=1}^r \mathfrak{g}_i$ with $\mathfrak{g}_i$ simple then $\mathfrak{z} = \oplus_{i=1}^r \mathfrak{z} \cap \mathfrak{g}_i$ so that $\mathfrak{n}^c = \oplus_{i=1}^r \mathfrak{n}^c \cap \mathfrak{g}_i$. If $\mathfrak{j}$ is a Cartan subalgebra containing $\mathfrak{a}$, if an order is chosen on $\mathfrak{j}$ as before, and if $\mathfrak{w}$ is a subspace of $\mathfrak{n}^c \cap \mathfrak{g}_i$ invariant under $\mathrm{Ad}\, p$ for $p \in P$ then the complexification of $\mathfrak{w}$ contains a vector belonging to the lowest weight of the representation $g \longrightarrow \mathrm{Ad}(g^{-1})$ of G on $\mathfrak{g}_i$. Since this vector is unique $\mathfrak{n}^c \cap \mathfrak{g}_i \subseteq \mathfrak{n}_{\beta_i}$ for some β_i in Q. The lowest weight is the negative of a dominant integral function so $\beta_i = \Sigma_{j=1}^p b_i^j \alpha_{,j}$ with $b_i^j \ge 0$. Thus there is a constant ν' so that, for all γ in Γ,

$$\min_{1 \le j \le p} \xi_{\alpha_{,j}}(a(\gamma)) \le \nu'$$

In any case Lemma 2.8 is now proved for percuspidal subgroups of rank one.

If G_0 is a maximal compact normal subgroup of G then in the proof of the lemma G may be replaced by $G_0\backslash G$ and Γ by $G_0\backslash\Gamma G_0$. In other words it may be supposed that G has no compact normal subgroup. Let Z be the centre of G. If we could show that $\Gamma\cap Z\backslash Z$ was compact we could replace G by $Z\backslash G$ and Γ by $Z\backslash\Gamma Z$ and assume that G has no centre. We will show by induction on the dimension of G that if $\Gamma\backslash G$ has finite volume then $\Gamma\cap Z\backslash Z$ is compact. This is certainly true if G is abelian and, in particular, if the dimension of G is one. Suppose then that G is not abelian and of dimension larger than one. Because of our assumptions the group G has a finite covering which is a product of simple groups. We may as well replace G by this group and Γ by its inverse image in this group. Let $G = \prod_{i=1}^{n} G_i$ where G_i is simple $1 \leq i \leq r$. We may as well assume that G_i is abelian for some i. Choose δ in Γ but not in Z. It follows from Corollary 4.4 in [1] that the centralizer of δ in Γ is not of finite index in Γ and hence that for some γ in Γ $\delta^{-1}\gamma^{-1}\delta\gamma = \varepsilon$ does not lie in the centre of G. Let $\varepsilon = \prod_{i=1}^{n}\varepsilon_i$ and suppose that ε_i does not lie in the centre of G_i for $1 \leq i \leq m$ where $1 \leq m < n$. It follows as in [20] that the projection of Γ on $G' = \prod_{i=1}^{m} G_i$ is discrete and that the volume of $\Gamma\cap G''\backslash G''$ is finite if $G'' = \prod_{i=m+1}^{n} G_i$. Since G'' contains a subgroup of Z which is of finite index in Z and otherwise satisfies the same conditions as G the proof may be completed by induction.

If G has no centre and no compact normal subgroup Γ is said to be reducible if there are two non-trivial closed normal subgroups G_1 and G_2 so that $G_1 \cap G_2 = \{1\}$, $G = G_1G_2$, and Γ is commensurable with the product of $\Gamma_1 = \Gamma\cap G_1$ and $\Gamma_2 = \Gamma \cap G_2$. Γ is irreducible when it is not reducible. Since if one of a pair of commensurable groups satisfies the basic assumption so does the other, it may be supposed when Γ is reducible that it is the product of Γ_1 and Γ_2. If we show that Γ_1 and Γ_2 satisfy the basic assumption we need only prove the lemma for irreducible groups. If ${}^*P$ is a cuspidal subgroup for Γ then

${}^*P = {}^*P_1\,{}^*P_2$ with ${}^*P_i = {}^*P \cap G_i$ and ${}^*N = {}^*N_1\,{}^*N_2$ with ${}^*N_i = {}^*N \cap G_i$. Since $\Gamma \cap {}^*N \backslash {}^*N$ is thus the product of $\Gamma \cap {}^*N_1 \backslash {}^*N_1$ and $\Gamma \cap {}^*N_2 \backslash {}^*N_2$ both factors are compact. Moreover if ${}^*S_i = {}^*S \cap G_i$ then $\Gamma_i \cap {}^*P_i \subseteq {}^*S_i$ and $\Gamma \cap {}^*P \subseteq {}^*S_1\,{}^*S_2$. If ${}^*A$ is a split component of $({}^*P, {}^*S)$ and ${}^*A_i$ is the projection of ${}^*A$ on G_i then ${}^*A_1\,{}^*A_2$ is a split component of ${}^*P$ and determines the split parabolic subgroup $({}^*P, {}^*S_1\,{}^*S_2)$. Since the measure of $\Gamma \cap {}^*S_1\,{}^*S_2 \backslash {}^*S_1\,{}^*S_2$ is clearly finite Lemma 2.7 implies that ${}^*S = {}^*S_1\,{}^*S_2$. It follows readily that $({}^*P_i, {}^*S_i)$ is a cuspidal subgroup for Γ_i, $i = 1, 2$. Once this is known it is easy to convince oneself that Γ_i, $i = 1, 2$, satisfies the basic assumption.

To make use of the condition that Γ is irreducible another lemma is necessary.

LEMMA 2.9. <u>Suppose</u> Γ <u>is irreducible</u>. <u>If</u> P <u>is a cuspidal subgroup for</u> Γ <u>and</u> $\alpha_{,1}, \ldots, \alpha_{,q}$ <u>are the linear functions on</u> $\mathfrak{a}$ <u>dual to the simple roots then</u> $\langle \alpha_{,i}, \alpha_{,j} \rangle > 0$ <u>for all</u> i <u>and</u> j.

To prove this it is necessary to show that if the first alternative of Lemma 2.6 obtains then Γ is reducible. Let F_1 and F_2 be the two subsets of that lemma and let P_1 and P_2 be the two cuspidal subgroups determined by them. We will show in a moment that $\mathfrak{n}$ is the Lie algebra generated by $\Sigma_{i=1}^{q} \mathfrak{n}_{\alpha_i,}$ and that if $i \in F_1$, $j \in F_2$ then $[\mathfrak{n}_{\alpha_i,}, \mathfrak{n}_{\alpha_j,}] = 0$. Thus $\mathfrak{n} = \mathfrak{n}_1 \oplus \mathfrak{n}_1$ if $\mathfrak{n}_i$ is the algebra generated by $\Sigma_{j \notin F_i} \mathfrak{n}_{\alpha_j}$. Moreover $\mathfrak{n}_i$ is the maximal normal subalgebra of $\mathfrak{p}_i \cap \mathfrak{g}^s$ containing only elements whose adjoints are nilpotent. The centralizer of $\mathfrak{a}_1$ is a fully reducible subalgebra of $\mathfrak{g}$ and lies in the normalizer of $\mathfrak{n}_1$. The kernel of the representation of this algebra on $\mathfrak{n}_1$ is a fully reducible subalgebra $\mathfrak{g}_2$. The normalizer $\mathfrak{g}'$ of $\mathfrak{g}_2$ is the sum of a fully reducible subalgebra $\mathfrak{g}_1$ and $\mathfrak{g}_2$. $\mathfrak{g}'$ contains $\mathfrak{n}_1$ and the centralizer of $\mathfrak{a}_1$ and thus contains $\mathfrak{p}_1$. Since $\mathfrak{p}_1$ is parabolic $\mathfrak{g}' = \mathfrak{g}$. Since $\mathfrak{g}_2$ contains $\mathfrak{n}_2$ and $\Gamma \cap N_i \neq \{1\}$ for $i = 1, 2$ it follows from Theorem 1' of [20] that Γ is reducible.

To begin the proof of the first of the above assertions we show that if α is in Q then $\langle \alpha, \alpha_{j,} \rangle > 0$ for some j. If this were not so then, since $\alpha = \Sigma_{j=1}^{q} m_j \alpha_{j,}$,

$$0 < \langle \alpha, \alpha \rangle = \Sigma_{j=1}^{q} m_j \langle \alpha_{j,}, \alpha \rangle \leq 0$$

Choose a Cartan subalgebra $\mathfrak{j}$ of $\mathfrak{g}$ containing $\mathfrak{a}$ and choose an order as before on the real linear functions on $\mathfrak{j}_c$. If α' is a positive root and the restriction α of α' to $\mathfrak{a}$ is neither zero nor $\alpha_{i,}$, $1 \leq i \leq q$, then for some j there is a β' in $Q'_{\alpha_{j,}}$ such that $\alpha' - \beta'$ is a positive root. Indeed if this were not so then, since $\beta' - \alpha'$ is not a root for any such β', we would have $(\alpha', \beta') \leq 0$. Consequently

$$\langle \alpha, \alpha_{i,} \rangle \leq 0, \ 1 \leq i \leq q$$

which is impossible. Let $\mathfrak{n}'$ be the algebra generated by $\Sigma_{j=1}^{q} \mathfrak{n}_{\alpha_{j,}}$; it is enough to show that $\mathfrak{n}'_c$, the complexification of $\mathfrak{n}'$, equals $\mathfrak{n}_c$. We suppose that this is not so and derive a contradiction. Order the elements of Q lexicographically according to the basis $\{\alpha_{1,}, \ldots, \alpha_{q,}\}$ and let α be a minimal element for which there is a root α' in Q'_{α} such that $X_{\alpha'}$, a root vector belonging to α', is not in $\mathfrak{n}'_c$. Choose a j and a β' in $Q'_{\alpha_{j,}}$ so that $\alpha' - \beta'$ is a root. The root vectors $X_{\alpha'}$ and $X_{\alpha'-\beta'}$ both belong to $\mathfrak{n}'_c$ and thus $X_{\alpha'}$ which is a complex multiple of $[X_{\beta'}, X_{\alpha'-\beta'}]$ does also. As for the second assertion we observe that if

$$i \in F_1, \ j \in F_2$$

and

$$\alpha' \in Q'_{\alpha_{i,}}, \ \beta' \in Q'_{\alpha_{j,}}$$

then $\alpha' - \beta'$ is neither a root nor zero. Moreover

$$0 = \Sigma_{\beta' \in Q'_{\alpha_{j,}}} (\alpha', \beta')$$

so each term is zero and for no β' in $Q'_{\alpha_{j,}}$ is $\alpha' + \beta'$ a root. This shows that

$$[\mathfrak{n}_{\alpha_{i}}, \mathfrak{n}_{\alpha_{j}}] = 0$$

Suppose that Γ is irreducible and that the assertion of Lemma 2.8 is not true for the percuspidal subgroup P. There is a sequence $\{\gamma_j\} \subseteq \Gamma$ and a k, $1 \leq k \leq p$, so that

$$\lim_{j\to\infty} \xi_{\alpha_{,k}}(a(\gamma_j)) = \infty$$

It may be supposed that $k = 1$. Let ${}^*P$ be the cuspidal subgroup belonging to P determined by $\{\alpha_{,i} \mid i \neq 1\}$. Let $\gamma_j = n_j a_j m_j k_j$ with n_j in ${}^*N$, a_j in ${}^*A$, m_j in ${}^*M$, and k_j in K. Replacing γ_j by $\delta_j\gamma_j$ with $\delta_j \in {}^*\Delta = \Gamma \cap {}^*S$ and choosing a subsequence if necessary we may assume that $\{n_j\}$ belongs to a fixed compact set and that $\{m_j\}$ belongs to a given Siegel domain associated to the percuspidal subgroup ${}^{\Psi}P' = {}^*N\backslash P' \cap {}^*S$ of ${}^*M$. P' is a percuspidal subgroup of G to which ${}^*P$ belongs. If ${}^{\Psi}A'$ is the split component of ${}^{\Psi}P'$ and $A' = A$ is the split component of P' then $a'(\gamma_j) = a_j{}^{\Psi}a'(m_j)$. There is a constant c so that

$$\xi_{\alpha_{,i}}(a'(\gamma_j)) \geq c\xi_{\alpha_{,i}}(a_j)$$

This follows immediately if $i = 1$ since

$$\xi_{\alpha_{,1}}(a'(\gamma_j)) = \xi_{\alpha_{,1}}(a')$$

and from Lemma 2.5 if $i > 1$. Since

$$\langle \alpha_{,i}, \alpha_{,1} \rangle > 0, \quad 1 \leq i \leq q$$

there is a positive constant r so that

$$\xi^{r}_{\alpha_{,i}}(a_j) \geq \xi_{\alpha_{,1}}(a_j)$$

However

$$\xi_{\alpha_{,1}}(a_j) = \xi_{\alpha_{,1}}(a(\gamma_j))$$

so

$$\lim_{j\to\infty} \xi_{\alpha_{,i}}(a'(\gamma_j)) = \infty, \quad 1 \leq i \leq p$$

which we know to be impossible.

The next lemma is a simple variant of a well known fact but it is best to give a proof since it is basic to this paper. Suppose P is a parabolic with split component $\mathfrak{a}$. Let $\mathfrak{j}$ be a Cartan subalgebra such that $\mathfrak{a} \subseteq \mathfrak{j} \subseteq \mathfrak{g}$ and choose an order on the real linear functions on $\mathfrak{j}_c$ as before. Let $\alpha_{,1}, \ldots, \alpha_{,q}$ be the linear functions on $\mathfrak{a}$ dual to the simple roots and let $\bar{\alpha}_{,i}$ be the linear function on $\mathfrak{j}$ which agrees with $\alpha_{,i}$ on $\mathfrak{a}$ and is zero on the orthogonal complement of $\mathfrak{a}$. There is a negative number d_i so that $d_i\bar{\alpha}_{,i}$ is the lowest weight of a representation ρ_i of G_0, the connected component of G, acting on the complex vector space V_i to the right.

LEMMA 2.10. *If* λ *is a linear function on* $\mathfrak{a}$ *such that there is a non-zero vector* v *in* V_i *with* $v\rho_i(a) = \xi_\lambda(a)$ *for all* a *in* A *then*

$$\lambda = d_i\alpha_{,i} + \Sigma_{j=1}^{q} n_j\alpha_j,$$

with $n_j \geq 0$, $1 \leq j \leq q$. *Moreover if* v_i *is a non-zero vector belonging to the lowest weight then*

$$\{g \in G_0 \mid v_i\rho_i(g) = \mu v_i \text{ with } \mu \in \mathbb{C}\}$$

in the intersection with G_0 *of the split parabolic subgroup* P_i *determined by* $\{\alpha_{j,} \mid j \neq i\}$.

Let

$$\mathfrak{a}_i = \{H \in \mathfrak{a} \mid \alpha_{j,}(H) = 0,\ j \neq i\}$$

and let Q_i' be the set of positive roots of $\mathfrak{v}$ which do not vanish on $\mathfrak{v}_i$. Set

$$\mathfrak{n}_i^- = \Sigma_{\alpha \epsilon Q_i'} \mathfrak{n}_\alpha^-$$

then

$$\mathfrak{g} = \mathfrak{n}_i^- + \mathfrak{v}_i + \mathfrak{m}_i + \mathfrak{n}_i$$

Let

$$V_i' = \{v \mid v\rho_i(X) = 0 \text{ for } X \epsilon \mathfrak{n}_i\}$$

If W is a subspace of V_i' invariant and irreducible under $\mathfrak{v}_i + \mathfrak{m}_i$ then the vector belonging to the lowest weight of the representation of $\mathfrak{v}_i + \mathfrak{m}_i$ on W must be a multiple of v_i and the lowest weight must be $d_i \bar{\alpha}^i$. Consequently V_i' is the set of multiples of v_i. Let $V_i^{(n)}$ be the linear space spanned by

$$\{v_i\rho_i(X_1) \dots \rho_i(X_k) \mid X_j \epsilon \mathfrak{g} \text{ and } k \leq n\}$$

and let ${}^{(n)}V_i$ be the linear space spanned by

$$\{v_i\rho_i(X_1) \dots \rho_i(X_k) \mid X_j \epsilon \mathfrak{n}_i^- \text{ and } k \leq n\}$$

We show by induction that $V_i^{(n)} \subseteq {}^{(n)}V_i$. This is certainly true for $n = 1$ since k may be zero. If $X_1, \dots, X_{n-1}$ belong to $\mathfrak{n}_i^-$ and X_n belongs to $\mathfrak{g}$ then

$$v_i\rho_i(X_1) \dots \rho_i(X_n)$$

is equal to

$$v_i\rho_i(X_1) \dots \rho_i(X_{n-2})\rho_i([X_{n-1}, X_n]) + v_i\rho_i(X_1) \dots \rho_i(X_{n-2})\rho_i(X_n)\rho_i(X_{n-1}) \ .$$

Applying induction to the two terms on the right we are finished. The first assertion of the lemma follows immediately. Let

$$P'_i = \{g \epsilon G_0 \mid v_i\rho_i(g) = \mu v_i \text{ with } \mu \epsilon \mathbb{C}\}$$

The intersection of P_i with G_0 is just the normalizer of $\mathfrak{n}_i$ in G_0. Thus it leaves V_i' invariant and is contained in P_i'. To complete the proof we need only show that $\mathfrak{p}_i'$ is contained in $\mathfrak{p}_i$. If $\mathfrak{m}_i'$ is a maximal fully reducible subalgebra of $\mathfrak{p}_i'$ containing $\mathfrak{a}_i + \mathfrak{m}_i$ then $\{X \in \mathfrak{m}_i' \mid v_i \rho_i(X) = 0\}$ is a normal subalgebra of $\mathfrak{m}_i'$ and its orthogonal complement in $\mathfrak{m}_i'$ with respect to the Killing form lies in $\mathfrak{a}_i + \mathfrak{m}_i$ because it commutes with $\mathfrak{a}_i$. Thus its orthogonal complement is $\mathfrak{a}_i$ and $[\mathfrak{a}_i, \mathfrak{m}_i'] = 0$ so $\mathfrak{m}_i' = \mathfrak{a}_i + \mathfrak{m}_i$. Let $\mathfrak{n}_i'$ be a maximal normal subalgebra of $\mathfrak{p}_i'$ such that ad X is nilpotent for all X in $\mathfrak{n}_i'$. Then $\mathfrak{n}_i'$ is contained in $\mathfrak{n}_i$ and $\mathfrak{p}_i' = \mathfrak{m}_i' + \mathfrak{n}_i'$. It follows that $\mathfrak{p}_i' = \mathfrak{p}_i$.

Before stating the next lemma we make some comments on the normalization of Haar measures. We suppose that the Haar measure on G is given. The Haar measure on K will be so normalized that the total volume of K is one. If P is a cuspidal subgroup the left-invariant Haar measure on P will be so normalized that

$$\int_G \phi(g)dg = \int_P \int_K \phi(pk)dpdk$$

Let ρ be one-half the sum of the elements of Q and if $a = \exp H$ belongs to A let $\omega(a) = \exp(-\rho(H))$. Let dH be the Lebesgue measure $\mathfrak{a}$ normalized so that the measure of a unit cube is one and let da be the Haar measure on A such that $d(\exp H) = dH$. Choose, as is possible, a Haar measure on S so that

$$\int_P \phi(p)dp = \int_S \int_A \phi(sa)\omega^2(a)dsda$$

Choose the invariant measure on $\Gamma \cap N \backslash N$ so that the volume of $\Gamma \cap N \backslash N$ is one and choose the Haar measure on N so that

$$\int_N \phi(n)dn = \int_{\Gamma \cap N \backslash N} \Sigma_{\delta \in \Gamma \cap N} \phi(\delta n)dn$$

Finally choose the Haar measure on M so that

$$\int_S \phi(s)ds = \int_N \int_M \phi(nm)dndm$$

LEMMA 2.11. *Let* P *be a percuspidal subgroup and* ω *a compact subset of* S. *There are constants* c *and* r *that for any* $t \leq 1$ *and any* g *in* G *the intersection of* Γg *and the Siegel domain* $\mathcal{S}(\omega, t)$ *associated to* $\mathfrak{P}$ *has at most* ct^{-r} *element*

It is easy to convince oneself that it is enough to prove the lemma when G i connected. In this case the representations ρ_i introduced before Lemma 2.10 are representations of G. Choose a norm on V_i so that $\rho_i(k)$ is unitary for all k in K. If $g = sa(g)k$ then

$$v_i\rho_i(g) = \xi_{\alpha,i}^{d_i}(a(g))v_i\rho_i(k)$$

so that

$$\| v_i\rho_i(g) \| = \xi_{\alpha,i}^{d_i}(a(g)) \| v_i \|$$

If T is a linear transformation then $\| T \|$ denotes as usual the norm of T. Choosing a basis v_{ij}, $1 \leq j \leq n_i$ for V_i so that

$$v_{ij}\rho_i(a) = \xi_{\lambda_{ij}}(a)v_{ij}$$

for all a in A we see that there is a constant c_1 so that, for all v in V_i and all a in A,

$$\| v\rho_i(a) \| \geq c_1(\min_{1\leq j\leq n_i} \xi_{\lambda_{ij}}(a)) \| v \|$$

Moreover, it follows from Lemma 2.10 that there is a constant s so that, for all a in $A^+(t, \infty)$,

$$\min_{1\leq j\leq n_i} \xi_{\lambda_{ij}}(a) \geq t^s \xi_{\alpha,i}^{d_i}(a)$$

Let c_2 be such that, for all s in ω and all v in V_i,

$$\| v\rho_i(s) \| \geq c_2 \| v \|$$

Suppose g and $g' = \gamma g$, with γ in Γ, both belong to $\mathcal{J}(\omega, t)$. Certainly

$$\|v_i\rho_i(g')\| = \xi_{\alpha,i}^{d_i}(a(g'))\|v_i\|$$

On the other hand

$$\|v_i\rho_i(\gamma g)\| \geq c_1c_2t^s\xi_{\alpha,i}^{d_i}(a(g))\|v_i\rho_i(\gamma)\|$$

and

$$\|v_i\rho_i(\gamma)\| = \xi_{\alpha,i}^{d_i}(a(\gamma))\|v_i\|$$

It follows from Lemma 2.8 that there are constants c_3 and c_4 and s_1 so that

$$\xi_{\alpha,i}(a(g')) \leq c_3t^{s_1}\xi_{\alpha,i}(a(\gamma))\xi_{\alpha,i}(a(g)) \leq c_4t^{s_1}\xi_{\alpha,i}(a(g))$$

Since $g = \gamma^{-1}g'$ the argument may be reversed. Thus there are constants c_5, c_6, and s_2 so that

$$c_5 > \xi_{\alpha,i}(a(\gamma)) > c_6t^{s_2}, \quad 1 \leq r \leq p$$

Let us estimate the order of

$$U(t) = \{\gamma = sak \mid s \in \omega_1,\ a \in {}^+A(c_6t^{s_2}, c_5),\ k \in K\}$$

with ω_1 a compact subset of S. There are certainly constants b_1, b_2 and r_1, r_2 so that ${}^+A(c_6t^{s_2}, c_5)$ is contained in $A^+(b_1t^{r_1}, b_2t^{r_2})$. Choose a conditionally compact open set in G such that $\gamma_1U \cap \gamma_2U \neq \phi$ implies $\gamma_1 = \gamma_2$; then b_1 can be so chosen that $\gamma \in U(t)$ implies

$$\gamma U \in \omega_2(t)A^+(b_1t^{r_1}, b_2t^{r_2})K$$

where

$$\omega_2(t) = \{s_1as_2a^{-1} \mid s_1 \in \omega_1,\ s_2 \in \omega_2,\ a \in A^+(b_1t^{r_1}, b_2t^{r_2})\}$$

and ω_2 is the projection of KU on S. Consequently the order of U(t) is at most a constant times the product of

$$\int_{A^+(b_1 t^{r_1}, b_2 t^{r_2})} \omega^2(a)da$$

and the volume of $\omega_2(t)$. A simple calculation, which will not be given here, now shows that the order of U(t) is bounded by a constant times a power of t. If it can be shown that for each g in $\mathfrak{S}(\omega, t)$ and each γ in Γ the number of elements δ in $\Delta = \Gamma \cap P$ such that $\delta\gamma g$ belongs to $\mathfrak{S}(\omega, t)$ is bounded by a constant independent of t, γ, and g then the lemma will be proved. If $\gamma g = sak$ then δs must be in ω. If there is no such δ the assertion is true; if there is one, say δ_0, then any other δ equals $\delta'\delta_0$ with $\delta'\omega \cap \omega \neq \phi$.

COROLLARY. Let P_1 and P_2 be percuspidal subgroups and let ${}^*P$ be a cuspidal subgroup belonging to P_2. Let $\mathfrak{S}_1$ be a Siegel domain associated to P_1, let ${}^\Psi\mathfrak{S}_2$ be a Siegel domain associated to ${}^\Psi P_2 = {}^*N\backslash P_2 \cap {}^*S$, let ω be a compact subset of ${}^*N$, and let b, s, and t be positive numbers. Let ${}^\Psi\mathfrak{a}_2$ be the split component of ${}^\Psi P_2$. There is a constant r, which depends only on G and s, and a constant c so that if $g \in \mathfrak{S}_1$, $\gamma \in \Gamma$, and $\gamma g = namk$ with n in ω, a in ${}^*A^+(t, \infty)$, m in ${}^\Psi\mathfrak{S}_2$, k in K, and $\eta(a) \leq b\eta^s({}^\Psi a_2(m))$ then

$$\eta(a_1(g)) \leq c\eta^r({}^\Psi a_2(m))$$

Moreover if ${}^*P = G$ the constant r can be taken to be 1.

If $\alpha_1, \dots, \alpha_p$, are the simple roots of $\mathfrak{a}_1$ then

$$\eta(a_1(g)) = \sup_{1 \leq i \leq p} \xi_{\alpha_i}(a_1(g))$$

similarly, if $\beta_1, \dots, \beta_q$, are the simple roots of ${}^\Psi\mathfrak{a}_2$,

$$\eta({}^\Psi a_2(m)) = \sup_{1 \leq i \leq q} \xi_{\beta_i}({}^\Psi a_2(m))$$

Suppose that

$$\mu \leq \xi_{\beta_{i,}}({}^{\Psi}a_2(m)), \quad 1 \leq i \leq q$$

for all m in ${}^{\Psi}\mathfrak{T}$. If m is given as in the lemma let $M = \eta({}^{\Psi}a_2(m))$; then

$$\log \mu \leq \xi_{\beta_{i,}}({}^{\Psi}a_2(m)) \leq \log M, \quad 1 \leq i \leq q$$

Since

$$\log t \leq \log \xi_{\alpha_{i,}}(a) \leq \log b + s \log M$$

and since $a_2(\gamma g) = a{}^{\Psi}a_2(m)$ there is a constant r_1, which depends only on G and s, and a constant r_2 so that

$$|\log \xi_{\alpha_{i,}}(a_2(\gamma g))| \leq r_1 \log M + r_2, \quad 1 \leq i \leq p$$

In particular there is a constant r_3, which depends only on G and s, and two positive constants c_1 and c_2 so that

$$\xi_{\alpha_{i,}}(a_2(\gamma g)) \geq c_1 M^{-r_3}$$

and

$$\xi_{\alpha_{,i}}(a_2(\gamma g)) \leq c_2 M^{r_3}$$

for $1 \leq i \leq p$. Choose u so that $uP_2u^{-1} = P_1$; then $a_1(u\gamma g u^{-1}) = a_2(\gamma g)$. Let v_i have the same significance as above except that the group P is replaced by P_1. Then there is a constant r_4, which depends only on G and s, and a constant c_3 so that

$$\xi_{\alpha_{,i}}^{d_i}(a_1(g)) \|v_i\| = \|v_i \rho_i(\gamma^{-1} u^{-1}(u\gamma g u^{-1})u)\| \geq c_3 M^{r_4} \xi_{\alpha_{,i}}^{d_i}(a_2(\gamma g)) \|v_i\|$$

Thus there is a constant r_5, which depends only on G and s, and a constant c_4 so that

$$\xi_{\alpha_{,i}}(a_1(g)) \le c_4 M^{r_5}$$

Appealing to Lemma 2.4 we see that $\xi_{\alpha_{,i}}(a_1(g))$ is bounded away from zero for $1 \le i \le p$. Since $\log \xi_{\alpha_{j,}}(a_1(g))$ is a linear combination of $\log \xi_{\alpha_{,i}}(a_1(g))$, $1 \le i \le p$ the first assertion of the lemma is proved.

To complete the proof of the lemma we have to show that if $\mathfrak{S}_1$ and $\mathfrak{S}_2$ are Siegel domains associated to P_1 and P_2 respectively then there is a constant c so that if g belongs to $\mathfrak{S}_1$ and γg belongs to $\mathfrak{S}_2$ then

$$\eta(a_1(g)) \le c\eta(a_2(\gamma g))$$

Using Lemma 2.10 as above we see that $\xi_{\alpha_{,i}}(a_1(g)a_2^{-1}(\gamma g))$ is bounded away from zer and infinity for $1 \le i \le p$. $\xi_{\alpha_{i,}}(a_1(g)a_2^{-1}(\gamma g))$ must be also.

The next lemma will not be needed until Section 5.

LEMMA 2.12. Suppose P and P' are two percuspidal subgroups and $\mathfrak{S}$ and $\mathfrak{S}'$ are associated Siegel domains. Let F and F' be two subsets, with the same number of elements, of the set of simple roots of $\mathfrak{f}$ and let ${}^*P$ and ${}^*P'$ be the cuspidal subgroups belonging to P and P' respectively determined by F and F'. If $0 \le b < 1$ there are constants t and t' so that if g belongs to $\mathfrak{S}$ and $\xi_\alpha(a(g)) > t$ when α does not belong to F and $\xi_\alpha^b(a(g)) > \xi_\beta(a(g))$ when β belongs to F and α does not, if g' belongs to $\mathfrak{S}'$ and satisfies the corresponding conditions, and if γ belongs to F and $\gamma g = g'$ then $\gamma\,{}^*P\gamma^{-1} = {}^*P'$. Moreover if $P = P'$ and, for some g in G, $g\,{}^*Pg^{-1} = {}^*P'$ and $g\,{}^*Sg^{-1} = {}^*S'$ then ${}^*P = {}^*P'$, ${}^*S = {}^*S'$.

Suppose, for the moment, merely that g belongs to $\mathfrak{S}$, g' belongs to $\mathfrak{S}'$, and $\gamma g = g'$. Choose u so that $u\gamma$ belongs to G_0 and so that $uP'u^{-1} = P$. Choosing v_i in V_i as above we see that

$$\xi_{\alpha_{,i}}^{d_i}(a(g))\,\|v_i\| = \xi_{\alpha_{,i}}^{d_i}(a(\gamma^{-1}u^{-1}))\,\|w_i\rho_i(u\gamma g)\| \ge c_i\xi_{\alpha_{,i}}^{d_i}(a(\gamma^{-1}u^{-1}))\xi_{\alpha_{,i}}^{d}(a'(g'))\,\|v_i\|$$

if w_i is such that

$$\xi_{\alpha_{,i}}^{d_i}(a(\gamma^{-1}u^{-1}))w_i = v_i\rho_i(\gamma^{-1}u^{-1})$$

Of course a similar inequality is valid if g and g' are interchanged. Since u may be supposed to lie in a finite set independent of γ we conclude as before that $a^{-1}(g)a'(g')$ lies in a compact set. Moreover, as in the proof of Lemma 2.11, γ^{-1} must belong to one of a finite number of left-cosets of Δ. Consequently w_i, $1 \leq i \leq p$, must belong to a finite subset of V_i and there must be a constant c so that

$$\| w_i\rho_i(ug'u^{-1}) \| \leq c\xi_{\alpha_{,i}}^{d} a'(g'))$$

Moreover, it follows from the proof of Lemma 2.10 that there are positive constants b and r so that if w_i is not a multiple of v_i then

$$\| w_i\rho_i(ug'u^{-1}) \| \geq b\xi_{\alpha_{i,}}(a'(g'))\xi_{\alpha_{,i}}^{d_i}(a'(g'))$$

Choose t' so large that bt' > c and choose t in an analogous fashion. If g and g' satisfy the conditions of the lemma then $\gamma^{-1}u^{-1}$ must belong to $\bigcap_{\alpha_{i,} \epsilon F} P_i$, where P_i is defined as in Lemma 2.10. It is easily seen that

$$^*P = \bigcap_{\alpha_{i,} \epsilon F} P_i$$

so $\gamma^{-1}u^{-1}$ belongs to $^*P$. Index the system of simple roots so that

$$\xi_{\alpha_{1,}}(a'(g')) \geq \xi_{\alpha_{2,}}(a'(g'))(\ldots \geq \xi_{\alpha_{p,}}(a'(g'))$$

There is an integer q so that $F' = \{\alpha_{q+1,}, \ldots, \alpha_{p,}\}$. If t' is very large then

$$\xi_{\alpha_{i,}}(a(g)) > \xi_{\alpha_{j,}}(a(g))$$

if $i \leq q < j$. Thus if $\beta_{1,}, \ldots, \beta_{p,}$ is the system of simple roots indexed so that

$$\xi_{\beta_1}(a(g)) \geq \xi_{\beta_2}(a(g)) \geq \ldots \geq \xi_{\beta_p}(a(g))$$

then

$$\{\beta_1, \ldots, \beta_q\} = \{\alpha_1, \ldots, \alpha_q\}$$

Since $\{\beta_{q+1}, \ldots, \beta_p\} = F$ the sets F and F' are equal and $u\,{}^*P'u^{-1} = {}^*P$. Then

$$\gamma^{-1}\,{}^*P'\gamma = \gamma^{-1}u^{-1}\,{}^*Pu\gamma = {}^*P$$

To prove the second assertion we observe that $({}^*P', {}^*S')$ belongs to (P, S) and to (gPg^{-1}, gSg^{-1}). We have proved while discussing the basic assumption that this implies that g belongs to ${}^*P'$.

The next lemma will not be needed until Section 6 when we begin to prove the functional equations for the Eisenstein series in several variables. Let P be a cuspidal subgroup of rank q with $\mathfrak{a}$ as split component. A set $\{\beta_1, \ldots, \beta_q\}$ of roots of $\mathfrak{a}$ is said to be a fundamental system if every other root can be written as a linear combination of $\beta_1, \ldots, \beta_q$ with integral coefficients all of the same sign. It is clear that if P_1 and P_2 are two cuspidal subgroups, g belongs to G, $\operatorname{Ad} g(\mathfrak{a}_1) = \mathfrak{a}_2$, and $B = \{\beta_1, \ldots, \beta_q\}$ is a fundamental system of roots for $\mathfrak{a}_2$ then

$$g^{-1}B = \{\beta_1 \circ \operatorname{Ad} g, \ldots, \beta_q \circ \operatorname{Ad} g\}$$

is a fundamental system. The Weyl chamber W_B associated to a fundamental system is

$$\{H \in \mathfrak{a} \mid \beta_i(H) > 0,\ 1 \leq i \leq q\}$$

so that

$$\operatorname{Ad}(g^{-1})W_B = W_{g^{-1}B}$$

It is clear that the Weyl chambers associated to two distinct fundamental systems are disjoint. The only fundamental system immediately at hand is the set of simple roots and the associated Weyl chamber is $\mathfrak{a}^+$. If P_1 and P_2 are as above we define $\Omega(\mathfrak{a}_1, \mathfrak{a}_2)$ to be the set of all linear transformations from $\mathfrak{a}_1$ to $\mathfrak{a}_2$ obtained by restricting $\mathrm{Ad}\, g$ to $\mathfrak{a}_1$ if g in G is such that $\mathrm{Ad}\, g(\mathfrak{a}_1) = \mathfrak{a}_2$. P_1 and P_2 are said to be associate if $\Omega(\mathfrak{a}_1, \mathfrak{a}_2)$ is not empty.

Suppose P_0 is a percuspidal subgroup and P is a cuspidal subgroup belonging to P_0. Let $\{a_{1,}, \ldots, a_{p,}\}$ be the set of simple roots for $\mathfrak{f}$ and suppose that P is determined by $\{a_{q+1,}, \ldots, a_{p,}\}$. If $1 \leq j \leq q$ let ${}^*P_j$ be the cuspidal subgroup determined by $\{a_{j,}, a_{q+1,}, \ldots, a_{p,}\}$. Suppose ${}^*\mathfrak{a}_j$ is contained in $\mathfrak{a}$. To prove the next lemma it is necessary to know that for each P and each j there is an element g in ${}^*M_j$ so that $\mathrm{Ad}\, g(\mathfrak{a} \cap {}^*\mathfrak{m}_j)$ is the split component of a cuspidal subgroup which belongs to $P_0 \cap {}^*M_j$ and so that if a is the unique simple root of $\mathfrak{a} \cap {}^*\mathfrak{m}_j$ then $a \circ \mathrm{Ad}\, g^{-1}$ is a negative root of $\mathrm{Ad}\, g(\mathfrak{a} \cap {}^*\mathfrak{m}_j)$. Unfortunately the only apparent way to show this is to use the functional equations for the Eisenstein series. Since the lemma is used to prove some of these functional equations a certain amount of care is necessary. Namely if $q > 1$ one needs only the functional equations of the Eisenstein series for the pairs $({}^*\Theta_j, {}^*M_j)$ and since the percuspidal subgroups have rank less than those for (Γ, G) one can assume them to be proved. On the other hand if $q = 1$ the lemma is not used in the proof of the functional equations. In any case we will take this fact for granted and prove the lemma. Everyone will be able to resolve the difficulty for himself once he has finished the paper.

LEMMA 2.13. Let P_0 be a percuspidal subgroup and let $F = \{P_1, \ldots, P_r\}$ be a complete family of associate cuspidal subgroups belonging to P_0. If P belongs to F and E is the collection of fundamental systems of roots of $\mathfrak{a}$ then $\mathfrak{a}$ is the closure of $\bigcup_{B \in E} W_B$. If $B \in E$ then there is a unique i, $1 \leq i \leq r$, and a unique s in $\Omega(\mathfrak{a}_i, \mathfrak{a})$ so that $s\,\mathfrak{a}_i^+ = W_B$.

Suppose as before that P is determined by $\{\alpha_{q+1}, \ldots, \alpha_p\}$. If $1 \le j \le q$ let g_j be one of the elements of ${}^*M_j$ whose existence was posited above. Denote the restriction of $\mathrm{Ad}\, g_j$ to $\mathfrak{a}$ by s_j and let $s_j(\mathfrak{a}) = \mathfrak{b}_j$. Denote the restriction of $\alpha_1, \ldots, \alpha_q$ to $\mathfrak{a}$ also by $\alpha_1, \ldots, \alpha_q$. Then $\alpha_j \circ s_j^{-1}$ restricted to $\mathfrak{b}_j \cap {}^*\mathfrak{m}_j$ is the unique simple root. Thus the simple roots $\beta_1, \ldots, \beta_q$ of $\mathfrak{b}_j$ can be so indexed that $\alpha_j \circ s_j^{-1} = -\beta_j$ and $\alpha_i \circ s_j^{-1} = \beta_i + b_{ij}\beta_j$ with $b_{ij} \ge 0$ if $i \ne j$. More conveniently, $\beta_j \circ s_j = -\alpha_j$ and $\beta_i \circ s_j = \alpha_i + b_{ij}\alpha_j$. To prove the first assertion it is enough to show that if H_0 belongs to $\mathfrak{a}$ and $\alpha(H_0) \ne 0$ for all roots α then there is some i and some s in $\Omega(\mathfrak{a}_i, \mathfrak{a})$ so that $s^{-1}(H_0)$ belongs to $\mathfrak{a}_i^+$. There is a point H_1 in $\mathfrak{a}^+$ so that the line through H_0 and H_1 intersects none of the sets

$$\{H \in \mathfrak{a} \mid \alpha(H) = \beta(H) = 0\}$$

where α and β are two linearly independent roots. If no such i and s exist let H_2 be the point closest to H_0 on the segment joining H_0 and H_1 which is such that the closed segment from H_1 to H_2 lies entirely in the closure of

$$\bigcup_{i=1}^{r} \bigcup_{s \in \Omega(\mathfrak{a}_i, \mathfrak{a})} s(\mathfrak{a}_i^+)$$

H_2 is not H_0. Let H_2 lie in the closure of $t\mathfrak{a}_k^+$ with t in $\Omega(\mathfrak{a}_k, \mathfrak{a})$. Replacing H_0 by $t^{-1}(H_0)$ and P by P_k if necessary it may be supposed that H_2 lies in the closure of $\mathfrak{a}^+$. Choose j so that

$$\alpha_\ell(H_2) > 0, \ 1 \le \ell \le q, \ \ell \ne j$$

and $\alpha_j(H_2) = 0$. Then $\alpha_j(H_0) < 0$ so that if H lies on the segment joining H_0 and H_2 and is sufficiently close to H_2 then $s_j H$ lies in $\mathfrak{b}_j^+$; this is a contradiction.

It is certainly clear that if B belongs to E then there is an i and an s in $\Omega(\mathfrak{a}_i, \mathfrak{a})$ so that $s\mathfrak{a}_i^+ = W_B$. Suppose that t belongs to $\Omega(\mathfrak{a}_k, \mathfrak{a})$ and

$t\mathfrak{a}_k^+ = W_B$. Then $s^{-1}t(\mathfrak{a}_k^+) = \mathfrak{a}_i^+$. If s is the restriction of Ad h to $\mathfrak{a}_i^+$ and t is the restriction of Ad g to $\mathfrak{a}_k^+$ then $h^{-1}gP_kg^{-1}h = P_i$. The previous lemma implies that $i = k$ and that $h^{-1}g$ belongs to P_i. Since the normalizer and centralizer of $\mathfrak{a}_i$ in P_i are the same it follows that $s^{-1}t$ is the identity.

If $\mathfrak{a}$ is as in the lemma the transformations $s_1, \dots, s_q$ just introduced will be called the reflections belonging, respectively, to $\alpha_1, \dots, \alpha_q$. We have proved that if $\mathfrak{a}$ and $\mathfrak{b}$ belong to $\{\mathfrak{a}_1, \dots, \mathfrak{a}_r\}$ then every element of $\Omega(\mathfrak{a}, \mathfrak{b})$ is a product of reflections; if s is the product of n but of no fewer reflections then n is called the length of s. Two refinements of this corollary will eventually be necessary; the first in the proof of Lemma 6.1 and the second in the proof of Lemma 7.4.

COROLLARY 1. Every s in $\Omega(\mathfrak{a}, \mathfrak{b})$ can be written as a product $s_n \dots s_2 s_1$ of reflections in such a way that if s_k lies in $\Omega(\mathfrak{a}_{i_k}, \mathfrak{a}_{j_k})$ and belongs to the simple root α_k of $\mathfrak{a}_{i_k}$ then $s_{k-1} \dots s_1(\mathfrak{a}_{i_1}^+)$ is contained in

$$\{H \in \mathfrak{a}_{i_k} \mid \alpha_k(H) > 0\}$$

Of course n is not necessarily the length of s. Let $W_B = s\,\mathfrak{a}^+$. Take a line segment joining a point in the interior of W_B to a point in the interior of $\mathfrak{b}^+$ which does not meet any of the sets $\{H \in \mathfrak{b} \mid \alpha(H) = \beta(H) = 0\}$ where α and β are two linearly independent roots. If the segment intersects only one Weyl chamber the result is obvious. The lemma will be proved by induction on the number, m, of Weyl chambers which it intersects. If m is greater than one let the segment intersect the boundary of $\mathfrak{b}^+$ at H_0. Index the simple roots $\beta_{1,}, \dots, \beta_{q,}$ of $\mathfrak{b}$ so that $\beta_{1,}(H_0) = 0$ and $\beta_{j,}(H_0) > 0$ if $j > 1$. Then if H belongs to $\mathfrak{b}^+$ the number $\beta_{1,}(sH)$ is negative so that if r is the reflection belonging to $\beta_{1,}$ the number $(-\beta_{1,} \circ r^{-1})(rsH)$ is positive. Let $t = rs$; if r belongs to $\Omega(\mathfrak{b}, \mathfrak{c})$ then t belongs to $\Omega(\mathfrak{a}, \mathfrak{c})$. Since there is a line segment connecting W_B and $r^{-1}(\mathfrak{c}^+)$

which meets only m-1 Weyl chambers, there is a line segment connecting $\mathfrak{i}^+$ and $t\,\mathfrak{a}^+ = rW_B$ which meets only m-1 Weyl chambers. If the corollary is true for t, say $t = s_{n-1} \cdots s_1$ and $s_n = r^{-1}$ then $s = s_n \cdots s_1$ and this product satisfies the conditions of the corollary.

Suppose $\mathfrak{a}_1, \ldots, \mathfrak{a}_r$ are, as in the lemma, split components of $P_1, \ldots, P_r$ respectively. Suppose that, for $1 \leq i \leq r$, S_i is a collection of m-dimensional affine subspaces of the complexification of $\mathfrak{a}_i$ defined by equations of the form $\alpha(H) = \mu$ where α is a root and μ is a complex number. If $\mathfrak{s}$ belongs to S_i and $\mathfrak{t}$ belongs to S_j we shall define $(\mathfrak{s}, \mathfrak{t})$ as the set of distinct linear transformations from $\mathfrak{s}$ to $\{H \mid -\overline{H} \in \mathfrak{t}\}$ obtained by restricting the elements of $\Omega(\mathfrak{a}_i, \mathfrak{a}_j)$ to $\mathfrak{s}$. Suppose that each $\mathfrak{s}$ in $S = \bigcup_{i=1}^r S_i$ is of the form $X(\mathfrak{s}) + \tilde{\mathfrak{s}}$ where $\tilde{\mathfrak{s}}$ is the complexification of a distinguished subspace of $\mathfrak{h}$ and the point $X(\mathfrak{s})$ is orthogonal to $\tilde{\mathfrak{s}}$; suppose also that for each $\mathfrak{s}$ in S the set $\Omega(\mathfrak{s}, \mathfrak{s})$ contains an element s_0 such that

$$s_0(X(\mathfrak{s}) + H) = -X(\mathfrak{s}) + H$$

for all H in $\tilde{\mathfrak{s}}$. Then if $r \in \Omega(\mathfrak{r}, \mathfrak{s})$ and $t \in \Omega(\mathfrak{s}, \mathfrak{t})$ the transformation ts_0r belongs to $\Omega(\mathfrak{r}, \mathfrak{t})$. Every element s of $\Omega(\mathfrak{s}, \mathfrak{t})$ defines an element of $\Omega(\tilde{\mathfrak{s}}, \tilde{\mathfrak{t}})$ in an obvious fashion. s is called a reflection belonging to the simple root α of $\tilde{\mathfrak{s}}$ if the element it defines in $\Omega(\tilde{\mathfrak{s}}, \tilde{\mathfrak{t}})$ is that reflection. It is easy to convince oneself of the following fact.

COROLLARY 2. Suppose that for every $\mathfrak{s}$ in S and every simple root α of $\tilde{S}$ there is a $\mathfrak{t}$ in S and a reflection in $\Omega(\mathfrak{s}, \mathfrak{t})$ which belongs to α. Then if $\mathfrak{s}$ and $\mathfrak{t}$ belong to S and s belongs to $\Omega(\mathfrak{s}, \mathfrak{t})$ there are reflections $r_n, \ldots, r_1$ so that if r_k belongs to $\Omega(\mathfrak{s}, \mathfrak{s}_k)$ and s_k in $\Omega(\mathfrak{s}_k, \mathfrak{s}_k)$ defines the identity in $\Omega(\tilde{\mathfrak{s}}_k, \tilde{\mathfrak{s}}_k)$ the transformation s equals the product $r_n s_{n-1} r_{n-1} \cdots r_2 s_1 r_1$.

As before the minimal value for n is called the length of s.

3. Cusp forms.

As usual the invariant measure on $\Gamma\backslash G$ is normalized by the condition that

$$\int_G \phi(g)dg = \int_{\Gamma\backslash G} \{\Sigma_\Gamma \phi(\gamma g)\}dg$$

If ϕ is a locally integrable function on $\Gamma\backslash G$, P is a cuspidal subgroup, and $T = N\Delta$, then

$$\phi^\wedge(g) = \int_{\Delta\backslash T} \phi(tg)dt = \int_{\Gamma\cap N\backslash N} \phi(ng)dn$$

is defined for almost all g. A function ϕ in $\mathfrak{L}(\Gamma\backslash G)$, the space of square integrable functions on $\Gamma\backslash G$, such that $\phi^\wedge(g)$ is zero for almost all g and all cuspidal subgroups except G itself will be called a cusp form. It is clear that the space of all cusp forms is a closed subspace of $\mathfrak{L}(\Gamma\backslash G)$ invariant under the action of G on $\mathfrak{L}(\Gamma\backslash G)$; it will be denoted by $\mathfrak{L}_0(\Gamma\backslash G)$. Before establishing the fundamental property of $\mathfrak{L}_0(\Gamma\backslash G)$ it is necessary to discuss in some detail the integral

$$(\lambda(f)\phi)(g) = \int_G \phi(gh)f(h)dh$$

when ϕ is a locally integrable function on $\Gamma\backslash G$ and f is a once continuously differentiable function on G with compact support.

Suppose P is a percuspidal subgroup of G and F is a subset of the set of simple roots of $\mathfrak{h}$. Let P_1 be the cuspidal subgroup belonging to P determined by the set F. Let

$$\phi_2(g) = \int_{\Gamma\cap N_1\backslash N_1} \phi(ng)dn$$

and let $\phi_1 = \phi - \phi_2$. Then $(\lambda(f)\phi)(g)$ equals

$$\int_G \phi(h)f(g^{-1}h) = \int_G \phi_1(h)f(g^{-1}h)dh + \int_G \phi_2(h)f(g^{-1}h)dh \tag{3.a}$$

The second integral will be allowed to stand. The first can be written as

$$\int_{N_1(\Gamma\cap N)\backslash G} \Sigma_{\delta\epsilon\Gamma\cap N_1\backslash\Gamma\cap N} \left\{\int_{\Gamma\cap N_1\backslash N_1} \phi_1(n\delta h)\Sigma_{\delta_1\epsilon\Gamma\cap N_1} f(g^{-1}\delta_1 n\delta h)dn\right\} dh$$

If we make use of the fact that

$$\phi_1(\delta nh) = \phi_1(nh)$$

and

$$\int_{\Gamma\cap N_1\backslash N_1} \phi_1(nh)dn = 0$$

this repeated integral can be written as

$$\int_{N_1(\Gamma\cap N)\backslash G} \left\{\int_{\Gamma\cap N_1\backslash N_1} \phi_1(nh)f(g,\ nh)dn\right\} dh$$

with

$$f(g,\ h) = \Sigma_{\delta\epsilon\Gamma\cap N} f(g^{-1}\delta h) - \Sigma_{\Gamma\cap N_1\backslash\Gamma\cap N} \int_{N_1} f(g^{-1}n\delta h)dn$$

$$= \int_{\Gamma\cap N_1\backslash N_1} \Sigma_{\delta\epsilon\Gamma\cap N}\{f(g^{-1}\delta h) - f(g^{-1}\delta nh)\}dn$$

It should be recalled that N_1 is a normal subgroup of N and $\Gamma\cap N_1$ a normal subgroup of $\Gamma\cap N$.

If $\mathfrak{S} = \mathfrak{S}(t,\ \omega)$ is a Siegel domain associated to P it is necessary to estimate $f(g,\ h)$ when g is in $\mathfrak{S}(t,\ \omega)$. It may be supposed that $(\Gamma\cap N)\omega$ contains N. Since $f(g,\ \delta h) = f(g,\ h)$ if $\delta\ \epsilon\ \Gamma\cap N$ we can take $h = n_1a_1m_1k_1$ with n_1 in $\omega\cap N$, a_1 in A, m_1 in M, and k_1 in K. Suppose $g = sak = a(a^{-1}sa)k = au$ with s in ω and a in $A^+(t,\ \infty)$; then u lies in a compact set U_1 which depends on ω and t. The integrand in the expression for $f(g,\ h)$ equals

$$\Sigma_{\delta\epsilon\Gamma\cap N}\{f(u^{-1}a^{-1}\delta ah_1) - f(u^{-1}a^{-1}\delta nah_1)\}$$

with $h_1 = a^{-1}h$. If ω_1 is a compact subset of N_1 such that $(\Gamma\cap N_1)\omega_1 = N_1$ it is enough to estimate this sum for n in ω_1. Let U be a compact set containing the support of f. If a given term of this sum is to be different from zero either $a^{-1}\delta ah_1$ or $a^{-1}\delta nah_1$ must belong to U_1U. Then either

$$(a^{-1}\delta aa^{-1}n_1 a)(a^{-1}m_1 a_1)$$

or

$$(a^{-1}\delta naa^{-1}n_1 a)(a^{-1}m_1 a_1)$$

belongs to $P \cap U_1 UK$. It follows that there is a compact set V in N depending only on $\mathfrak{r}$ and U so that $a^{-1}\delta a$ belongs to V. Choose a conditionally compact open set V_1 in N so that if δ belongs to $\Gamma \cap N$ and $\delta V_1 \cap V_1$ is not empty then $\delta = 1$; there is a compact set V_2 in N so that $a^{-1}V_1 a$ is contained in V_2 if a belongs to $A^+(t, \infty)$. If $a^{-1}\delta a$ belongs to V then δV_1 is contained in aVV_2a^{-1}. Consequently the number of terms in the above sum which are different from zero is at most a constant times the measure of aVV_2a^{-1} and a simple calculation shows that this is at most a constant times $\omega^{-2}(a)$. Finally there is a compact subset ω_2 of AM so that every term of the above sum vanishes unless $m_1 a_1$ belongs to $a\omega_2$.

If $\{X_i\}$ is a basis of $\mathfrak{g}$ there is a constant μ so that $|\lambda(X_i)f(g)| \leq \mu$ for all i and g. If $X \in \mathfrak{g}$ then $\lambda(X)f(g)$ is defined to be the value of $\frac{df}{dt}(g \text{ expt } X)$ at $t = 0$. Then

$$|f(u^{-1}a^{-1}\delta ah_1) - f(u^{-1}a^{-1}\delta nah_1)|$$

is less than or equal to

$$\int_0^1 |\lambda(\mathrm{Ad}(h_1^{-1})\mathrm{Ad}(a^{-1})X)f(u^{-1}a^{-1}\delta \text{ expt } X ah_1)|dt$$

if $n = \exp X$. Since n lies in a fixed compact set so does X. Moreover

$$h_1 = a^{-1}n_1 aa^{-1}m_1 a_1 k_1$$

lies in a compact set depending only on $\mathfrak{r}$ and U. Consequently the right hand side is less than a constant, depending only on $\mathfrak{S}$, U, and μ, times the largest eigenvalue of $\mathrm{Ad}(a^{-1})$ on N_1. In conclusion there is a constant c, depending only on $\mathfrak{S}$, U, and μ, so that for all g in $\mathfrak{S}$ and all h

$$|f(g, h)| \leq c\omega^{-2}(a(g))\{\min_{\alpha_{i,} \notin F} \xi_{\alpha_{i,}}(a(g))\}^{-1}$$

Moreover the first integral in expression for $\lambda(f)\phi(g)$ is equal to

$$\int_{a\omega_2 \times K} \omega^2(b) \left\{ \int_{N_1(\Gamma\cap N)\backslash N} \left\{ \int_{\Gamma \cap N_1 \backslash N_1} \phi_1(n_1 nbmk) f(g, n_1 nbmk) dn_1 \right\} dn \right\} dbdmdk$$

or, as is sometimes preferable,

$$\int_{a\omega_2 \times K} \omega^2(b) \left\{ \int_{\Gamma\cap N\backslash N} \phi_1(nbmk) f(g, nbmk) dn \right\} dbdmdk$$

The absolute value of the first integral is at most

$$c\omega^{-2}(a(g)) \left\{ \min_{\alpha_{i,} \notin F} \xi_{\alpha_{i,}}(a(g)) \right\}^{-1} \int_{a\omega_2 \times K} \omega^2(b) \left\{ \int_{\Gamma\cap N\backslash N} |\phi_1(nbmk)| dn \right\} dbdmdk$$

If ω_3 is a compact subset of N such that $(\Gamma \cap N)\omega_3 = N$ this expression is at most

(3.b) $$c\omega^{-2}(a(g)) \left\{ \min_{\alpha_{i,} \in F} \xi_{\alpha_{i,}}(a(g)) \right\}^{-1} \int_{\omega_3 a\omega_2 K} |\phi_1(h)| dh$$

For the same reasons the absolute value of the second integral is at most

(3.c) $$c\omega^{-2}(a(g)) \left\{ \min_{\alpha_i \notin F} \xi_{\alpha_i}(a(g)) \right\}^{-1} \int_{\omega_3 a\omega_2 K} |\phi(h)| dh$$

LEMMA 3.1. Let ϕ belong to $\mathcal{L}_0(\Gamma\backslash G)$, let f be a once continuously differentiable function with compact support, and let P be a percuspidal subgroup. If $\mathfrak{S} = \mathfrak{S}(t, \omega)$ is a Siegel domain associated to P there is a constant c depending only on $\mathfrak{S}$ and f so that for g in $\mathfrak{S}$

$$|\lambda(f)\phi(g)| \leq c_1\omega^{-1}(a(g))\eta^{-1}(a(g)) \|\phi\|$$

Here $\|\phi\|$ is the norm of ϕ in $\mathcal{L}(\Gamma\backslash G)$ and if a belongs to A then

$$\eta(a) = \max_{1 \leq i \leq p} \xi_{\alpha_{i,}}(a)$$

It is enough to establish the inequality on each

$$\mathcal{S}_i = \{g \in \mathcal{S} \mid \xi_{\alpha_{i,}}(a(g)) \geq \xi_{\alpha_{j,}}(a(g)),\ 1 \leq j \leq p\}$$

For simplicity take $i = 1$. In the above discussion take $F = \{\alpha_{j,} \mid j \neq 1\}$. The second term in (3.a) is zero so to estimate $\lambda(f)\phi(g)$ we need only estimate (3.b). The integral is at most

$$\left\{\int_{\omega_3 a \omega_2 K} dh\right\}^{\frac{1}{2}} \left\{\int_{\omega_3 a \omega_2 K} |\phi(h)|^2 dh\right\}^{\frac{1}{2}}$$

Since $\omega_3 a \omega_2 K$ is contained in a fixed Siegel domain $\mathcal{S}(t', \omega')$ for all a in $A^+(t, \infty)$ the second integral is at most a constant times $\|\phi\|^2$. If ω_4 and ω_5 are compact subsets of A and M respectively so that ω_2 is contained in $\omega_4\omega_5$ the first integral is at most

$$\left\{\int_{\omega_3} dn\right\} \left\{\int_{\omega_4} \omega^2(ab)db\right\} \left\{\int_{\omega_5} dm\right\}$$

Since

$$\min_{\alpha_{i,} \notin F} \xi_{\alpha_i}(a) = \xi_{\alpha_{1,}}(a) = \eta(a(g))$$

if $g = sak$ is in $\mathcal{S}_1$ the lemma follows.

It is a standard fact that $\lambda(f)$ is a bounded linear operator on $L(\Gamma\backslash G)$. It is readily seen to leave $\mathcal{L}_0(\Gamma\backslash G)$ invariant.

COROLLARY. <u>If</u> f <u>is once continuously differentiable with compact support then the restriction of</u> $\lambda(f)$ <u>to</u> $\mathcal{L}_0(\Gamma\backslash G)$ <u>is a compact operator</u>.

Since $\omega^{-1}(a)\eta^{-1}(a)$ is square integrable on any Siegel domain the corollary follows immediately from Ascoli's lemma, the above lemma, and the fact that $\Gamma\backslash G$ is covered by a finite number of Siegel domains. The significance of the corollary is seen from the following lemma.

LEMMA 3.2. *Let G be a locally compact group and π a strongly continuous unitary representation of G on the separable Hilbert space $\mathfrak{H}$. Suppose that for any neighbourhood U of the identity in G there is an integrable function f on G with support in U such that*

$$f(g) \geq 0,\ f(g) = f(g^{-1}),\ \int_G f(g)dg = 1$$

and $\pi(f)$ is compact; then $\mathfrak{H}$ is the orthogonal direct sum of countably many invariant subspaces on each of which there is induced an irreducible representation of G. Moreover no irreducible representation of G occurs more than a finite number of times in $\mathfrak{H}$.

Of course $\pi(f)$ is defined by

$$\pi(f)v = \int_G f(g)\pi(g)vdg$$

if v belongs to V. Consider the families of closed mutually orthogonal subspaces of V which are invariant and irreducible under the action of G. If these families are ordered by inclusion there will be a maximal one. Let the direct sum of the subspaces in some maximal family be $\mathfrak{M}$. In order to prove the first assertion it is necessary to show that $\mathfrak{M}$ equals $\mathfrak{H}$. Suppose the contrary and let $\mathfrak{M}'$ be the orthogonal complement of $\mathfrak{M}$ in $\mathfrak{H}$. Choose a v in $\mathfrak{M}'$ with $\|v\| = 1$ and choose U so that $\|v - \pi(g)v\| < \frac{1}{2}$ if g is in U. Choose f as in the statement of the lemma. Then $\|\pi(f)v - v\| < \frac{1}{2}$ so that $\pi(f)v \neq 0$. The restriction of $\pi(f)$ to $\mathfrak{M}'$ is self-adjoint and thus has a non-zero eigenvalue μ. Let $\mathfrak{M}'_\mu$ be the finite-dimensional space of eigenfunctions belonging to the eigenvalue μ. Choose from the family of non-zero subspaces of $\mathfrak{M}'_\mu$ obtained by intersecting $\mathfrak{M}'_\mu$ with closed invariant subspaces of $\mathcal{W}'$ a minimal one $\mathcal{W}'_0$. Take the intersection $\mathcal{V}_0$ of all closed invariant subspaces of $\mathcal{W}'$ containing $\mathcal{W}'_0$. Since $\mathcal{V}_0 \neq \{0\}$ a contradiction will result if it is shown that $\mathcal{V}_0$ is irreducible. If $\mathcal{V}_0$ were not then it would be the orthogonal direct sum of two closed invariant subspaces $\mathcal{V}_1$ and $\mathcal{V}_2$. Since $\mathcal{V}_i \cap \mathcal{W}'_\mu$ is contained in $\mathcal{V}_0 \cap \mathcal{W}'_\mu = \mathcal{W}'_0$ for $i = 1$ and 2 the space $V_i \cap W'_\mu$ is either $\{0\}$ or $\mathcal{W}'_0$. But $\pi(f)\mathcal{V}_i \subseteq \mathcal{V}_i$ so

$$\mathcal{W}'_0 = (\mathcal{V}_1 \cap \mathcal{W}'_\mu) \oplus (\mathcal{V}_2 \cap \mathcal{W}'_\mu)$$

and, consequently, $\mathcal{V}_i \cap \mathcal{W}'_\mu = \mathcal{W}'_\mu$ for i equal to 1 or 2. This is impossible. The second assertion follows from the observation that if some irreducible representation occurred with infinite multiplicity then, for some f, $\pi(f)$ would have a non-zero eigenvalue of infinite multiplicity.

Before proceeding to the next consequence of the estimates (3.b) and (3.c) we need a simple lemma.

LEMMA 3.3. *Let* $\mathfrak{S}^{(1)}, \ldots, \mathfrak{S}^{(m)}$ *be Siegel domains, associated to the percuspidal subgroups* $P^{(1)}, \ldots, P^{(m)}$ *respectively, which cover* $\Gamma\backslash G$. *Suppose* c *and* r *are real numbers and* $\phi(g)$ *is a locally integrable function on* $\Gamma\backslash G$ *such that*

$$|\phi(g)| \leq c\eta^r(a^{(i)}(g))$$

if g *belongs to* $\mathfrak{S}^{(i)}$. *If* ${}^*P$ *is a cuspidal subgroup and*

$${}^*\phi^\wedge(a, m, k) = \int_{\Gamma \cap {}^*N \backslash {}^*N} \phi(namk^{-1})dn$$

for a *in* ${}^*A$, m *in* ${}^*M$, *and* k *in* K *then there is a constant* r_1, *which does not depend on* ϕ, *so that for any compact set* C *in* ${}^*A$, *any percuspidal subgroup* ${}^\Psi P$ *of* ${}^*M$, *and any Siegel domain* ${}^\Psi S$ *associated to* ${}^\Psi P$ *there is a constant* c_1, *which does not depend on* ϕ, *so that*

$$|{}^*\phi^\wedge(a, m, k)| \leq c_1 \eta^{r_1}({}^\Psi a(m))$$

if a *belongs to* C *and* m *belongs to* ${}^\Psi\mathfrak{S}$. *In particular if* ${}^*P = G$ *then* r_1 *can be taken equal to* r.

If ω is a compact subset of ${}^*N$ so that $(\Gamma \cap {}^*N)\omega = {}^*N$ then

$$|{}^*\phi^\wedge(a, m, k)| \leq \sup_{n\epsilon\omega} |\phi(namk^{-1})|$$

If $g = namk^{-1}$ choose γ in Γ so that γg belongs to $\mathfrak{S}^{(i)}$ for some i. According to the corollary to Lemma 2.11 there is a constant r_2 so that for any C, ${}^\Psi P$, and ${}^\Psi\mathfrak{S}$ there is a constant c_2 so that

$$\eta(a^{(i)}(\gamma g)) \leq c_2 \eta({}^\Psi a(m))^{r_2}$$

Since $\eta(a^{(i)}(g))$ is bounded below on $\mathfrak{S}^{(i)}$ for each i, it can be supposed for the first assertion that $r \geq 0$. Then take $r_1 = rr_2$ and $c_1 = cc_2^r$. If ${}^*P$ is G the lemma also asserts that if $\mathfrak{S}$ is any Siegel domain associated to a percuspidal subgroup P then there is a constant c_1 so that $|\phi(g)| \leq c_1\eta^r(a(g))$ on $\mathfrak{S}$. Given g in $\mathfrak{S}$ again choose γ in Γ so that γg belongs to $\mathfrak{S}^{(i)}$ for some i. The corollary to Lemma 2.11 asserts that there is a number c_2 independent of i and g so that

$$c_2^{-1} \leq \eta^{-1}(a(g))\eta(a^{(i)}(g)) \leq c_2$$

Take $c_1 = cc_2^r$ if $r \geq 0$ and take $c_1 = cc_2^{-r}$ if $r < 0$.

LEMMA 3.4. *Suppose* $\mathfrak{S}^{(1)}, \ldots, \mathfrak{S}^{(m)}$ *are Siegel domains, associated to percuspidal subgroups, which cover* $\Gamma\backslash G$. *Suppose that* $\phi(g)$ *is a locally integrable function on* $\Gamma\backslash G$ *and that there are constants* c *and* r *so that* $|\phi(g)| \leq c\eta^r(a^{(i)}(g))$ *if* g *belongs to* $S^{(i)}$. *Let* U *be a compact subset of* G, *let* μ *be a constant, let* $\{X_i\}$ *be a basis of* $\mathfrak{g}$, *and let* f(g) *be a once continuously* differentiable function on G with support in U such that $|\lambda(X_i)f(g)| \le \mu$ for all g and i. If $\mathfrak{S}$ is a Siegel domain associated to the percuspidal subgroup P and if k is a non-negative integer there is a constant c_1, depending on c, r, U, μ, $\mathfrak{S}$, and k but not on ϕ or f, so that

$$|\lambda^k(f)\phi(g) - \lambda^k(f)\phi_i^\wedge(g)| \le c_1 \eta^{r-k}(a(g))$$

on

$$\mathfrak{S}_i = \{g \in \mathfrak{S} \,|\, \xi_{\alpha_{i,}}(a(g)) \ge \xi_{\alpha_{j,}}(a(g)),\ 1 \le j \le p\}$$

In accordance with our notational principles

$$\phi_i^\wedge(g) = \int_{\Gamma \cap N_i \backslash N_i} \phi(ng)dn$$

if P_i is the percuspidal subgroup belonging to P determined by $\{\alpha_{j,} \,|\, j \ne i\}$. The assertion of the lemma is certainly true for $k = 0$. The proof for general k will proceed by induction. For simplicity take $i = 1$. Since

$$(\lambda^k(f)\phi)_i^\wedge = \lambda^k(f)\phi_i^\wedge$$

it will be enough to show that if there is a constant s so that for any Siegel domain $\mathfrak{S}$ associated to P there is a constant c' such that $|\phi_1^\wedge(g)| \le c'\eta^s(a(g))$ on $\mathfrak{S}_1$ then for any $\mathfrak{S}_1$ there is a constant c_1' so that

$$|\lambda(f)\phi(g) - \lambda(f)\phi_1^\wedge(g)) \le c_1'\eta^{s-1}(a(g))$$

on S_1. Of course it will also have to be shown that the constants c_1' do not depend on f or ϕ. Indeed we apply this assertion first to ϕ with $s = r$ and then in general to $\lambda^k(f)\phi$ with $s = r-k$. Since

$$\lambda(f)\phi(g) - \lambda(f)\phi_1^\wedge(g)$$

is nothing but the first term on the right side of (3.a) it can be estimated by means of (3.b). Thus

$$|\lambda(f)\phi(g) - \lambda(f)\phi_1^{\wedge}(g)| \leq c_2\omega^{-2}(a(g))\xi^{-1}_{\alpha_{1,}}(a(g))\int_{\omega_3 a\omega_2 K} |\phi_1(h)|dh$$

if g belongs to $\mathfrak{S}_1$. First observe that if g belongs to $\mathfrak{S}_1$ then

$$\eta(a(g)) = \xi_{\alpha_{1,}}(a(g))$$

There is a Siegel domain $\mathfrak{S}'$ so that when $a = a(g)$ and g belongs to $\mathfrak{S}$ the set $\omega_3 a\omega_2 K$ belongs to $\mathfrak{S}'$. Let ω_4 and ω_5 be compact subsets of A and M respectively so that ω_2 is contained in $\omega_4\omega_5$; then the integral is less than or equal to a constant, which does not depend on ϕ, times

$$\int_{\omega_4} \omega^2(ab)\eta^s(ab)db$$

which is certainly less than a constant times $\omega^2(a)\eta^s(a)$.

COROLLARY. <u>Suppose</u> V <u>is a finite-dimensional subspace of</u> $\mathcal{L}_0(\Gamma\backslash G)$ <u>invariant under</u> $\lambda(f)$ <u>for</u> f <u>continuous with compact support and such that</u> $f(kgk^{-1}) = f(g)$ <u>for all</u> g <u>in</u> G <u>and all</u> k <u>in</u> K. <u>Then given any real number</u> r <u>and any Siegel domain</u> $\mathfrak{S}$ <u>associated to a percuspidal subgroup</u> P <u>there is a constant</u> c <u>so that, for all</u> ϕ <u>in</u> V <u>and all</u> g <u>in</u> $\mathfrak{S}$,

$$|\phi(g)| \leq c\eta^r(a(g))\,\|\phi(g)\|$$

Since for a given t there are constants c_1 and r_1 so that $\omega^{-1}(a) \leq c_1\eta^{r_1}(a)$ for a in $A^+(t, \infty)$ the corollary will follow from Lemmas 3.1 and 3.3 if it is shown that there is a once continuously differentiable function f_0 satisfying the conditions of the lemma so that $\lambda(f_0)\phi = \phi$ for all ϕ in V. Let $\{\phi_1, \ldots, \phi_n\}$ be an orthonormal basis for V and let U be a neighbourhood of the identity in G so that

$$\|\lambda(g)\phi_i - \phi_i\| < (2n)^{-1}$$

if g belongs to U and $1 \leq i \leq n$. Then, for any ϕ in $\mathcal{H}$,

$$\|\lambda(g)\phi - \phi\| \leq \tfrac{1}{2}\|\phi\|$$

if g is in U. Choose f to be a non-negative function, once continuously differentiable with support in u, such that $\int_G f(g)dg = 1$ and $f(kgk^{-1}) = f(g)$ for all g in G and all k in K. Then the restriction of $\lambda(f)$ to V is invertible. Thus there is a polynomial p with no constant term so that $p(\lambda(f))$ is the identity on V. In the group algebra $p(f)$ is defined; set $f_0 = p(f)$. If V was not a space of square integrable functions but a space of continuous functions and otherwise satisfied the conditions of the lemma then a simple modification of the above argument would show the existence of the function f_0.

If P is a cuspidal subgroup then the pair M, Θ satisfies the same conditions as the pair G, Γ. It will often be convenient not to distinguish between functions on $\Theta\backslash M$, $T\backslash S$, and $AT\backslash P$. Also every function ϕ on G defines a function on $P \times K$ by $\phi(p, k) = \phi(pk^{-1})$. Since $G = PK$ functions on G may be identified with functions on $P \times K$ which are invariant under right translation by (k, k) if k belongs to $K \cap P$. If V is a closed invariant subspace of $\mathcal{L}(\Theta\backslash M)$ let $\mathcal{E}(V)$ be the set of measurable functions Φ on $AT\backslash G$ such that $\Phi(mg)$ belongs to V as a function of m for each fixed g in G and

$$\int_{\Theta\backslash M\times K} |\Phi(mk)|^2 dmdk = \|\Phi\|^2 < \infty$$

If H belongs to $\mathfrak{a}_c$, the complexification of $\mathfrak{a}$, and Φ belongs to $\mathcal{E}(V)$ consider the function

$$\exp(\langle H(h), H\rangle + \rho(H(h)))\Phi(h)$$

on G. If g belongs to G it is not difficult to see that there is another function

$\Phi_1(h)$ in $\mathcal{E}(V)$ so that

$$\exp(\langle H(hg), H\rangle + \rho(H(hg)))\Phi(hg) = \exp(\langle H(h), H\rangle + \rho(H(h)))\Phi_1(h)$$

Φ_1 depends on Φ, g, and H. If we set $\Phi_1 = \pi(g, H)\Phi$ then $\pi(g, H)$ is a bounded linear transformation from $\mathcal{E}(V)$ to $\mathcal{E}(V)$, $\pi(g_1 g_2, H) = \pi(g_1, H)\pi(g_2, H)$, and $\pi(1, H) = I$. In fact it is easy to see that $\pi(g, H)$ is a strongly continuous representation of G on $\mathcal{E}(V)$ for each H in $\mathfrak{a}_c$. The representation is unitary if H is purely imaginary. If f is a continuous function on G with compact support then $\pi(f, H)$ can be defined as usual by

$$\pi(f, H)\Phi = \int_G f(g)\pi(g, H)\Phi dg$$

It is readily seen that for almost all g

$$\exp(\langle H(g), H\rangle + \rho(H(g)))(\pi(f, H)\Phi)(g)$$

is equal to

$$\int_G \exp(\langle H(gh), H\rangle + \rho(H(gh)))\Phi(gh)f(h)dh$$

If F is a finite set of irreducible representations of K let W be the space of functions on K spanned by the matrix elements of the representations in F. W will be called an admissible subspace of the space of functions on K. Let $\mathcal{E}(V, W)$ be the space of functions Φ in $\mathcal{E}(V)$ such that, for almost all g, $\Phi(gk)$ belongs to W, that is, agrees with an element of W except on a set of measure zero. With no loss it may be assumed that it always belongs to W. $\mathcal{E}(V, W)$ is just the space of functions Φ in $\mathcal{E}(V)$ such that the space spanned by $\{\lambda(k)\Phi \mid k \in K\}$ is finite dimensional and contains only irreducible representations of K equivalent to those in F. If f is a continuous function on G with compact support and $f(kgk^{-1}) = f(g)$ for all g and k then $\pi(f, H)$ leaves $\mathcal{E}(V, W)$ invariant.

Suppose ${}^*P$ is a cuspidal subgroup belonging to P. If Φ belongs to $\mathcal{E}(V)$ define a function on ${}^*M \times K$ by $\Phi({}^*m, k) = \Phi({}^*pk^{-1})$ if ${}^*p$ in ${}^*P$ projects onto

${}^*m$. Since $G = {}^*PK$ this defines an isomorphism of $\mathcal{E}(V)$ with a space of functions on ${}^*M \times K$. Indeed let ${}^{\Psi}P = {}^*N\backslash P \cap {}^*S$ then ${}^{\Psi}P$ is a cuspidal subgroup of ${}^*M$ and ${}^{\Psi}M$ is the same as M. Also ${}^{\Psi}P \times K$ is a cuspidal subgroup of ${}^*M \times K$ for the group ${}^*\Theta \times \{1\}$. If L is the space of square integrable functions on K then the image of $\mathcal{E}(V)$ is the set of all functions in $\mathcal{E}(V \otimes L)$ which are invariant under right translations by (k^*, k) where k belongs to $K \cap {}^*P$ and ${}^*k$ is the projection of k on ${}^*M$. Denote the group of such elements by ${}^*K_0$ and let ${}^*K$ be the projection of $K \cap {}^*P$ on ${}^*M$. The group ${}^*K$ plays the same role for ${}^*M$ as K does for G. Suppose Φ belongs to $\mathcal{E}(V, W)$ then

$$\Phi({}^*m\,{}^*k_1, kk_2) = \Phi({}^*m\,{}^*k_1 k_2^{-1} k^{-1})$$

for fixed ${}^*m$ and k this function belongs to the space of functions on ${}^*K \times K$ of the form $\phi(k_1 k_2^{-1})$ with ϕ in W. A typical element of W is of the form σ_{ij}, that is, the matrix element of a representation in F. Since

$$\sigma_{ij}(k_1 k_2^{-1}) = \Sigma_{\ell}\sigma_{i\ell}(k_1)\sigma_{\ell j}(k_2^{-1})$$

it belongs to the space W^* if W^* is the space of functions on ${}^*K \times K$ spanned by the matrix elements of those irreducible representations of ${}^*K \times K$ obtained by taking the tensor product of an irreducible representation of $K \cap {}^*P$ contained in the restriction to $K \cap {}^*P$, which is isomorphic to ${}^*K$, of one of the representations in F with a representation of K contragredient to one of the representations in F. Thus the image of $\mathcal{E}(V, W)$ is contained in $\mathcal{E}(V \otimes L, W^*)$; indeed it is readily seen to be contained in $\mathcal{E}(V \otimes W, W^*)$ and to be the space of all functions in $\mathcal{E}(V \otimes W, W^*)$ invariant under right translation by elements of ${}^*K_0$. On occasion it will be convenient to identify $\mathcal{E}(V, W)$ with this subspace.

Since the representation of M on $\mathcal{L}(\Theta\backslash M)$ is strongly continuous there is associated to each element X in the centre $\mathfrak{Z}'$ of the universal enveloping algebra of $\mathfrak{m}$ a closed operator $\lambda(X)$ on $\mathcal{L}(\Theta\backslash M)$. Indeed if π is any strongly continuous representation of M on a Hilbert space $\mathcal{L}$ there is associated to each X in $\mathfrak{Z}'$ a closed operator $\pi(X)$. If $\mathcal{L}$ is irreducible then

$$\mathcal{L} = \oplus_{j=1}^{n} \mathcal{L}_j$$

where each $\mathcal{L}_j$ is invariant and irreducible under the action of M_0, the connected component of M. The restriction of $\pi(X)$ to $\mathcal{L}_j$ is equal to a multiple, $\xi_j(X)I$, of the identity. The map $X \longrightarrow \xi_j(X)$ is a homomorphism of $\mathfrak{Z}'$ into the complex numbers. Let us say that the representation belongs to the homomorphism ξ_j. Suppose the closed invariant subspace V is a direct sum $\oplus V_i$ of closed, mutually orthogonal subspaces V_i each of which is invariant and irreducible under the action of M. As we have just remarked each V_i is a direct sum

$$\oplus_{j=1}^{n_i} V_{ij}$$

of subspaces invariant and irreducible under the action of M_0. Suppose V_{ij} belongs to the homomorphism ξ_{ij}. V will be called an admissible subspace of $\mathcal{L}(\Theta\backslash M)$ if V is contained in $\mathcal{L}(\Theta\backslash M)$ and there is only a finite number of distinct homomorphisms in the set $\{\xi_{ij}\}$.

LEMMA 3.5. _If_ V _is an admissible subspace of_ $\mathcal{L}_0(\Theta\backslash M)$ _and_ W _is an admissible subspace of the space of functions on_ K _then_ $\mathcal{E}(V, W)$ _is finite dimensional_.

In the discussion above take $^*P$ equal to P. Then $\mathcal{E}(V, W)$ is isomorphic to a subspace of $\mathcal{E}(V \otimes W, W^*)$. It is readily seen that $V \otimes W$ is an admissible subspace of $\mathcal{L}_0(\Theta \times \{1\}\backslash M \times K)$ and that W^* is an admissible subspace of the space of functions on $^*K \times K$. Since it is enough to show that $\mathcal{E}(V \otimes W, W^*)$ is finite dimensional we have reduced the lemma to the case that P and M are equal to G. Suppose $V = \oplus V_i$. If $V' = \oplus V_i$ where the second sum is taken over those V_i which contain vectors transforming according to one of the representations in F

then $\mathcal{E}(V, W) = \mathcal{E}(V', W)$. In other words it can be supposed that each V_i contains vectors transforming under K according to one of the representations in F. For each i let

$$V_i = \oplus_{j=1}^{n_i} V_{ij}$$

where each V_{ij} is invariant and irreducible under the action of G_0, the connected component of G, and belongs to the homomorphism ξ_{ij}. It is known ([10], Theorem 3) that there are only a finite number of irreducible unitary representations of G_0 which belong to a given homomorphism of $\mathfrak{Z}$, the centre of the universal enveloping algebra of $\mathfrak{g}$, and which contain vectors transforming according to a given irreducible representation of $K \cap G_0$. Thus there is a finite set E of irreducible representations of G_0 so that for each i there is a j so that the representation of G_0 on V_{ij} is equivalent to one of the representations in E. As a consequence of Lemma 3.2 applied to G_0 there are only a finite number of V_i. It is known however (cf. [10], Theorem 4) that for each i the space of functions in V_i transforming according to one of the representations in F is finite dimensional. This completes the proof of the lemma. Since $\mathcal{E}(V, W)$ is finite-dimensional it follows from the proof of the corollary to Lemma 3.4 that it can be considered as a space of continuous functions.

Suppose $\phi(g)$ is a continuous function on $T\backslash G$ such that for each g in G the function $\phi(mg)$ on $\Theta\backslash M$ belongs to V and the function $\phi(gk)$ on K belongs to W. For each a in A consider the function $\phi(sak)$ on $T\backslash S \times K$ or on $AT\backslash P \times K$. If k_0 belongs to $K \cap P = K \cap S$ then $\phi(sk_0^{-1}ak_0k) = \phi(sak)$ since $sk_0^{-1}ak_0a^{-1}s^{-1}$ is in N. Thus it defines a function $\Phi'(a)$ on $AT\backslash G$ which is seen to belong to $\mathcal{E}(V, W)$. The space of all such functions ϕ for which $\Phi'(\cdot)$, which is a function on A with values in $\mathcal{E}(V, W)$, has compact support will be called $\mathcal{D}(V, W)$.

LEMMA 3.6. *Suppose* V *is an admissible subspace of* $\mathcal{L}_0(\Theta\backslash M)$ *and* W *is an admissible subspace of functions on* K. *If* ϕ *belongs to* $\mathcal{H}(V, W)$ *then* $\Sigma_{\Delta\backslash\Gamma}\,\phi(\gamma g)$ *is absolutely convergent; its sum* $\phi^\wedge(g)$ *is a function on* $\Gamma\backslash G$. *If* $\mathfrak{S}_0$ *is a Siegel domain associated to a percuspidal subgroup* P_0 *and if* r *is a real number there is a constant* c *so that* $|\phi^\wedge(g)| \leq c\eta^r(a_0(g))$ *for* g *in* $\mathfrak{S}_0$.

There is one point in Section 6 where we will need a slightly stronger assertion than that of the lemma. It is convenient to prove it at the same time as we prove the lemma.

COROLLARY. *Let* $\phi(g)$ *be a function on* $T\backslash G$ *and suppose that there is a constant* t *so that* $\phi(namk) = 0$ *unless* a *belongs to* $A^+(t, \infty)$. *Let* $P_1, \ldots, P_m$ *be percuspidal subgroups to which* P *belongs and suppose that there are Siegel domains* ${}^\Psi\mathfrak{S}_1, \ldots, {}^\Psi\mathfrak{S}_m$ *associated to* ${}^\Psi P_i = N\backslash P_i \cap S$ *which cover* $\Theta\backslash M$. *Suppose that there is a constant* s *so that given any constant* r_1 *there is a constant* c_1 *so that, for* $1 \leq i \leq m$,

$$|\phi(namk)| \leq c_1\eta^s(a)\eta^{r_1}({}^\Psi a_i(m))$$

if m *belongs to* ${}^\Psi\mathfrak{S}_i$. *Finally suppose that there are constants* u *and* b *with* $1 \geq b > 0$ *so that* $\phi(namk) = 0$ *if* $\eta(a) > u$ *and the projection of* m *on* $\Theta\backslash M$ *belongs to the projection on* $\Theta\backslash M$ *of*

$$\{m \in {}^\Psi\mathfrak{S}_i \,|\, \eta({}^\Psi a_i(m)) < \eta^b(a)\}$$

for some i. *Then*

$$\Sigma_{\Delta\backslash\Gamma}\,\phi(\gamma g) = \phi^\wedge(g)$$

is absolutely convergent and if $\mathfrak{S}_0$ *is a Siegel domain associated to a percuspidal subgroup* P_0 *and* r *is a real number there is a constant* c *so that* $|\phi^\wedge(g)| \leq c\eta^r(a_0(g))$ *for* g *in* $\mathfrak{S}_0$.

It is a consequence of Lemma 3.5 and the corollary to Lemma 3.4 that the function of the lemma satisfies the conditions of the corollary. Let ω be a compact subset of N such that $(\Gamma \cap N)\omega = N$. If g is in $\mathfrak{S}_0$ let U be the set of all elements γ in Γ such that $\gamma g = namk$ with n in ω, a in $A^+(t, \infty)$, m in ${}^\Psi\mathfrak{S}_i$ for some i, and k in K. Since any left coset of Δ in Γ contains an element γ so that $\gamma g = namk$ with n in ω, a in A, m in ${}^\Psi\mathfrak{S}_i$ for some i, and k in K and since $\phi(namk) = 0$ unless a belongs to $A^+(t, \infty)$ it is enough to estimate

$$\Sigma_{\gamma \in U} |\phi(\gamma g)|$$

We first estimate the number of elements in $U(v)$ which is the set of all γ in U such that $\gamma g = namk$ with n in ω, a in $A^+(t, \infty)$, m in ${}^\Psi\mathfrak{S}_i$ for some i and such that $\eta({}^\Psi a_i(m)) \leq v$, and k in K. Suppose ${}^\Psi\mathfrak{S}_i = {}^\Psi\mathfrak{S}_i({}^\Psi\omega_i, {}^\Psi t_i)$ and let $\omega_i = \omega\,{}^\Psi\omega_i$. If γ belongs to $U(v)$ then, for some i, $\gamma g = n_i a a_i k_i$ with n_i in ω_i, a in $A^+(t, \infty)$, a_i in ${}^\Psi A_i^+({}^\Psi t_i, \infty)$ and such that $\eta(a_i) \leq v$, and k in K. Since a_i is considered as an element of ${}^\Psi A_i$ the number $\eta(a_i)$ is the maximum of $\xi_\alpha(a_i)$ as α varies over the simple roots of ${}^\Psi\mathfrak{n}_i$. Consider the point aa_i in A_i. Let $\alpha_{1}, \ldots, \alpha_{q}$ be the simple roots of $\mathfrak{h}$ which vanish on $\mathfrak{n}$; then

$$\xi_{\alpha_j}(aa_i) = \xi_{\alpha_j}(a_i) \geq {}^\Psi t_i$$

for $1 \leq j \leq q$. If $j > q$ then

$$\xi_{\alpha_j}(a_i) = \prod_{k=1}^{q} \xi_{\alpha_k}^{\delta_k}(a_i)$$

with $\delta_k \leq 0$; thus if $\delta = \Sigma_{k=1}^{q} \delta_k$ then

$$\xi_{\alpha_j}(aa_i) = \xi_{\alpha_j}(a)\xi_{\alpha_j}(a_i) \geq t\eta^\delta(a_i) \geq tv^\delta$$

Consequently γg is contained in the Siegel domain $\mathfrak{S}_i(\omega_i, tv^\delta)$ associated to P_i if $tv^\delta \leq \min\{{}^\Psi t_i, \ldots, {}^\Psi t_m\}$. In any case it follows from Lemma 2.11 that there are constants c_2 and r_2 which are independent of g so that $U(v)$ has at most $c_2 v^{r_2}$ elements. If $\phi(namk)$ is not zero either $\eta(a) \leq u$ or $\eta({}^{\Psi}a_i(m)) \geq \eta^b(a)$, where $\eta(a)$ is the maximum of $\xi_\alpha(a)$ as α varies over the simple roots of $\mathfrak{n}$. Consequently given any number r_1 there is a constant c_1' so that

$$|\phi(namk)| \leq c_1' \eta^{r_1}({}^{\Psi}a_i(m))$$

If $N(g)$ is the largest integer such that $\gamma g = namk$ with n in ω, a in $A^+(t, \infty)$, m in ${}^{\Psi}\mathfrak{S}_i$ for some i, k in K, and $\phi(\gamma g) \neq 0$, implies $\eta({}^{\Psi}a_i(m)) \geq N(g)$ then

$$\Sigma_{\gamma \in u} |\phi(\gamma g)| \leq c_1' c_2 \Sigma_{n=N(g)}^{\infty} (n+1)^{r_2} n^{r_1}$$

which in turn is at most

$$-c_1' c_2 2^{r_2} (r_1 + r_2 - 1)(N(g)+1)^{r_1 + r_2 + 1}$$

if $N(g) > 1$, $r_1 < 0$, $r_2 > 0$, and $r_1 + r_2 + 1 < 0$. Since the corollary to Lemma 2.11 implies that there are positive constants c_3 and r_3 so that

$$N(g) + 1 \geq c_3 \eta^{r_3}(a_0(g))$$

the lemma and corollary are proved.

Let P be a cuspidal subgroup and let $\phi(g)$ be a measurable function on $\Gamma \backslash G$. Suppose that given any Siegel domain ${}^{\Psi}\mathfrak{S}$ associated to a percuspidal subgroup ${}^{\Psi}P$ of M and any compact subset C of A there are constants c and r so that

$$|\phi(namk)| \leq c\eta^r({}^{\Psi}a(m))$$

if a belongs to C and m belongs to ${}^{\Psi}\mathfrak{S}$. If V is an admissible subspace of $\mathcal{L}_0(\Theta \backslash M)$ and W is an admissible subspace of the space of functions on K and if ψ belongs to $\mathfrak{H}(V, W)$ then

$$\int_{T\backslash G} \psi(g)\bar{\phi}(g)dg$$

is convergent. If it vanishes for all choices of V and W and all ψ then we say that the cuspidal component of ϕ is zero.

LEMMA 3.7. *Let* $\mathfrak{S}^{(1)}, \ldots, \mathfrak{S}^{(m)}$ *be Siegel domains, associated to the percuspidal subgroups* $P^{(1)}, \ldots, P^{(m)}$ *respectively, which cover* $\Gamma\backslash G$. *Suppose that* $\phi(g)$ *is a continuous function on* $\Gamma\backslash G$ *and there are constants* c *and* r *so that*

$$|\phi(g)| \leq c\eta^{r}(a^{(i)}(g))$$

if g *belongs to* $\mathfrak{S}^{(i)}$. *If the cuspidal component of*

$$\phi^{\wedge}(g) = \int_{\Gamma\cap N\backslash N} \phi(ng)dn$$

is zero for every cuspidal subgroup P *then* $\phi(g)$ *is identically zero.*

It is a consequence of Lemma 3.3 that it is meaningful to speak of the cuspidal component of $\phi^{\wedge}$ being zero. The lemma will be proved by induction on the rank of the percuspidal subgroups of G. If they are of rank 0 so that $\Gamma\backslash G$ is compact then ϕ is itself a cusp form. It follows from Lemma 3.2 and the corollary to Lemma 3.1 that the subspace of $\mathcal{L}(\Gamma\backslash G)$ spanned by the spaces $\mathcal{E}(V, W)$ with V an admissible subspace of $\mathcal{L}_0(\Gamma\backslash G)$ and W an admissible subspace of the space of functions on K is dense in $\mathcal{L}(\Gamma\backslash G)$. Since in this case $\vartheta(V, W) = \mathcal{E}(V, W)$ and $\phi^{\wedge}(g) = \phi(g)$ when $P = G$ the assumptions of the lemma imply that ϕ is orthogonal to every element of $\mathcal{L}(\Gamma\backslash G)$ and is consequently zero.

If the rank of the percuspidal subgroups of G is p suppose that the lemma is true when the percuspidal subgroups are of rank less than p. Let ${}^{*}P$ be a cuspidal subgroup and consider

$$^{*}\phi^{\wedge}(a, m, k) = \int_{\Gamma \cap {}^{*}N \backslash {}^{*}N} \phi(namk^{-1})dn$$

According to Lemma 3.3 $^{*}\phi^{\wedge}(a, m, k)$ is for each fixed a in $^{*}A$ a function on $^{*}\Theta \times \{1\} \backslash {}^{*}M \times K$ which satisfies the given conditions on its rate of growth on Siegel domains of $^{*}M \times K$. If $^{\Psi}P$ is a cuspidal subgroup of $^{*}M$ there is a cuspidal subgroup P to which $^{*}P$ belongs so that $^{\Psi}P = {}^{*}N \backslash P \cap {}^{*}S$. Then

$$\begin{aligned} {}^{\Psi}({}^{*}\phi^{\wedge})^{\wedge}(a, m, k) &= \int_{{}^{*}\Theta \cap {}^{\Psi}N \backslash {}^{\Psi}N} {}^{*}\phi^{\wedge}(a, nm, k)dn \\ &= \int_{{}^{*}\Theta \cap {}^{\Psi}N \backslash {}^{\Psi}N} dn \left\{ \int_{\Gamma \cap {}^{*}N \backslash {}^{*}N} \phi(n_1 namk^{-1})dn_1 \right\} dn \\ &= \int_{\Gamma \cap N \backslash N} \phi(namk^{-1})dn \end{aligned}$$

so that

$$(3.d) \qquad {}^{\Psi}({}^{*}\phi^{\wedge})^{\wedge}(a, m, k) = \phi^{\wedge}(amk^{-1})$$

Suppose that V' is an admissible subspace of $\mathcal{L}_0(\Theta \times \{1\} \backslash M \times K)$ and W' is an admissible subspace of the space of functions on $^{*}K \times K$. As in the remarks preceding Lemma 3.5 $^{*}K$ is the projection on $^{*}M$ of $K \cap {}^{*}P$. If ψ belongs to $\mathcal{V}(V', W')$ then

$$\int_{{}^{\Psi}T \times \{1\} \backslash {}^{*}M \times K} \psi(m, k)\bar{\phi}^{\wedge}(amk^{-1})dmdk$$

is equal to

$$\int_{{}^{\Psi}T \times \{1\} \backslash {}^{*}M \times K} \left\{ \int_{{}^{*}K_0} \psi(mk_0, kk_0)dk_0 \right\} \bar{\phi}^{\wedge}(amk^{-1})dmdk$$

This equality will be referred to as (3.e). Suppose $\zeta(a)$ is a continuous function on $^{*}A$ with compact support then we can define a function $\xi(g)$ on $T \backslash G$ by setting

$$\xi(namk^{-1}) = \zeta(a) \int_{{}^{*}K_0} \psi(mk_0, kk_0)$$

If F' is the set of irreducible representations of ${}^*K \times K$ whose matrix elements span W' let F be a finite set of irreducible representations of K which contains the representations contragredient to the irreducible representations of K occurring in the restrictions of the representations of F' to K. If W is the space of functions on K spanned by the matrix elements of the representation in F then, for each g in G, $\xi(gk)$ is a function in W. It is also easy to see that there is an admissible subspace V of $\mathcal{L}_0(\Theta\backslash M)$ so that V' is contained in $V \otimes W$ if F is suitably chosen; then $\xi(g)$ belongs to $\mathcal{V}(V, W)$. Consequently

$$\int_{{}^*A} \omega^2(a)\zeta(a) \left\{ \int_{{}^\Psi T\times\{1\}\backslash {}^*M\times K} \psi(m, k)\overline{\phi}^\wedge(amk^{-1})dmdk \right\} da$$

is equal to

$$\int_{T\backslash G} \xi(g)\overline{\phi}^\wedge(g)dg = 0$$

Since $\zeta(a)$ is arbitrary we conclude that the left side of (3.e) is zero and hence that for each a in ${}^*A$ the function ${}^*\phi^\wedge(a, m, k)$ on ${}^*M \times K$ satisfies the conditions of the lemma. By the induction assumption ${}^*\phi^\wedge(a, m, k)$, and hence ${}^*\phi^\wedge(g)$, is identically zero if the rank of ${}^*P$ is positive.

Suppose $f_1, \ldots, f_\ell$ are once continuously differentiable functions on G with compact support. Let $\phi_1 = \lambda(f_1) \ldots \lambda(f_k)\phi$. It follows from Lemma 3.4 that there is a constant c_1 so that

$$|\phi_1(g)| \leq c_1\eta^{r-\ell}(a^{(i)}(g))$$

if g belongs to $\mathcal{F}^{(i)}$, $1 \leq i \leq m$. Let ℓ be some fixed integer greater than r so that $\phi_1(g)$ is bounded and hence square integrable on $\Gamma\backslash G$. If P is a cuspidal subgroup different from G then

$$\phi_1^\wedge = \lambda(f_1) \ldots \lambda(f_k)\phi^\wedge = 0$$

so that ϕ_1 is a cusp form. $f_1, \ldots, f_\ell$ can be so chosen that $f_j(kgk^{-1}) = f(g)$ for all g and all k and for $1 \leq j \leq \ell$ and $\phi_1(h)$ is arbitrarily close to $\phi(h)$ for any given h in G. Consequently if it can be shown that ϕ_1 is identically zero for all such $f_1, \ldots, f_\ell$ it will follow that ϕ is identically zero. Suppose V is an admissible subspace of $\mathcal{L}_0(\Gamma\backslash G)$, W is an admissible subspace of the space of functions on K, and ψ belongs to $\mathcal{E}(V, W)$; then

$$\int_{\Gamma\backslash G} \psi(g)\overline{\phi}_1(g)dg = \int_{\Gamma\backslash G} \lambda(f_\ell^*) \ldots \lambda(f_1^*)\psi(g)\overline{\phi}(g)dg = 0$$

since $\lambda(f_\ell^*) \ldots \lambda(f_1^*)\psi$ also belongs to $\mathcal{E}(V, W)$. The functions f_j^* are defined by $f_j^*(g) = \overline{f}_j(g^{-1})$. Since, as follows from Lemma 3.2, the space spanned by the various $\mathcal{E}(V, W)$ is dense in $\mathcal{L}_0(\Gamma\backslash G)$ the function ϕ_1 must be identically zero.

We also see from the above proof that if $\phi(g)$ satisfies the first condition of the lemma and if the cuspidal component of $\phi^\wedge$ is zero for all cuspidal subgroups of rank at least q then $\phi^\wedge$ is identically zero for all cuspidal subgroups of rank at least q. Let us now prove a simple variant of the above lemma which will be used in Section 4.

COROLLARY. <u>Suppose that</u> ϕ <u>belongs to</u> $\mathcal{L}(\Gamma\backslash G)$ <u>and that if</u> P <u>is any cuspidal subgroup,</u> V <u>an admissible subspace of</u> $\mathcal{L}_0(\Theta\backslash M)$, W <u>an admissible subspace of the space of functions of</u> K, <u>and</u> ψ <u>an element of</u> $\mathcal{V}(V, W)$ <u>then</u>

$$\int_{\Gamma\backslash G} \psi^\wedge(g)\overline{\phi}(g)dg = 0$$

<u>The function</u> ϕ <u>is then zero.</u>

It is enough to show that if f is a once continuously differentiable function on G with compact support such that $f(kgk^{-1}) = f(g)$ for all g and k then $\lambda(f)\phi$ is identically zero. Let $\phi_1 = \lambda(f)\phi$ then, if ψ belongs to $\mathcal{V}(V, W)$ and $\psi_1 = \lambda(f^*)\psi$,

$$\int_{T\backslash G} \psi(g)\overline{\phi}_1^\wedge(g)dg = \int_{T\backslash G} \psi_1(g)\overline{\phi}^\wedge(g)dg$$

$$= \int_{\Delta\backslash G} \psi_1(g)\overline{\phi}(g)dg$$

$$= \int_{\Gamma\backslash G} \psi_1^\wedge(g)\overline{\phi}(g)dg$$

$$= 0$$

since ψ_1 also belongs to $\mathcal{V}(V, W)$. If we can obtain a suitable estimate on ϕ_1 we can conclude from the lemma that ϕ_1 is identically zero. But $\lambda(f)\phi(g)$ is equal to

$$\int_G \phi(h)f(g^{-1}h)dh = \int_{\Gamma\backslash G} \phi(h)\{\Sigma_\Gamma f(g^{-1}\gamma h)\}dh$$

Consequently $|\lambda(f)\phi(g)|$ is at most

$$\left\{\int_{\Gamma\backslash G} |\phi(h)|^2 dh\right\}^{\frac{1}{2}} \left\{\int_{\Gamma\backslash G} dh\right\}^{\frac{1}{2}} \left\{\sup_{h\in G} \Sigma_\Gamma |f(g^{-1}\gamma h)|\right\}$$

Let U be the support of f and suppose that for all h in G the set $\{\gamma | \gamma h \in gU\}$ has at most $N(g)$ elements; then the above expression is less than or equal to a constant times $N(g)$. Let $\mathfrak{S}_0 = \mathfrak{S}_0(\omega, t)$ be a Siegel domain associated to the percuspidal subgroup P_0; at the cost of increasing the size of $\mathfrak{S}_0$ it may be supposed that $\Delta_0\omega = S_0$. Let ω_1 and ω_2 be compact subsets of S_0 and A_0 respectively so that KU is contained in $\omega_1\omega_2K$. Choose a number t' so that $A_0^+(t', \infty)$ contains the product of ω_2 and $A_0^+(t, \infty)$ and let $\mathfrak{S}'_0 = \mathfrak{S}_0(\omega, t')$. Every element γ' of Γ such that $\gamma' h$ belongs to gU can be written as a product $\delta\gamma$ so that γh belongs to $\mathfrak{S}'_0$ and $\delta\omega \cap \omega a_0 \omega_1 a_0^{-1}$ is not empty if $a_0 = a_0(g)$. It follows from Lemma 2.11 that the number of choices for γ is bounded independent of h. The condition on δ is that $a_0^{-1}\delta a_0$ is contained in $a_0^{-1}\omega a_0 \omega_1 a_0^{-1}\omega^{-1} a_0$. But the union over all a_0 in $A_0^+(t, \infty)$ of these sets is contained in a compact set. We conclude first of all that the projection of δ on $M = M\backslash S$ must belong to a fixed compact set and therefore must be one of a finite number of points. Consequently δ

can be written as a product $\delta_1\delta_2$ where δ_2 is one of a finite set of points, δ_1 belongs to $\Gamma \cap N_0$, and $a_0^{-1}\delta_1 a_0$ belongs to a fixed compact subset of N_0. The discussion preceding Lemma 3.1 shows that the number of choices for δ_1 is at most a constant times $\omega^{-2}(a_0)$. Thus, on $\mathcal{S}_0$, $N(g)$ can be taken as a constant times $\omega^{-2}(a_0(g))$. The required estimate is now established.

4. Eisenstein Series.

Let P be a cuspidal subgroup of G, let V be an admissible subspace of $\mathcal{L}(\Theta\backslash M)$, and let W be the space of functions on K spanned by the matrix elements of some representation of K. It will follow from Lemma 4.3 that $\mathcal{E}(V, W)$ is finite-dimensional and thus by the argument used in Section 3 that every element of $\mathcal{E}(V, W)$ in continuous. We assume then that $\mathcal{E}(V, W)$ is a finite-dimensional space of continuous functions. If Φ is an element of $\mathcal{E}(V, W)$ and H belongs to $\mathfrak{a}_c$, the series

$$\Sigma_{\Delta\backslash\Gamma} \exp(\langle H(\gamma g), H\rangle + \rho(H(\gamma g)))\Phi(\gamma g) \tag{4.a}$$

is called an Eisenstein series.

LEMMA 4.1. *The series (4.a) converges uniformly absolutely on compact subsets of the Cartesian product of*

$$\mathfrak{U} = \{H \in \mathfrak{a}_c \mid \mathrm{Re}\,\alpha_{i,}(H) > \langle \alpha_{i,}, \rho\rangle, \ 1 \leq i \leq \text{rank } P\}$$

and G_0. *If the sum is* $E(g, \Phi, H)$ *then* $E(g, \Phi, H)$ *is infinitely differentiable as a function of* g *and* H *and is analytic as a function of* H *for each fixed* g. *Moreover if* P_0 *is a percuspidal subgroup of* G *and* $\mathfrak{S}_0$ *a Siegel domain associated to* P_0 *there is a locally bounded function* $c(H)$ *on* $\mathfrak{U}$ *which depends only on the real part of* H *so that, for* g *in* $\mathfrak{S}_0$,

$$|E(g, \Phi, H)| \leq c(H)\exp(\langle H_0(g), \mathrm{Re}\,H\rangle + 2\rho(H_0(g)) - \rho(H'_0(g)))$$

where $H'_0(g)$ *is the projection of* $H_0(g)$ *on* $\mathfrak{a}$.

Let $\mathfrak{D}$ be the universal enveloping algebra of $\mathfrak{g}$. The map

$$Y \longrightarrow \frac{df}{dt}(g \exp t Y) = \lambda(Y)f(g)$$

of $\mathfrak{g}$ into the space of left-invariant vector fields on G can be extended to an isomorphism $X \longrightarrow \lambda(X)$ of $\mathfrak{b}$ with the algebra of left-invariant differential operators on G and the map

$$Y \longrightarrow \frac{df}{dt}(\exp(-tY)g) = \lambda'(Y)$$

of $\mathfrak{g}$ into the space of right-invariant vector fields on G can be extended to an isomorphism $X \longrightarrow \lambda'(X)$ of $\mathfrak{b}$ with the algebra of right invariant differential operators on G. If f is an infinitely differentiable function on G with compact support and if

$$F(g, \Phi, H) = \exp(\langle H(g), H\rangle + \rho(H(g)))\Phi(g)$$

with Φ in $\mathcal{E}(V, W)$, then as we have observed above

$$\lambda(f)F(g, \Phi, H) = \exp(\langle H(g), H\rangle + \rho(H(g)))F(g, \pi(f, H)\Phi, H) \tag{4.b}$$

It is easily verified that if $\phi(g)$ is any locally integrable function on G then

$$\lambda(X)\lambda(f)\phi(g) = \lambda(\lambda'(X)f)\phi(g)$$

Arguing as in the corollary to Lemma 3.3 we see that for a given H_0 there is an infinitely differentiable function f_0 with compact support so that $f_0(kgk^{-1}) = f_0(g)$ for all g in G and all k in K and so that $\pi(f_0, H_0)$ is the identity on $\mathcal{E}(V, W)$. For H close to H_0, $\pi(f_0, H)$ is non-singular and we see from (4.b) that for any such H

$$\lambda(X)F(g, \Phi, H) = \exp(\langle H(g), H\rangle + \rho(H(g)))F(g, \pi(X, H)\Phi, H)$$

if we define $\pi(X, H)$ to be $\pi(\lambda'(X)f_0, H)\pi^{-1}(f_0, H)$. Of course $\pi(X, H)$ is independent of the choice of f_0. The map $(X, \Phi) \longrightarrow \pi(X, H)\Phi$ can be extended to a linear map of $B \otimes \mathcal{E}(V, W)$ into $\mathcal{E}(V)$. If $\mathfrak{b}_m$ is the space spanned by

$$\{X_1 \dots X_k \mid k \leq m, X_i \in \mathfrak{g}, 1 \leq i \leq k\}$$

then $\mathfrak{b}_m$ is invariant under the adjoint group of G. If $k \in K$ and ϕ is a differentiable function then

$$\lambda(\mathrm{Ad}\,k(X))\lambda(k)\phi(g) = \lambda(k)\lambda(X)\phi(g)$$

so that the map of $\mathfrak{b}_m \otimes \mathcal{E}(V, W)$ into $\mathcal{E}(V)$ commutes with K. If W_1 is the space of functions on K which is spanned by the matrix elements of the representation of K in $\mathfrak{b}_m \otimes W$ and if the degree of X is at most m then $\pi(X, H)\Phi$ belongs to $\mathcal{E}(V, W_1)$. Consequently the second assertion of the lemma follows immediately from the first.

To prove the last assertion we will estimate the series

$$\Sigma_{\Delta\backslash\Gamma} |\exp(\langle H(\gamma g), H\rangle + \rho(H(\gamma g)))\Phi(\gamma g)|$$

which equals

$$\Sigma_{\Delta\backslash\Gamma} \exp(\langle H(\gamma g), \mathrm{Re}\,H\rangle + \rho(H(\gamma g)))\,|\Phi(\gamma g)| \tag{4.c}$$

so that it may as well be supposed that H is real. To prove the first assertion it is enough to show that the second series is uniformly convergent on compact subsets of $\mathfrak{U} \times G$. It follows from Lemma 2.5 that if C is a compact subset of G there is a constant μ so that

$$\alpha_{,i}(H(\gamma g)) \leq \mu, \quad 1 \leq i \leq q$$

for γ in Γ and g in C. q is of course the rank of P. If C_1 is a compact subset of $\mathfrak{U}$ and if H_0 is such that $\alpha_{,i}(H_0) \leq \mathrm{Re}\,\alpha_{,i}(H)$ for all H in C_1 and $1 \leq i \leq q$ then

$$|\exp(\langle H(\gamma g), H\rangle + \rho(H(\gamma g)))| \leq c \exp(\langle H(\gamma g), H_0\rangle + \rho(H(\gamma g)))$$

for all H in C_1 and all g in C. c is some constant depending on μ. To prove the first assertion it is then enough to prove that the series (4.c) converges uniformly for H_0 fixed and for g in a compact subset of G.

Given H_0 choose $f_0(g)$ as above so that $\pi(f_0, H)$ is the identity on $\mathcal{E}(V, W)$ then

$$F(\gamma g, \Phi, H_0) = \int_G F(\gamma g, \Phi, H_0) f_0(g^{-1}h)dh$$

Let C_2 be the support of f_0 and let $C_3 = CC_2$; then if g belongs to C the series on the right is dominated by

$$M\Sigma_{\Delta\backslash\Gamma}\int_{C_3} |F(\gamma h, \Phi, H_0)| dh$$

if $M = \sup_{h\epsilon G} |f(h)|$. If the number of elements in $\{\gamma | \gamma g \epsilon C_3\}$ is less than or equal to N for all g in G and if C_4 is the projection of C_3 on $\Gamma\backslash G$ the sum above is at most N times

$$\int_{C_4} \Sigma_{\Delta\backslash\Gamma} |F(\gamma h, \Phi, H_0)| dh \leq \int_{C_5} |F(h, \Phi, H_0)| dh$$

where C_5 is the projection on $T\backslash G$ of ΓC_3. To prove the first assertion it has merely to be shown that the integral on the right is finite. Before doing this we return to the last assertion. If H is in a sufficiently small neighbourhood of H_0 then $\pi(f_0, H)$ is non-singular on $\mathcal{E}(V, W)$ and if $\Phi \epsilon \mathcal{E}(V, W)$ then

$$2 \| \pi(f, H_0)\Phi \| \geq \| \Phi \|$$

Given Ψ in $\mathcal{E}(V, W)$ and H in this neighbourhood choose Φ so that $\pi(f_0, H)\Phi = \Psi$. Then

$$|F(\gamma g, \Psi, H)| \leq \int_G |F(\gamma h, \Phi, H)| |f_0(g^{-1}h)| dh$$

so that to estimate the series (4. c) and establish the last assertion it will be enough to show that there is a locally bounded function $c_1(H)$ on $\mathcal{U}$ so that for g in $\mathcal{Y}_0$, Φ in $\mathcal{E}(V, W)$, and H real and in $\mathcal{U}$

(4.d) $$\Sigma_{\Delta\backslash\Gamma}\int_G |F(h, \Phi, H)|\,|f_0(g^{-1}\gamma^{-1}h)|\,dh$$

is at most

$$c_1(H)\,\|\Phi\|\exp(\langle H_0(g), H\rangle + 2\rho(H_0(g)) - \rho(H'_0(g)))$$

The expression (4.d) equals

$$\int_{\Delta\backslash G} |F(h, \Phi, H)|\{\Sigma_\Gamma |f_0(g^{-1}\gamma^{-1}h)|\}dh \leq c\omega^{-2}(a_0(g))\int_{C(g)} |F(h, \Phi, H)|dh$$

if g is in $\mathcal{Y}_0$. The set $C(g)$ is the projection on $T\backslash G$ of $\Gamma g C_2$ and c is some constant. The inequality is a consequence of the estimate used to prove the corollary to Lemma 3.7. Lemma 2.10 can be used to prove that $\Gamma g C_2$ is contained in

$$\{s \exp(H + H'_0(g))k \mid s \in S,\ k \in K,\ H \in {}^+\mathfrak{n}(-\infty, \mu)\}$$

where μ is some constant. The integral is at most $\exp(\langle H'_0(g), H\rangle - \rho(H'_0(g)))$ times

$$\int_{{}^+\mathfrak{n}(-\infty,\mu)} \exp(\langle X, H\rangle - \rho(X))\,|dX|\int_{\Theta\backslash M\times K} |\Phi(mk)|\,dmdk$$

The second integral is at most $\mu(\Theta\backslash M)^{\frac{1}{2}}\|\Phi\|$ and the first is a constant times

$$\prod_{i=1}^{q}\{(\alpha_{i,}(H) - \langle\alpha_{1,}, \rho\rangle)^{-1}\exp\mu\alpha_{1,}(H)\}$$

This completes the proof of both the first and the last assertion.

Two remarks should now be made. The first is that if C is a compact subset of $\Gamma\backslash G$ and ε is a positive number there is a constant c and a point H_0 in $\mathfrak{n}$ so that if

$$\mathrm{Re}(\alpha_{i,}(H)) > \langle\alpha_{i,}, \rho\rangle + \varepsilon$$

for $1 \leq i \leq q$ and Φ is in $\mathcal{E}(V, W)$ then, for g in C,

$$|E(g, \Phi, H)| \leq c\|\Phi\| \exp\langle H_0, \operatorname{Re} H\rangle$$

The second is that if X belongs to $\mathfrak{Z}$, then

$$\lambda(X)E(g, \Phi, H) = E(g, \pi(X, H)\Phi, H)$$

Both statements have been essentially proved in the course of proving the above lemma.

We can in particular choose V and W to be the space of constant functions on M and K respectively. It is clear that if $\Phi(g) \equiv 1$ and H is real then

$$E(g, \Phi, H) \geq F(g, \Phi, H)$$

This observation will allow us to prove a variant of Lemma 4.1 which will be used in the proof of the functional equations for the Eisenstein series in several variables. Suppose that ${}^*P$ is a cuspidal subgroup belonging to P and $\phi(g)$ a function on ${}^*A\,{}^*T\backslash G$. ${}^\dagger\mathfrak{a}$, the orthogonal complement of ${}^*\mathfrak{a}$ in $\mathfrak{a}$, can be regarded as the split component of ${}^\dagger P = {}^*N\backslash P \cap {}^*S$. It is contained in ${}^\dagger\mathfrak{h}$, the orthogonal complement of ${}^*\mathfrak{a}$ in $\mathfrak{h}$, which in turn can be regarded as the split component of the percuspidal subgroups of ${}^*M$. Suppose that there is a point ${}^\dagger H$ in ${}^\dagger\mathfrak{a}$ so that if ${}^\dagger\mathfrak{S}_0$ is a Siegel domain associated to the percuspidal subgroup ${}^\dagger P_0$ of ${}^*M$ then

$$|\phi(mk)| \leq c \exp(\langle {}^\dagger H_0(m), {}^\dagger H\rangle + \rho({}^\dagger H_0'(m))$$

if m belongs to ${}^\dagger\mathfrak{S}_0$ and k belongs to K. ${}^\dagger H_0'(m)$ is the projection of ${}^\dagger H_0(m)$ on ${}^\dagger\mathfrak{a}$. Suppose ${}^*H$ belongs to ${}^*\mathfrak{a}_c$. Let us verify that the series

$$\Sigma_{{}^*\Delta\backslash\Gamma} \exp(\langle {}^*H(\gamma g), {}^*H\rangle + \rho({}^*H(\gamma g))\phi(\gamma g)$$

converges absolutely if $H = {}^*H + {}^\dagger H$ belongs to $\mathfrak{A}$. Suppose that $P_{01}, \ldots, P_{0r}$

are percuspidal subgroups of G to which ${}^*P$ belongs and ${}^{\Psi}\mathfrak{S}_1, \dots, {}^{\Psi}\mathfrak{S}_r$ are Siegel domains of ${}^*M$, associated to the groups ${}^{\Psi}P_{01}, \dots, {}^{\Psi}P_{0r}$ respectively, such that $\bigcup_{i=1}^r {}^{\Psi}\mathfrak{S}_i$ covers ${}^*\Theta \backslash {}^*M$. Let $P_1, \dots, P_r$ be the cuspidal subgroups with the split component $\mathfrak{a}$ belonging to $P_{01}, \dots, P_{0r}$ respectively. The function $|\phi(g)|$ is bounded by a constant multiple of

$$\Sigma_{i=1}^r \Sigma_{\Delta_i \backslash {}^*\Delta} \exp(\langle H_i(\delta g), {}^{\Psi}H\rangle + \rho({}^{\Psi}H'_{0i}(\delta g))$$

which equals

$$\Sigma_{i=1}^r \Sigma_{{}^{\Psi}\Delta_i \backslash {}^*\Theta} \exp(\langle {}^{\Psi}H_i(\theta m), {}^{\Psi}H\rangle + \rho(H'_{0i}(\theta)))$$

if $g = namk$ with n in ${}^*N$, a in ${}^*A$, m in ${}^*M$, and k in K and if ${}^{\Psi}H'_{0i}(g)$ is the projection of $H_{0i}(g)$ on ${}^{\Psi}\mathfrak{a}$. Since

$$\langle H_i(g),\ H\rangle + \rho(H_i(g)) = \langle {}^*H(g),\ {}^*H\rangle + \rho({}^*H(g)) + \langle H_i(g),\ {}^{\Psi}H\rangle + \rho({}^{\Psi}H'_{0i}(\delta g))$$

the assertion is seen to follow from the lemma. The assertion has now to be refined slightly.

Suppose that in Lemma 2.10 the parabolic group is a percuspidal subgroup. If s belongs to $\Omega(\mathfrak{h}, \mathfrak{h})$ then λ can be taken to be the linear function defined by $\lambda(H) = d_i \alpha_{,i}(sH)$. We infer from the lemma that $\alpha_{,i}(H) - \alpha_{,i}(sH)$ is non-negative on $\mathfrak{h}^+$. It will be seen eventually that if $\mathfrak{a}$ and $\mathfrak{b}$ are distinguished subspaces of $\mathfrak{h}$ then $\Omega(\mathfrak{a}, \mathfrak{b})$ is the set of linear transformations from $\mathfrak{a}$ to $\mathfrak{b}$ obtained by restricting those elements of $\Omega(\mathfrak{h}, \mathfrak{h})$ which take $\mathfrak{a}$ onto $\mathfrak{b}$ to $\mathfrak{a}$. It follows readily that if H belongs to $\mathfrak{a}^+$ and s belongs to $\Omega(\mathfrak{a}, \mathfrak{b})$ then $H - sH$ belongs to ${}^+\mathfrak{h}$.

Suppose that ${}^*P$ and P are as before but that the function $\phi(g)$ on ${}^*A\,{}^*T\backslash G$ satisfies

$$|\phi(mk)| \le c\Sigma_{i=1}^n \Sigma_{s\in {}^{\Psi}\Omega(\mathfrak{a}, \mathfrak{a}_i)} \exp(\langle {}^{\Psi}H_0(m),\ s({}^{\Psi}H)\rangle + \rho(H'_0(m))$$

Here $\mathfrak{a}_1, \ldots, \mathfrak{a}_n$ are the distinguished subspaces of $\mathfrak{f}$ such that ${}^{\Psi}\Omega(\mathfrak{a}, \mathfrak{a}_i)$, which is the set of all linear transformations from $\mathfrak{a}$ to $\mathfrak{a}_i$ induced by elements of $\Omega(\mathfrak{f}, \mathfrak{f})$ that leave each point of ${}^*\mathfrak{a}$ fixed, is not empty. Combining the result of the previous paragraph with the convexity of the exponential function we see that

$$\sum_{{}^*\Delta\backslash\Gamma} \exp(\langle {}^*H(\gamma g), {}^*H\rangle + \rho({}^*H(\gamma g)))\phi(\gamma g)$$

converges if ${}^*H + {}^{\Psi}H$ belongs to the convex hull of

$$\bigcup_{i=1}^n \bigcup_{s \in {}^{\Psi}\Omega(\mathfrak{a}, \mathfrak{a}_i)} s^{-1}(\mathfrak{a}_i)$$

There is no need to be explicit about the sense in which the convergence is uniform.

For the further study of Eisenstein series some facts about differential operators on G must be reviewed. In [9] it has been shown that $\mathfrak{Z}$, the centre of $\mathfrak{G}$, is isomorphic to the algebra $\mathfrak{J}$ of polynomials on $\mathfrak{j}_c$ invariant under the Weyl group Ω of $\mathfrak{g}_c$. Let this isomorphism take X in $\mathfrak{Z}$ to p_X. For our purposes the form of the isomorphism is of some importance. If P is a split parabolic subgroup of G with A as a split component and if α is in Q let

$$\mathfrak{n}_\alpha^- = \{X \in \mathfrak{g} \mid [H, X] = -\alpha(H)X \text{ for all } H \text{ in } \mathfrak{a}\}$$

If $\mathfrak{n}^- = \sum_{\alpha \in Q} \mathfrak{n}_\alpha^-$ then $\mathfrak{g}_c = \mathfrak{n}_c + \mathfrak{a}_c + \mathfrak{m}_c + \mathfrak{n}_c^-$. If the universal enveloping algebras of $\mathfrak{n}, \mathfrak{a}, \mathfrak{m}, \mathfrak{n}^-$ are $\mathfrak{N}, \mathfrak{A}, \mathfrak{M}, \mathfrak{N}^-$ respectively then the map

$$X_1 \otimes X_2 \otimes X_3 \otimes X_4 \longrightarrow X_1X_2X_3X_4$$

extends to a vector space isomorphism of $\mathfrak{N} \otimes \mathfrak{A} \otimes \mathfrak{M} \otimes \mathfrak{N}^-$ with $\mathfrak{G}$. Identify the image of $1 \otimes \mathfrak{A} \otimes \mathfrak{M} \otimes 1$ with $\mathfrak{A} \otimes \mathfrak{M}$. If X belongs to $\mathfrak{Z}$ then X is congruent modulo $\mathfrak{n}_c\mathfrak{G}$ to a unique element X_1 in $\mathfrak{A} \otimes \mathfrak{M}$, say $X = X_1 + X_2$. If $\mathfrak{Z}'$ is the centre of $\mathfrak{M}$ it is clear that X_1 belongs to $\mathfrak{A} \otimes \mathfrak{Z}'$. The advantage of this decomposition for us rests on the fact that if X belongs to $\mathfrak{Z}$ then $\lambda(X) = \lambda'(X')$ if X' is the result of applying to X that anti-automorphism of $\mathfrak{L}$ which sends Y in $\mathfrak{g}$ to $-Y$. Thus, if $\phi(g)$ is a function on $N\backslash G$, $\lambda(X)\phi(g) = \lambda'(X_1')\phi(g)$. Let $\mathfrak{j}_c = \mathfrak{n}_i \oplus \mathfrak{j}_c'$ where $\mathfrak{j}_c'$ is the Cartan subalgebra of $\mathfrak{m}_c$. There is of course an isomorphism of $\mathfrak{Z}'$ with the algebra $\mathfrak{J}'$ of polynomials on $\mathfrak{j}_c'$ invariant under the Weyl group of $\mathfrak{m}_c$. Let $X \longrightarrow p_X$ be that isomorphism of $\mathfrak{A}$ with the algebra of polynomials on $\mathfrak{n}_c$ which assigns to Y in $\mathfrak{n}$ the polynomial $p_Y(H) = \langle H, H\rangle + \rho(H)$. Since $\mathfrak{j}_c$ is the direct sum of $\mathfrak{n}_c$ and $\mathfrak{j}_c'$ a polynomial on either of the latter defines a polynomial on $\mathfrak{j}_c$. If $X = \Sigma X_i \otimes Y_i$ belongs to $\mathfrak{A} \otimes \mathfrak{Z}'$ let $p_X = \Sigma p_{X_i} p_{Y_i}$. The image of $\mathfrak{A} \otimes \mathfrak{Z}'$ is the set $\mathfrak{J}_1$ of all polynomials on $\mathfrak{j}_c$ invariant under the Weyl group Ω' of $\mathfrak{n}_c + \mathfrak{m}_c$. If X belongs to $\mathfrak{Z}$ and $X = X_1 + X_2$ as above then $p_X = p_{X_1}$. $\mathfrak{J}_1$ is a finite module over $\mathfrak{J}$ and so is the set of all polynomials on $\mathfrak{j}_c$. If $X \longrightarrow \xi(X)$ is a homomorphism of $\mathfrak{J}$ or $\mathfrak{J}'$ into the complex numbers there is a point Z in $\mathfrak{j}_c$ or in $\mathfrak{j}_c'$ respectively so that $\xi(X) = p_X(Z)$.

If P is a cuspidal subgroup and V is an admissible subspace of $\mathfrak{L}(\Theta\backslash M)$ then V can be written as a direct sum,

$$\oplus_{i=1}^{r} V_i$$

where V_i is closed and invariant under the action of the connected component of M and $\lambda(X)\phi = p_X(Z_i')\phi$ if ϕ belongs to V_i and X belongs to $\mathfrak{Z}'$. $\mathfrak{Z}_i'$ is some point in $\mathfrak{j}_c'$. Although V_i is not admissible we can still define $\mathcal{E}(V_i, W)$ and $\mathcal{E}(V, W) = \oplus_{i=1}^{r} \mathcal{E}(V_i, W)$. If Φ belongs to $\mathcal{E}(V_i, W)$, X belongs to $\mathfrak{Z}$, and $X = X_1 + X_2$ as above then

$$\begin{aligned}
\lambda(X)F(g, \Phi, H) &= \lambda'(X_1')F(g, \Phi, H) \\
&= \Sigma_j \lambda'(U_j')\exp(\langle H(g), H\rangle + \rho(H(g)))p_{Y_j}(Z_i)\Phi(g) \\
&= \Sigma_j p_{U_j}(H)p_{Y_j}(Z_i')F(g, \Phi, H) \\
&= p_X(Z_i)F(g, \Phi, H)
\end{aligned}$$

if $X_1 = \Sigma_j U_j \otimes Y_j$ and $Z_i = H + Z_i'$. Thus

$$\lambda(X)E(g, \Phi, H) = p_X(Z_i)E(g, \Phi, H) \tag{4.e}$$

LEMMA 4.2. <u>Let</u> P <u>be a cuspidal subgroup of</u> G; <u>let</u> ϕ <u>be an infinitely differentiable function on</u> $N\backslash G$; <u>and suppose that there is an integer</u> ℓ <u>and a</u> Z <u>in</u> $\mathfrak{j}_c$ <u>so that, for all</u> X <u>in</u> $\mathfrak{Z}$, $(\lambda(X) - p_X(Z))^{\ell}\phi = 0$. <u>Let</u> $k = [\Omega : \Omega']$. <u>If</u> $\{p_j\}$ <u>is a basis for the polynomials on</u> $\mathfrak{h}$ <u>of degree at most</u> $k^2\ell$, <u>if</u> $\{Z_1, \ldots, Z_t\}$ <u>is a set of representatives of the orbits of</u> Ω' <u>in</u> ΩZ, <u>and</u> $Z_i = H_i + Z_i'$ <u>with</u> H_i <u>in</u> $\mathfrak{h}_c$ <u>and</u> Z_i' <u>in</u> $\mathfrak{j}_c'$ <u>then there are unique functions</u> ϕ_{ij} <u>on</u> $NA\backslash G$ <u>so that</u>

$$(\lambda'(X') - p_X(Z_i'))^{k^2\ell}\phi_{ij} = 0$$

<u>if</u> X <u>belongs to</u> $\mathfrak{Z}'$ <u>and</u>

$$\phi(g) = \Sigma_{i=1}^{t} \exp(\langle H(g), H_i\rangle + \rho(H(g)))\{\Sigma p_j(H(g))\phi_{ij}(g)\}$$

If $\{Y_1, \ldots, Y_u\}$ generate $\mathfrak{N} \otimes \mathfrak{Z}'$ over the image of $\mathfrak{Z}$ and if $\{X_1, \ldots, X_v\}$ generate $\mathfrak{Z}$ the linear space $\mathfrak{M}$ spanned by

$$\{\lambda'(X_1')^{a_1} \ldots \lambda'(X_v')^{a_v}\lambda'(Y_j')\phi \mid 1 \leq a_i \leq \ell,\ 1 \leq j \leq u\}$$

is finite-dimensional and is invariant under $\lambda'(X')$ for X' in $\mathfrak{A} \otimes \mathfrak{Z}'$. Since $\mathfrak{A} \otimes \mathfrak{Z}'$ is commutative one has a representation of this algebra on $\mathfrak{M}$. Let $\mathfrak{K}$ be a set of representatives for the left-cosets of Ω' in Ω and if $s \in \Omega$ and p is a polynomial on $\mathfrak{j}_c$ let $p^s(W) = p(sW)$ for W in $\mathfrak{j}_c$. If X belongs to $\mathfrak{A} \otimes \mathfrak{Z}'$ the polynomial

$$p(U) = \prod_{s \in K}(U - p_X^s)$$

has coefficients in $\mathfrak{J}$; by means of the isomorphism between $\mathfrak{J}$ and $\mathfrak{Z}$ it defines a polynomial q with coefficients in $\mathfrak{Z}$ and $q(X) = 0$. If $p(U, Z)$ is the polynomial

$$\prod_{s \in K} (U - p_X(sZ))$$

with constant coefficients then, restricted to $\mathfrak{M}$,

$$(p(\lambda'(X'), Z) - \lambda'(p(X')))^{k\ell} = 0$$

so

$$\prod_{s \in \mathfrak{K}} (\lambda'(X') - p_X(sZ))^{k\ell} = 0$$

From this it follows immediately that $\mathfrak{M}$ is the direct sum of spaces $\mathfrak{M}_1, \ldots, \mathfrak{M}_t$ with

$$\mathfrak{M}_i = \{\psi \in \mathfrak{M} | (\lambda'(X') - p_X(Z_i))^{k^2\ell} = 0 \text{ for all } X \text{ in } \mathfrak{a} \otimes \mathfrak{Z}'\}$$

Then ϕ can be written uniquely as $\Sigma_{i=1}^t \phi_i'$ with ϕ_i' in $\mathfrak{M}_i$. Suppose ψ belongs to $\mathfrak{M}_i$ for some i. If g is fixed in G let $\psi(a, g)$ be the function $\psi(ag)$ on A. If X belongs to $\mathfrak{a}$ then

$$(\lambda(X) - p_X(H_i))^{k^2\ell} \psi(a, g) = 0$$

This implies that

$$\psi(\exp H, g) = \exp(\langle H, H_i \rangle + \rho(H)) \Sigma_j \psi_j'(g) p_j(H)$$

where the functions $\psi_j'(g)$ are uniquely determined and infinitely differentiable. If $a' = \exp H'$ let

$$\exp(\langle H-H', H_i \rangle + \rho(H-H')) p_m(H-H') = \Sigma_j \tau_{jm}(a') \exp(\langle H, H_i \rangle + \rho(H)) p_j(H)$$

Since $\psi(a'^{-1}a, a'g) = \psi(ag)$ we have

$$\Sigma_j \tau_{mj}(a') \psi_j'(a'g) = \psi_m'(g)$$

Consequently

$$\psi_j(g) = \Sigma_m \tau_{jm}(a(g))\psi'_m(g)$$

is a function on $A\backslash G$ and

$$\psi(g) = \Sigma_m \psi'_m(g)p_m(0) = \exp(\langle H(g), H\rangle + \rho(H(g)))\{\Sigma_j \psi_j(g)p_j(H(g))\} \quad .$$

Since the functions $\psi_j(g)$ are readily seen to be functions on $N\backslash G$ the lemma follows.

Two remarks should be made in connection with this lemma. The first is just that if ϕ is a function on $T\backslash G$ then, for all i and j, ϕ_{ij} will be a function on $AT\backslash G$. For the second suppose that $\ell = 1$ and suppose that there is a subset $\{Z_1, \ldots, Z_u\}$ of $\{Z_1, \ldots, Z_t\}$ so that ϕ_{ij} is identically zero unless $1 \leq i \leq u$. Suppose moreover that for $1 \leq i \leq u$ there is a unique element s_i in $\mathfrak{K}$ such that $s_i Z = Z_i$ and that

$$H_i = H_j, \quad 1 \leq j \leq t$$

implies $Z_i = Z_j$. Referring back to the proof we see that

$$p(\lambda'(X'), Z) - \lambda'(p(X')) = 0$$

so

$$\prod_{s \in \mathfrak{K}} (\lambda'(X') - p_X(sZ)) = 0$$

If X belongs to $\mathfrak{a}$ we see also that

$$\prod_{i=1}^{r} (\lambda(X) - p_X(H_i))\phi = 0$$

Hence

$$\lambda(X)\phi'_i = p_X(H_i)\phi'_i$$

and

(4.f) $$\phi(g) = \Sigma_{i=1}^{u} \exp(\langle H(g), H_i\rangle + \rho(H(g)))\phi_i(g)$$

where $\phi_i(g)$ is a function on $NA\backslash G$ such that

$$\lambda'(X')\phi_i = p_X(Z_i')\phi_i$$

is X belongs to $\mathfrak{Z}'$.

If ℓ is a fixed integer, $Z_1, \ldots, Z_m$ points in $\mathfrak{j}_c$, and $\sigma_1, \ldots, \sigma_n$ irreducible representations of K let

$$\mathfrak{F}(Z_1, \ldots, Z_m; \sigma_1, \ldots, \sigma_n; \ell)$$

be the set of infinitely differentiable functions ϕ on $\Gamma\backslash G$ such that

$$\prod_{i=1}^{m}(\lambda(X) - p_X(Z_i))^{\ell}\phi = 0$$

for every X in $\mathfrak{Z}$, $\{\lambda(k)\phi \mid k \in K\}$ spans a finite-dimensional space such that the restriction of $\lambda(k)$, $k \in K$, to this space contains only irreducible representations equivalent to one of $\sigma_1, \ldots, \sigma_n$, and there is a constant r so that for any Siegel domain $\mathfrak{S}$, associated to a percuspidal subgroup P, there is a constant c so that

$$|\phi(g)| \leq c\eta^r(a(g))$$

for g in $\mathfrak{S}$. The following lemma is essentially the same as one stated in [14].

LEMMA 4.3. The space

$$\mathfrak{F}(Z_1, \ldots, Z_m; \sigma_1, \ldots, \sigma_n; \ell)$$

is finite-dimensional.

There is no loss of generality in assuming that Z_i and Z_j do not belong to the same orbit under Ω unless $i = j$. Then

$$\mathfrak{F}(Z_1, \ldots, Z_m; \sigma_1, \ldots, \sigma_n; \ell)$$

is the direct sum of

$$\mathfrak{F}(Z_i, \sigma_1, \ldots, \sigma_n; \ell), \ 1 \leq i \leq m$$

In other words it can be assumed that $m = 1$. Let

$$\mathfrak{F}(Z, \sigma_1, \ldots, \sigma_n; \ell) = \mathfrak{F}$$

The first step is to show that the set $\mathfrak{F}_0$ of all functions ϕ in $\mathfrak{F}$ such that

$$\int_{\Gamma\cap N\backslash N} \phi(ng)dn \equiv 0$$

for all cuspidal subgroups except G itself is finite-dimensional. From Lemma 3.4 we see that $\mathfrak{F}_0 \cap \mathfrak{L}_0(\Gamma\backslash G)$ is finite-dimensional. Consequently to prove that $\mathfrak{F}_0$ is finite-dimensional it would be enough to show that $\mathfrak{F}_0$ is contained in $\mathfrak{L}_0(\Gamma\backslash G)$. If s is a real number let $\mathfrak{F}_0(s)$ be the set of functions in $\mathfrak{F}_0$ such that for any Siegel domain $\mathfrak{S}$ there is a constant c so that

$$|\phi(g)| \leq c\eta^s(a(g))$$

for g in $\mathfrak{S}$. Since

$$\mathfrak{F}_0 = \bigcup_{s\in\mathbb{R}} \mathfrak{F}_0(s)$$

it must be shown that $\mathfrak{F}_0(s)$ is contained in $\mathfrak{L}_0(\Gamma\backslash G)$. This is certainly true if $s = 0$ and if it is true for s_1 it is true for all s less than s_1. If it is not true in general let s_0 be the least upper bound of all the s for which it is true. If f is once continuously differentiable with compact support and, for all g and k,

$$f(kgk^{-1}) = f(g)$$

then $\lambda(f)$ takes $\mathfrak{F}$ and $\mathfrak{F}_0$ into themselves. Indeed according to Lemma 3.3 if ϕ belongs to $\mathfrak{F}_0(s_0 + \frac{1}{2})$ then $\lambda(f)\phi$ belongs to $\mathfrak{F}_0(s_0 - \frac{1}{2})$ and hence to $\mathfrak{F}_0 \cap \mathfrak{L}_0(\Gamma\backslash G)$. There is a sequence $\{f_n\}$ of such functions so that $\lambda(f_n)\phi$ converges uniformly to ϕ on compact sets. Since $\{\lambda(f_n)\phi\}$ belongs to $\mathfrak{F}_0 \cap \mathfrak{L}_0(\Gamma\backslash G)$

which is finite-dimensional so does ϕ. This is a contradiction. We have in particular proved the lemma if the percuspidal subgroups of G are of rank 0 so that we can use induction on the rank of the percuspidal subgroups of G. To complete the proof it will be enough to show that the range of the map $\phi \longrightarrow \phi^\wedge$ where

$$\phi^\wedge(g) = \int_{\Gamma \cap N \backslash N} \phi(ng)dn$$

is finite-dimensional for every cuspidal subgroup of rank one. According to the previous lemma there is a finite set $\{Z_1, \ldots, Z_t\}$ of elements of $\mathfrak{z}_c$ so that if $Z_i = H_i + Z_i'$ then $\phi^\wedge(g)$ may be written as

$$\Sigma_{i=1}^{t} \exp(\langle H(g), H^i\rangle + \rho(H(g)))\{\Sigma_j p_j(H(g))\phi_{ij}(g)\}$$

where the ϕ_{ij} are functions on $AT\backslash G$. We shall show that, for each i and j, ϕ_{ij} lies in a certain finite-dimensional space. Consider ϕ_{ij} as a function on $\Theta \times \{1\}\backslash M \times K$. The percuspidal subgroups here have rank one less than for G. It will be enough to show that there are points $W_1, \ldots, W_u$ in $\mathfrak{z}_c'$, representations $\tau_1, \ldots, \tau_v$ of $N\backslash N(K \cap P) \times K$, and an integer ℓ' so that ϕ_{ij} belongs to

$$\mathfrak{I}(W_1, \ldots, W_u; \tau_1, \ldots, \tau_v; \ell'\}$$

This follows almost immediately from Lemma 4.2 and Lemma 3.3.

Observe that if ϕ belongs to

$$\mathfrak{I}(Z_1, \ldots, Z_m; \sigma_1, \ldots, \sigma_n; \ell)$$

and ψ belongs to

$$\mathfrak{I}(Z_1, \ldots, Z_m; \sigma_1, \ldots, \sigma_n; \ell) \cap \mathfrak{L}_0(\Gamma\backslash G) = \mathfrak{I}_0$$

then, by the corollary to Lemma 3.3,

$$\int_{\Gamma\backslash G} \psi(g)\overline{\phi}(g)dg$$

is defined. Thus there is a unique ϕ' in $\mathfrak{J}_0$ so that $\phi - \phi'$ is orthogonal to $\mathfrak{J}_0$; ϕ' is called the cuspidal component of ϕ. It is easy to see that if V is any admissible subspace of $\mathcal{L}_0(\Gamma\backslash G)$ and W is any admissible subspace of the space of continuous functions on K and ψ belongs to $\mathcal{E}(V, W)$ then

$$\int_{\Gamma\backslash G} \psi(g)\overline{\phi}(g)dg = \int_{\Gamma\backslash G} \psi(g)\overline{\phi}'(g)dg$$

These two lemmas will now be used to study the Eisenstein series. Suppose $P^{(1)}$ and $P^{(2)}$ are two cuspidal subgroups and $V^{(1)}$ is an admissible subspace of $\mathcal{L}(\Theta^{(1)}\backslash M^{(1)})$. As before write $V^{(1)}$ as $\Sigma_{i=1}^{r} V_i^{(1)}$ where $\lambda(X)\phi = p_X(Z_i^{(1)})\phi$ if ϕ belongs to $V_i^{(1)}$ and X belongs to $\mathfrak{Z}^{(1)}$. $Z_i^{(1)}$ is some point in $\mathfrak{j}_c^{(1)}$. Because we have two cuspidal subgroups it is notationally convenient to replace the prime that has been used earlier by the superscript (1) or (2). If Φ belongs to $\mathcal{E}(V_{i_0}^{(1)}, W)$ and $H^{(1)}$ in $\mathfrak{a}_c^{(1)}$ satisfies the conditions of Lemma 4.1 consider

$$\int_{\Gamma\cap N^{(2)}\backslash N^{(2)}} E(ng, \Phi, H^{(1)})dn = \int_{\Delta^{(2)}\backslash T^{(2)}} E(tg, \Phi, H^{(1)})dt$$

which is equal to

$$\int_{\Delta^{(2)}\backslash T^{(2)}} \Sigma_{\Delta^{(1)}\backslash\Gamma} \exp(\langle H^{(1)}(\gamma tg), H^{(1)}\rangle + \rho(H^{(1)}(\gamma tg)))\Phi(\gamma tg)dt$$

Replace the sum by a sum over double cosets to obtain

$$\Sigma_{\Delta^{(1)}\backslash\Gamma/\Delta^{(2)}} \int_{\Delta^{(2)}\cap\gamma^{-1}\Delta^{(1)}\gamma\backslash T^{(2)}} \exp(\langle H^{(1)}(\gamma tg), H^{(1)}\rangle + \rho(H^{(1)}(\gamma tg)))\Phi(\gamma tg)dt$$

The terms of this sum will be considered individually.

If

$$\Phi(g, H, \gamma) = \int_{\Delta^{(2)}\cap\gamma^{-1}\Delta^{(1)}\gamma\backslash T^{(2)}} \exp(\langle H^{(1)}(\gamma tg), H^{(1)}\rangle + \rho(H^{(1)}(\gamma tg)))\Phi(\gamma tg)dt$$

and if $W_1, \ldots, W_t$ is a set of representatives of the orbits of $\Omega^{(2)}$ in $\Omega(H^{(1)} + Z_{i_0}^{(1)})$ and $W_j = H_j^{(2)} + W_j^{(2)}$ we can write

$$\Phi(g, H, \gamma) = \Sigma_{j=1}^{t} \exp(\langle H^{(2)}(g), H_j^{(2)}\rangle + \rho(H^{(2)}(g))\{\Sigma_k p_k(H^{(2)}(g))\phi_{jk}(g)\}$$

Setting

$$\phi_{j,k}(m, k) = \phi_{j,k}(mk^{-1})$$

we obtain functions on

$$\Theta^{(2)} \times \{1\}\backslash M^{(2)} \times K$$

There are irreducible representations $\tau_1, \ldots, \tau_n$ of

$$(N^{(2)}\backslash N^{(2)}(K \cap P^{(2)})) \times K$$

and an integer ℓ so that $\phi_{j,k}$ belongs to

$$\mathfrak{J}(Z_j^{(2)}, \tau_1, \ldots, \tau_n; \ell)$$

Let $\phi'_{j,k}$ be the cuspidal component of $\phi_{j,k}$. If V is an admissible subspace of

$$\mathcal{L}_0(\Theta^{(2)} \times \{1\}\backslash M^{(2)} \times K)$$

and W' an admissible subspace of the space of continuous functions on

$$(N^{(2)}\backslash N^{(2)}(K \cap P^{(2)})) \times K$$

and if ψ belongs to $\mathcal{E}(V, W')$ then

$$\int_{\Theta^{(2)}\backslash M^{(2)}\times K} \Phi(\exp H^{(2)} mk^{-1}, H^{(1)}; \gamma)\bar{\psi}(m, k)dmdk$$

is equal to

$$\Sigma_{j=1}^{t} \exp(\langle H^{(2)}, H_j^{(2)}\rangle + \rho(H^{(2)}))\left\{\Sigma p_k(H_j^{(2)})\int_{\Theta^{(2)}\backslash M^{(2)}\times K} \phi'_{j,k}(m, k)\bar{\psi}(m, k)dmdk\right\}$$

The first integral is an analytic function of $H^{(1)}$ on

$$\mathfrak{U}^{(1)} = \{H^{(1)} \in \mathfrak{a}_c^{(1)} \mid \operatorname{Re} \alpha^1_{i,}(H^{(1)}) > \langle \alpha^1_{i,}, \rho\rangle,\ 1 \leq i \leq q^{(1)}\}$$

if $q^{(1)}$ is the rank of $P^{(1)}$. It vanishes identically if it vanishes identically on some open subset of $\mathfrak{U}^{(1)}$. If s_1 and s_2 belong to $\mathfrak{k}$, a set of representatives for the left cosets of $\Omega^{(2)}$ in Ω, and s belongs to $\Omega^{(2)}$ then the equation

$$ss_1(H^{(1)} + Z^{(1)}_{i_0}) = s_2(H^{(1)} + Z^{(1)}_{i_0})$$

is satisfied on all of $\mathfrak{U}^{(1)}$ or on a proper subspace of $\mathfrak{U}^{(1)}$. Let $\mathfrak{U}_1$ be the open set of points $H^{(1)}$ in $\mathfrak{U}^{(1)}$ such that for any s, s_1, and s_2,

$$ss_1(H^{(1)} + Z^{(1)}_{i_0}) = s_2(H^{(1)} + Z^{(1)}_{i_0})$$

only if this equation holds identically. On this set of points the number t above is constant. We can then choose fixed elements $s_1, \ldots, s_t$ in Ω and take, for $H^{(1)}$ in $\mathfrak{U}_1$,

$$W_j = s_j(H^{(1)} + Z^{(1)}_{i_0})$$

It is readily seen that

$$\Sigma_k p_k(H_j^{(2)}) \int_{\Theta^{(2)}\backslash M^{(2)} \times K} \phi'_{j,k}(m, k)\overline{\psi}(m, k)\,dm\,dk$$

is a continuous function on $\mathfrak{U}_1$. It vanishes unless $Z_j^{(2)}$ is one of a finite number of points. Since $Z_j^{(2)}$ is a linear function of $H^{(1)}$ it will be a constant if this integral does not vanish identically. Then $s_j(\mathfrak{U}_1)$ will be contained in $\mathfrak{U}^{(2)}$. Since s_j is non-singular this can only happen if the rank of $P^{(2)}$ is at least as great as the rank of $P^{(1)}$. We conclude that the cuspidal component of $\Phi(g, H^{(1)}, \gamma)$ and thus of

$$\int_{\Gamma \cap N^{(2)} \backslash N^{(2)}} E(ng, \Phi, H^{(1)})\,dn$$

is zero if the rank of $P^{(2)}$ is less than the rank of $P^{(1)}$.

We now treat the case that $V^{(1)}$ is contained in $\mathcal{L}_0(\Theta^{(1)}\backslash M^{(1)})$. It will be shown later that if the rank of $P^{(2)}$ is greater than the rank of $P^{(1)}$ then the cuspidal component of $\Phi(g, H^{(1)}, \gamma)$ vanishes identically. Anticipating this result we consider the case of equal rank. Let $s_1, \dots, s_m$ be the elements of $\mathfrak{k} = \{s_1, \dots, s_n\}$ such that $s_j(\mathfrak{a}^{(1)}) = \mathfrak{a}^{(2)}$. Let $H_j^{(2)}$ now be the projection of $s_j(H^{(1)} + Z_{i_0}^{(1)})$ on $\mathfrak{a}_c^{(2)}$. If $1 \leq j_1 \leq m$ and $m < j_2 \leq n$ the equation $H_{j_1}^{(2)} = H_{j_2}^{(2)}$ can not be identically satisfied. Let $\mathfrak{U}_2$ be the set of points $H^{(1)}$ in $\mathfrak{U}_1$ such that $H_{j_1}^{(2)} \neq H_{j_2}^{(2)}$ if $1 \leq j_1 \leq m$ and $m < j_2 \leq n$ and such that

$$s_{j_1}(H^{(1)} + Z_{i_0}^{(1)}) = s_{j_2}(H^{(1)} + Z_{i_0}^{(1)})$$

or

$$s_{j_1}(H^{(1)}) = s_{j_2}(H^{(1)}), \quad 1 \leq j_1, j_2 \leq m$$

only if this equation holds identically on $\mathfrak{a}^{(1)}$. Suppose $H^{(1)}$ belongs to $\mathfrak{U}_2$ and

$$s_{j_1}(H^{(1)}) = s_{j_2}(H^{(1)}), \quad 1 \leq j_1, j_2 \leq m$$

then $s_{j_1} s_{j_2}^{-1}$ belongs to $\Omega^{(2)}$ so that $j_1 = j_2$. According to the remark following Lemma 3.7 $\phi_{j,k} = \phi'_{j,k}$ and then according to the remark following the proof of Lemma 4.2

$$\Phi(g, H, \gamma) = \Sigma_{j=1}^{m} \exp(\langle H^{(2)}(g), s_j H^{(1)} \rangle + \rho(H^{(2)}(g)))\phi_j(g)$$

Grouping together those s_j which determine the same element of $\Omega(\mathfrak{a}^{(1)}, \mathfrak{a}^{(2)})$ we can write the right hand side as

$$\Sigma_{s \in \Omega(\mathfrak{a}^{(1)}, \mathfrak{a}^{(2)})} \exp(\langle H^{(2)}(g), sH^{(1)} \rangle + \rho(H^{(2)}(g)))\Phi_s(g; \gamma)$$

$\Phi_s(mk^{-1}; \gamma)$ belongs to

$$\mathcal{L}_0(\Theta^{(2)} \times \{1\} \backslash M^{(2)} \times K)$$

This sum is of course zero if $P^{(1)}$ and $P^{(2)}$ are not associate.

In general for any Φ in $\mathcal{E}(V^{(1)}, W)$ we see that, for $H^{(1)}$ in $\mathfrak{U}_2$,

$$\int_{\Gamma \cap N^{(2)} \backslash N^{(2)}} E(ng, \Phi, H^{(1)}) dn$$

is equal to

$$\sum_{\Delta^{(2)} \backslash \Gamma / \Delta^{(1)}} \sum_{s \in \Omega(\mathfrak{a}^{(1)}, \mathfrak{a}^{(2)})} \exp(\langle H^{(2)}(g), sH^{(1)} \rangle + \rho(H^{(2)}(g))) \Phi_s(g; \gamma)$$

In order to simplify the statements of our conclusions let us introduce the notion of a simple admissible subspace. Let P be a cuspidal subgroup and let $\mathfrak{j}_c = \mathfrak{a}_c + \mathfrak{j}'_c$ where $\mathfrak{j}'_c$ is the Cartan subalgebra of $\mathfrak{m}_c$. If Z' is a point in $\mathfrak{j}'_c$ and $Z'_1, \ldots, Z'_r$ is the orbit of Z' under those elements in the group of automorphisms of $\mathfrak{g}_c$ generated by the adjoint group of $\mathfrak{g}_c$ and G which normalize both $\mathfrak{j}_c$ and $\mathfrak{a}_c$ then the sum V of all closed subspaces of $\mathcal{L}_0(\Theta \backslash M)$ which are invariant and irreducible under the action of the connected component of M and which belong to one of the characters $X \longrightarrow p_X(Z'_i)$ of $\mathfrak{Z}'$ will be called a simple admissible subspace of $\mathcal{L}_0(\Theta \backslash M)$. Since V is invariant under M it is an admissible subspace. A simple admissible subspace of the space of continuous functions on K is the space of functions spanned by the matrix elements of an irreducible representation of K. If $P^{(1)}$ and $P^{(2)}$ are two associate cuspidal subgroups and $Z^{(1)}$ is a point of $\mathfrak{j}_c^{(1)}$ let $Z^{(2)}$ be the image of $Z^{(1)}$ under some element of Ω which takes $\mathfrak{a}^{(1)}$ onto $\mathfrak{a}^{(2)}$. If $V^{(1)}$ and $V^{(2)}$ are the simple admissible subspaces defined by $Z^{(1)}$ and $Z^{(2)}$ respectively then $V^{(1)}$ and $V^{(2)}$ are said to be associate. As a convention two associate admissible subspaces will always be simple. It will be enough to state the results for simple admissible subspaces because every admissible subspace is contained in a finite sum of simple admissible subspaces. In particular if $V^{(1)}$ and W are simple admissible subspaces and

$V^{(2)}$ is the simple admissible subspace associate to $V^{(1)}$, if Φ belongs to $\mathcal{E}(V^{(1)}, W)$, and if $H^{(2)}$ belongs to $U^{(2)}$ then

(4.g) $$\int_{\Delta^{(2)} \cap \gamma^{-1} \Delta^{(1)} \gamma \backslash T^{(2)}} \exp(\langle H^{(1)}(\gamma t g), H^{(1)} \rangle + \rho(H^{(1)}(\gamma t g)))\Phi(\gamma t g)\,dt$$

is equal to

(4.h) $$\sum_{s \in \Omega(\mathfrak{a}^{(1)}, \mathfrak{a}^{(2)})} \exp(\langle H^{(2)}(g), sH^{(1)} \rangle + \rho(H^{(2)}(g)))(N_\gamma(s, H^{(1)})\Phi)(g)$$

Here $N_\gamma(s; H^{(1)})$ is for each $H^{(1)}$ in $\mathfrak{U}_2$ and each s a linear transformation from $\mathcal{E}(V^{(1)}, W)$ to $\mathcal{E}(V^{(2)}, W)$; it is analytic as a function of $H^{(1)}$.

It is necessary to establish the formula (4.h) on all of $\mathfrak{U}$. To do this it is enough to show that all but one of the terms in (4.h) vanish identically on $\mathfrak{U}$. Choose some s_0 in $\Omega(\mathfrak{a}^{(1)}, \mathfrak{a}^{(2)})$; since $\mathfrak{U}_2$ is connected the corresponding term of (4.h) will vanish identically if it vanishes for real values of the argument. If $H^{(1)}$ is real and in $\mathfrak{U}_2$ then

$$\langle s_0 H^{(1)}, sH^{(1)} \rangle < \langle s_0 H^{(1)}, s_0 H^{(1)} \rangle$$

if s belongs to $\Omega(\mathfrak{a}^{(1)}, \mathfrak{a}^{(2)})$ but does not equal s_0. In (4.h) take

$$g = \exp a(s_0(H^{(1)})mk$$

where a is a positive real number, $\exp a(s_0 H^{(1)})$ belongs to some split component $A^{(2)}$ of $P^{(2)}$, m belongs to $M^{(2)}$, and k belongs to K and replace $H^{(1)}$ by $bH^{(1)}$ where b is a positive real number such that $bH^{(1)}$ belongs to $\mathfrak{U}_2$. Then multiply by

$$\exp(-ab\langle s_0 H^{(1)}, s_0 H^{(1)} \rangle - a\rho(s_0 H^{(1)}))$$

and take the limit as a approaches infinity. The result is

$$N_\gamma(s_0; H^{(1)})\Phi(mk)$$

On the other hand if the same substitution is effected in (4.g) the result is bounded by a constant times

$$\int_{\Delta^{(2)}\cap\gamma^{-1}\Delta^{(1)}\gamma\backslash T^{(2)}} \exp(\langle H^{(1)}(h(t,\ \gamma)),\ bH^{(1)}\rangle + \rho(H^{(1)}(h(t,\ \gamma))))dt$$

with $h(t,\ \gamma) = \gamma t \exp a(s_0 H^{(1)})mk$, because $\Phi(g)$ is a bounded function. Of course this integral is finite and it equals the sum over $\Delta^{(2)}\cap\gamma^{-1}\Delta^{(1)}\gamma\backslash\Delta^{(2)}/\Gamma\cap N^{(2)}$ of

$$\int_{\delta^{-1}\gamma^{-1}\Delta^{(1)}\gamma\delta\cap N^{(2)}\backslash N^{(1)}} \exp(\langle H^{(1)}(h(n,\ \gamma\delta)),\ bH^{(1)}\rangle + \rho(H^{(1)}(h(n,\ \gamma\delta))))dn$$

Choose u in the connected component of G so that $u^{-1}P^{(2)}u$ and $P^{(1)}$ both belong to the percuspidal subgroup P. Suppose that split components $\mathfrak{a}^{(1)}$, $\mathfrak{a}^{(2)}$ and $\mathfrak{a}$ have been chosen for $P^{(1)}$, $P^{(2)}$, and P respectively so that $\mathrm{Ad}(u^{-1})\mathfrak{a}^{(2)}$ and $\mathfrak{a}^{(1)}$ are both contained in $\mathfrak{a}$. $\mathfrak{p}$ contains a subalgebra $\mathfrak{b}$ such that $\mathfrak{a}\subseteq\mathfrak{b}\subseteq\mathfrak{g}_s$ and $\mathfrak{b}$ is a maximal subalgebra of $\mathfrak{g}_s$ such that $\{\mathrm{Ad}\,H \mid H\in\mathfrak{b}\}$ is diagonalizable. By Bruhat's lemma [12] $\gamma\delta u$ can be written as pvp_1 where p belongs to P, p_1 to $u^{-1}P^{(2)}u$ and v belongs to the normalizer of $\mathfrak{b}$. Then each integral above is the product of

$$\exp(\langle H^{(1)}(p),\ bH^{(1)}\rangle + \rho(H^{(1)}(p)))$$

and the integral over

$$\delta^{-1}\gamma^{-1}\Delta^{(1)}\gamma\delta\cap N^{(1)}\backslash N^{(1)}$$

of

$$\exp\langle H^{(1)}(vp_1u^{-1}nu\exp a(\mathrm{Ad}(u^{-1})(s_0H^{(1)}))u^{-1}mk),\ bH^{(1)} + H_\rho\rangle$$

if H_ρ is such that $\langle H,\ H_\rho\rangle = \rho(H)$ for H in h. Let $N_0 = u^{-1}N^{(2)}u$ and replace the integral by an integral over

$$t^{-1}p^{-1}\Delta^{(1)}pt\cap N_0\backslash N_0$$

Now $v^{-1}p^{-1}\Delta^{(1)}pv \cap N$ is contained in $v^{-1}p^{-1}S^{(1)}pv \cap N_0$ and both these groups are unimodular so the integral is a product of

$$\mu(v^{-1}p^{-1}\Delta^{(1)}pv \cap N_0 \backslash v^{-1}p^{-1}S^{(1)}pv \cap N_0)$$

and an integral over $v^{-1}S^{(1)}v \cap N_0 \backslash N_0$ since $p^{-1}S^{(1)}p = S^{(1)}$. If $p_1 = n_1 m_1 a_1$ with n_1 in N_0, m_1 in $u^{-1}M^{(2)}u$, and a_1 in $u^{-1}A^{(2)}u$ the integrand is

$$\exp\left\langle H^{(1)}(vn\exp(a\,\mathrm{Ad}(u^{-1})(s_0H^{(1)}))m_1a_1u^{-1}mk),\ bH^{(1)} + H\rho\right\rangle$$

If H belongs to $\mathfrak{b}$ then

$$n \longrightarrow \exp(-H)n\exp H = \xi(n)$$

defines a map of $v^{-1}S^{(1)}v \cap N_0 \backslash N_0$ onto itself; let $dn = \exp \rho_1(H)d\xi(n)$. Then this integral is the product of

$$\exp(ab\left\langle \mathrm{Ad}(vu^{-1})(s_0H^{(1)}),\ H^{(1)}\right\rangle + a\rho(\mathrm{Ad}(vu^{-1})(s_0H^{(1)})) + a\rho_1(\mathrm{Ad}(u^{-1})(s_0H^{(1)}))$$

and

$$\int_{v^{-1}S^{(1)}v\cap N_0\backslash N_0} \exp\left\langle H^{(1)}(vnm_1a_1u^{-1}mk),\ bH^{(1)} + H_\rho\right\rangle dn$$

This integral is of course independent of a. If $\mathrm{Ad}(vu^{-1})(s_0H^{(1)})$ does not equal $H^{(1)}$ choose b so large that

$$\rho(\mathrm{Ad}(vu^{-1})(s_0H^{(1)})) + \rho_1(\mathrm{Ad}(u^{-1})(s_0H^{(1)})) - \rho(s_0H^{(1)})$$

is less than

$$b\left\langle s_0H^{(1)},\ s_0H^{(1)}\right\rangle - b\left\langle \mathrm{Ad}(vu^{-1})(s_0H^{(1)}),\ H^{(1)}\right\rangle$$

Then the result of multiplying (4.g) by

$$\exp(-ab\left\langle s_0H^{(1)},\ s_0H^{(1)}\right\rangle - a\rho(s_0H^{(1)}))$$

and taking the limit as a approaches infinity is zero. Thus if $N_\gamma(s_0, H^{(1)})$ is not identically zero there is some δ in $\Delta^{(2)}$ so that $\mathrm{Ad}(vu^{-1})$ maps $\mathfrak{n}^{(2)}$ onto $\mathfrak{n}^{(1)}$ and is equal to the inverse of s_0 on $\mathfrak{n}^{(2)}$. If $\gamma\delta u = pvp_1$ then $\gamma u = pv(p_1 u^{-1}\delta^{-1}u)$ and v can be chosen to depend only on γ. Thus there is at most one term of (4.h) which does not vanish identically. Before summarizing the conclusions reached so far let us make some remarks which are useful for calculating the transformations $N_\gamma(s, H^{(1)})$ explicitly. If $N_\gamma(s, H^{(1)})$ does not vanish identically let $\gamma = p(vu^{-1})(up_1u^{-1})$; simplifying we can write $\gamma = n_1 a_1 w n_2$ with n_1 in $N^{(1)}$, a_1 in $A^{(1)}$, n_2 in $N^{(2)}$, and with w such that $\mathrm{Ad}\,w$ takes $\mathfrak{n}^{(2)}$ onto $\mathfrak{n}^{(1)}$ and is inverse to s on $\mathfrak{n}^{(2)}$. Then (4.g) equals the product of

$$\exp(\langle H^{(1)}(a_1), H^{(1)}\rangle + \rho(H^{(1)}(a_1)))$$

and the sum over $(\Gamma \cap N^{(2)})(\Delta^{(2)} \cap \gamma^{-1}\Delta^{(1)}\gamma)\backslash \Delta^{(2)}$ of

$$\int_{\delta^{-1}\gamma^{-1}\Delta^{(1)}\gamma\delta \cap N^{(2)}\backslash N^{(2)}} \exp(\langle H^{(1)}(wn_2\delta ng), H^{(1)}\rangle + \rho(H^{(1)}(wn_2\delta ng))\Phi(wn_2\delta ng)dn$$

Although we will not press the point here it is not difficult to see that the sum is finite and that $\delta^{-1}\gamma^{-1}\Delta^{(1)}\gamma\delta \cap N^{(2)}$ is equal to $\delta^{-1}\gamma^{-1}(N^{(1)} \cap \Gamma)\gamma\delta \cap N^{(2)}$. Consider the linear transformation on $\mathcal{E}(V^{(1)}, W)$ which sends Φ to Φ' with $\Phi'(g)$ equal to the product of

$$\exp(-\langle H^{(2)}(g), sH^{(1)}\rangle - \rho(H^{(2)}(g)))$$

and

$$\int_{w^{-1}N^{(1)}w \cap N^{(2)}\backslash N^{(2)}} \exp(\langle H^{(1)}(wng), H^{(1)}\rangle + \rho(H^{(1)}(wng)))\Phi(wng)dn$$

Φ' is a function on $A^{(2)}N^{(2)}\backslash G$. Considered as a function on $M^{(2)} \times K$, it is a function on $w^{-1}\Theta^{(1)}w\backslash M^{(2)}$. Since the sum is finite $w^{-1}\Theta^{(1)}w$ and $\Theta^{(2)}$ are commensurable. We can define the subspace $V^{(2)}(w)$ of $\mathcal{I}_0(w^{-1}\Theta^{(1)}w\backslash M^{(2)})$ associate to $V^{(1)}$ and Φ' belongs to $\mathcal{E}(V^{(2)}(w), W)$. Denote the linear transformation from $\mathcal{E}(V^{(1)}, W)$ to $\mathcal{E}(V^{(2)}(w), W)$ by $B(w, H^{(1)})$. Let

$$\mu(\delta^{-1}\gamma^{-1}(N^{(1)} \cap \Gamma)\gamma\delta \cap N^{(2)} \backslash \delta^{-1}\gamma^{-1}N^{(1)}\gamma\delta \cap N^{(2)})$$

which is independent of δ, equal μ; then

$$N_\gamma(s, H^{(1)})\Phi(m, k)$$

equals

$$\sum_{(\Gamma\cap N^{(2)})(\Delta^{(2)}\cap\gamma^{-1}\Delta^{(1)}\gamma)\backslash\Delta^{(2)}} \mu \exp(\langle H^{(1)}(a_1), H^{(1)}\rangle + \rho(H^{(1)}(a_1)))B(w, H^{(1)})\Phi(\bar{\delta}m, k)$$

if $\bar{\delta}$ is the projection of δ on $M^{(2)}$. The sum is a kind of Hecke operator. However this representation of the linear transformations will not be used in this paper.

If s belongs to $\Omega(\mathfrak{a}^{(1)}, \mathfrak{a}^{(2)})$ let

$$N(s, H^{(1)}) = \sum_{\Delta^{(1)}\backslash\Gamma/\Delta^{(2)}} N_\gamma(s, H^{(1)})$$

LEMMA 4.4. *Suppose* $P^{(1)}$ *and* $P^{(2)}$ *are two cuspidal subgroups and suppose* $V^{(1)}$ *is a simple admissible subspace of* $\mathcal{L}_0(\Theta^{(1)}\backslash M^{(1)})$ *and* W *is a simple admissible subspace of the space of continuous functions on* K. *If* $P^{(1)}$ *and* $P^{(2)}$ *are not associate then the cuspidal component of*

$$\int_{\Gamma\cap N^{(2)}\backslash N^{(2)}} E(ng, \Phi, H^{(1)})dn$$

is zero; however if $P^{(1)}$ *and* $P^{(2)}$ *are associate then*

$$\int_{\Gamma\cap N^{(2)}\backslash N^{(2)}} E(ng, \Phi, H^{(1)})dn$$

is equal to

$$\sum_{s\in\Omega(\mathfrak{a}^{(1)}, \mathfrak{a}^{(2)})} \exp(\langle H^{(2)}g, sH^{(1)}\rangle + \rho(H^{(2)}(g)))N(s, H^{(1)})\Phi(g)$$

where $N(s, H^{(1)})$ *is for each* $H^{(1)}$ *in* $\mathfrak{A}^{(1)}$ *a linear transformation from*

$\mathcal{E}(V^{(1)}, W)$ to $\mathcal{E}(V^{(2)}, W)$ which is analytic as a function of $H^{(1)}$.

This lemma is not yet completely proved; the proof will come eventually. First however let us establish some properties of the functions $N(s, H^{(1)})$.

LEMMA 4.5. (i) There is an element $H_0^{(1)}$ in $\mathfrak{a}^{(1)}$ and a constant $c = c(\varepsilon)$ so that, for all s in $\Omega(\mathfrak{a}^{(1)}, \mathfrak{a}^{(1)})$,

$$\|N(s, H^{(1)})\| \leq c \exp\langle H_0^{(1)}, \operatorname{Re} H^{(1)}\rangle$$

for all $H^{(1)}$ in $\mathfrak{A}^{(1)}$ with $\alpha_i(H^{(1)}) > \langle \alpha_{i,}, \rho\rangle + \varepsilon$.

(ii) Let F be a subset of the simple roots of $\mathfrak{h}$ and let

$$^*\mathfrak{a} = \{H \in \mathfrak{h} \mid \alpha(H) = 0 \text{ if } \alpha \in F\}$$

Suppose $^*\mathfrak{a}$ is contained in $\mathfrak{a}^{(1)}$ and $\mathfrak{a}^{(2)}$ and s in $\Omega(\mathfrak{a}^{(1)}, \mathfrak{a}^{(2)})$ leaves eac point of $^*\mathfrak{a}$ fixed. Let $^*P^{(1)}$ and $^*P^{(2)}$ be the unique cuspidal subgroups belonging to $P^{(1)}$ and $P^{(2)}$ respectively with $^*\mathfrak{a}$ as a split component. If $^*P^{(1)}$ and $^*P^{(2)}$ are not conjugate under Γ then $N(s, H^{(1)}) \equiv 0$; if $^*P^{(1)} = {}^*P^{(2)} = {}^*P$ then $N(s, H^{(1)}$ is the restriction to $\mathcal{E}(V^{(1)}, W)$ of $N({}^{\dagger}s, {}^{\dagger}H^{(1)})$.

(iii) If s belongs to $\Omega(\mathfrak{a}^{(1)}, \mathfrak{a}^{(2)})$ then $N(s, H^{(1)})$ is analytic on the conve hull of $\mathfrak{A}^{(1)}$ and $-s^{-1}(\mathfrak{A}^{(2)})$ and

$$N(s, H^{(1)}) = N^*(s^{-1}, -s\overline{H}^{(1)})$$

Let us start with part (ii). First of all we have to explain the notation. If $^*P^{(1)} = {}^*P^{(2)} = {}^*P$ let

$$^{\dagger}P^{(i)} = {}^*N\backslash P^{(i)} \cap {}^*S, \ i = 1, 2$$

$^{\dagger}P^{(i)} \times K$ is a cuspidal subgroup of $^*M \times K$ with split component $^{\dagger}\mathfrak{a}^{(i)}$ if $\mathfrak{a}^{(i)}$ is the orthogonal sum of $^*\mathfrak{a}$ and $^{\dagger}\mathfrak{a}^{(i)}$. The restriction of s to $^{\dagger}\mathfrak{a}^{(1)}$ defines an element $^{\dagger}s$ of $\Omega(^{\dagger}\mathfrak{a}^{(1)}, {}^{\dagger}\mathfrak{a}^{(2)})$ so $^{\dagger}P^{(1)}$ and $^{\dagger}P^{(2)}$ are associate. As we remarked in Section 3 the space $\mathcal{E}(V^{(i)}, W)$ can be identified with a subspace of $\mathcal{E}(V^{(i)} \times W, W^*)$. Although the subspace W^* is not simple it is a sum of simple admissible subspaces so that if ${}^{\Psi}H^{(1)}$ belongs to ${}^{\Psi}\mathcal{U}^{(1)}$, which is defined in the obvious manner, the linear transformation $N({}^{\Psi}s, {}^{\Psi}H^{(1)})$ from $\mathcal{E}(V^{(1)} \otimes W, W^*)$ to $\mathcal{E}(V^{(2)} \otimes W^*, {}^*W \otimes W^*)$ is still defined. If $H^{(1)}$ belongs to $\mathcal{U}^{(1)}$ and $H^{(1)} = {}^*H^{(1)} + {}^{\Psi}H^{(1)}$ then ${}^{\Psi}H^{(1)}$ belongs to ${}^{\Psi}\mathcal{U}^{(1)}$ and part (ii) of the lemma asserts that $N(s, H^{(1)})$ is the restriction to $\mathcal{E}(V^{(1)}, W)$ of $N({}^{\Psi}s, {}^{\Psi}H^{(1)})$. To prove it we start from the formula

$$N(s, H^{(1)}) = \Sigma_{\Delta^{(1)}\backslash\Gamma/\Delta^{(2)}} N_{\gamma}(s, H^{(1)}) \tag{4.i}$$

If $N_{\gamma}(s, H^{(1)})$ is not zero we know that $\gamma = p_1 v p_2$ with p_1 in $P^{(1)}$, p_2 in $P^{(2)}$, and v such that the restriction of $\mathrm{Ad}(v)$ to $\mathfrak{n}^{(2)}$ is the inverse of s. We are here considering $\mathfrak{n}^{(1)}$ and $\mathfrak{n}^{(2)}$ as subsets of $\mathfrak{g}$. Let ${}^*\mathfrak{n}^{(1)}$ and ${}^*\mathfrak{n}^{(2)}$ be the image of ${}^*\mathfrak{n}$ in $\mathfrak{n}^{(1)}$ and $\mathfrak{n}^{(2)}$ respectively. $\mathrm{Ad}\, v$ takes ${}^*\mathfrak{n}^{(1)}$ to ${}^*\mathfrak{n}^{(2)}$ so that positive roots go to positive roots. Thus

$$v({}^*P^{(2)})v^{-1} = ({}^*P^{(1)})$$

so

$$\gamma({}^*P^{(2)})\gamma^{-1} = {}^*P^{(1)}$$

which proves the first assertion. If ${}^*P^{(1)} = {}^*P^{(2)} = {}^*P$ then $\gamma\, {}^*P \gamma^{-1} = {}^*P$ so γ belongs to ${}^*P$. The sum defining $N(s, H^{(1)})$ may be replaced by a sum over a set of representatives of the cosets $\Delta^{(1)}\backslash {}^*\Delta/\Delta^{(2)}$. Moreover if γ_1 and γ_2 belong to ${}^*\Delta$ and $\delta_1\gamma_1\delta_2 = \gamma_2$ with δ_1 in $\Delta^{(1)}$, δ_2 in $\Delta^{(2)}$ then project on

$${}^*\Theta = ({}^*\Delta \cap {}^*N)\backslash {}^*\Delta$$

to obtain ${}^*\delta_1 {}^*\gamma_1 {}^*\delta_2 = {}^*\gamma_2$ with ${}^*\delta_i$ in

$${}^{\Psi}\Delta^{(i)} = ({}^*\Delta \cap {}^*N)\backslash\Delta^{(i)}, \ i = 1,\ 2$$

Conversely if ${}^{*}\delta_1 {}^{*}\gamma_1 {}^{*}\delta_2 = {}^{*}\gamma_2$ then there is a δ in ${}^{*}\Delta \cap {}^{*}N \subseteq \Delta^{(1)}$ so that $\delta\delta_1\gamma_1\delta_2 = \gamma_2$. Finally if H belongs to $\mathfrak{a}^{(2)}$ and γ belongs to ${}^{*}P$ then

$$\exp(\langle H, sH^{(1)}\rangle + \rho(H))N_\gamma(s, H^{(1)})\Phi(mk^{-1})$$

with m in ${}^{*}M$ and k in K, is equal to

$$\int_{\Delta^{(2)}\cap\gamma^{-1}\Delta^{(1)}\gamma\backslash T^{(2)}} \exp(\langle H^{(1)}(\gamma t \exp H\, mk^{-1}), H^{(1)} + H_\rho\rangle)\Phi(\gamma t \exp H\, mk^{-1})dt$$

Since $\Delta^{(2)} \cap \gamma^{-1}\Delta^{(1)}\gamma$ contains ${}^{*}N \cap \Gamma$ and

$$\mu({}^{*}N \cap \Gamma \backslash {}^{*}N) = 1$$

the integral is the product of

$$\exp(\langle {}^{*}H, s({}^{*}H^{(1)})\rangle + \rho({}^{*}H))$$

and

$$\int_{{}^{\Psi}\Delta^{(2)}\cap {}^{*}\gamma^{-1}({}^{\Psi}\Delta^{(1)})^{*}\gamma\backslash {}^{\Psi}T^{(2)}} \exp(\langle {}^{\Psi}H({}^{\Psi}\gamma {}^{\Psi}t \exp {}^{\Psi}H\, mk^{-1}), {}^{\Psi}H^{(1)} + H_\rho\rangle)\Phi({}^{\Psi}\gamma {}^{\Psi}t \exp {}^{\Psi}H\, mk^{-1})d^{\Psi}t$$

Here $H = {}^{*}H + {}^{\Psi}H$ with ${}^{*}H$ in ${}^{*}\mathfrak{a}$ and ${}^{\Psi}H$ in ${}^{\Psi}\mathfrak{a}^{(2)}$. This integral equals

$$\exp(\langle {}^{\Psi}H, s({}^{\Psi}H^{(1)})\rangle + \rho({}^{\Psi}H)\rangle)N_{{}^{*}\gamma}({}^{\Psi}s, {}^{\Psi}H^{(1)})\Phi(m, k)$$

Thus $N_\gamma(H^{(1)}, s)$ is the restriction of $N_{{}^{*}\gamma}({}^{\Psi}H^{(1)}, s)$. Substituting in (4.i) we obtain the result.

Before proving the rest of the lemma we should comment on the formulation of part (ii). Suppose P and $\gamma_0 P \gamma_0^{-1} = P'$ are two conjugate cuspidal subgroups. Then $\gamma_0 S \gamma_0^{-1} = S'$ and we may suppose that split components A and A' for (P, S) and (P', S') respectively have been so chosen that $\gamma_0 A \gamma_0^{-1} = A'$. Every function ϕ on $AT\backslash G$ defines a function $\phi' = D\phi$ on $A'T'\backslash G$ by

$$\phi'(g) = \omega(a'(\gamma_0))\phi(\gamma_0^{-1}g)$$

Let us verify that

$$\int_{\Theta\backslash M\times K} |\phi(mk)|^2 dmdk = \int_{\Theta'\backslash M'\times K} |\phi'(m'k)|^2 dm'dk$$

Since we may suppose that $M' = \gamma_0 M \gamma_0^{-1}$ the right side is equal to

$$\omega^2(a'(\gamma_0)) \int_{\Theta\backslash M\times K} |\phi(m\gamma_0^{-1}k)|^2 \mu(m) dmdk$$

which equals

$$\omega^2(a'(\gamma_0)) \int_{M\times K} |\phi(mk)|^2 \mu(m) dmdk$$

if

$$\mu(m) = \frac{d(\gamma_0 m \gamma_0^{-1})}{dm}$$

The map $n \longrightarrow \gamma_0 n \gamma_0^{-1}$ of N to N' is measure preserving since $\Gamma \cap N$ is mapped to $\Gamma \cap N'$ and $\Gamma \cap N\backslash N$ and $\Gamma \cap N'\backslash N'$ both have measure one. Since the map $H \longrightarrow \mathrm{Ad}\,\gamma_0(H)$ of $\mathfrak{n}$ to $\mathfrak{n}'$ is an isometry the map $a \longrightarrow \gamma_0 a \gamma_0^{-1}$ of A to A' is measure preserving. If $\psi(g)$ is a continuous function on G with compact support then

$$\int_G \psi(g) dg$$

is equal to

$$\int_N dn \int_A \omega^2(a) da \int_M dm \int_K dk\, \psi(\gamma_0 namk)$$

which equals

$$\int_{N'} dn' \int_{A'} \omega^2(a') da' \int_M dm \int_K dk\, \psi(n'a'\gamma_0 m \gamma_0^{-1} \gamma_0 k)$$

The latter integral is in turn equal to

$$\omega^{-2}(a'(\gamma_0)) \int_{N'} dn' \int_{A'} \omega^2(a') da' \int_M dm \int_K dk \{\psi(n'a'm'k) \mu^{-1}(\gamma_0^{-1} m' \gamma_0)\}$$

We conclude that

$$\mu(m) \equiv \omega^{-2}(a'(\gamma_0))$$

and the assertion is verified. In the same way if ϕ is a function on $\Theta\backslash M$ and $\phi' = D\phi$ is defined by

$$\phi'(m') = \omega(a'(\gamma_0))\phi(\gamma_0^{-1} m \gamma_0)$$

then

$$\int_{\Theta\backslash M} |\phi(m)|^2 dm = \int_{\Theta'\backslash M'} |\phi'(m')|^2 dm'$$

The map D takes $\mathcal{L}_0(\Theta\backslash M)$ to $\mathcal{L}_0(\Theta'\backslash M')$. If V is an admissible subspace of $\mathcal{L}(\Theta\backslash M)$ and W is an admissible subspace of the space of functions on K then D takes $\mathcal{E}(V, W)$ to $\mathcal{E}(V', W)$ if $V' = DV$. If Φ belongs to $\mathcal{E}(V, W)$ let

$$D(H)\Phi = \exp(-\langle H'(\gamma_0), H\rangle)D\Phi$$

Then

$$E(g, \Phi, H) = \Sigma_{\Delta\backslash\Gamma} \exp(\langle H(\gamma g), H\rangle + \rho(H(\gamma g)))\Phi(\gamma g)$$

or

$$\Sigma_{\Delta'\backslash\Gamma} \exp(\langle H(\gamma_0^{-1}\gamma g), H\rangle + \rho(H(\gamma_0^{-1}\gamma g)))\Phi(\gamma_0^{-1}\gamma g)$$

If $g = n'a'm'k$ then

$$\gamma_0^{-1} g = (\gamma_0^{-1} n' \gamma_0)(\gamma_0^{-1} a' \gamma_0)(\gamma_0^{-1} m' \gamma_0)\gamma_0^{-1} k$$

so that

$$H(\gamma_0^{-1} g) = H'(g) + H(\gamma_0^{-1})$$

In particular $H(\gamma_0^{-1}) = -H'(\gamma_0)$. Consequently the sum equals

$$\Sigma_{\Delta'\backslash\Gamma} \exp(\langle H'(\gamma g), H\rangle + \rho(H'(\gamma g)))(D(H)\Phi)(g) = E(g, D(H)\Phi, H)$$

Thus the theory of Eisenstein series is the same for both cuspidal subgroups. This is the reason that only the case that the cuspidal subgroups P_1^* and P_2^* are equal is treated explicitly in the lemma. Finally we remark that if ϕ belongs to $\mathfrak{D}(V, W)$ and ϕ' is defined by $\phi'(\gamma_0 g) = \phi(g)$ then ϕ' belongs to $\mathfrak{D}(V, W)$ and

$$\Sigma_{\Delta\backslash\Gamma} \phi(\gamma g) = \Sigma_{\Delta'\backslash\Gamma} \phi'(\gamma g)$$

Part (iii) and the improved assertion of Lemma 4.4 will be proved at the same time by means of Fourier integrals. Suppose P is a percuspidal subgroup, V is an admissible subspace of $\mathcal{L}_0(\Theta\backslash M)$, and W is an admissible subspace of the space of functions on K. If $\phi(g)$ belongs to $\mathfrak{D}(V, W)$ then, for each a in A, let $\Phi'(a)$ be that element of $\mathcal{E}(V, W)$ whose value at (m, k) is $\phi(amk^{-1})$. If q is the rank of P and H belongs to $\mathfrak{a}_c$ let

$$\Phi(H) = \int_{\mathfrak{a}} \Phi'(\exp X)\exp(-\langle X, H\rangle - \rho(X))dX$$

$\Phi(H)$, which is a meromorphic function on $\mathfrak{a}_c$, will be called the Fourier transform of ϕ. By the inversion formula

$$\phi(g) = \left(\frac{1}{2\pi}\right)^q \int_{Re(H)=Y} \exp(\langle H(g), H\rangle + \rho(H(g)))\Phi(H, g)\,|dH|$$

if Y is any point in $\mathfrak{a}$ and $\Phi(H, g)$ is the value of $\Phi(H)$ at g. In the following ϕ will be chosen to be infinitely differentiable so that this integral is absolutely convergent. If

$$\alpha(Y) > \langle \alpha, \rho\rangle$$

for every simple root of $\mathfrak{a}$ then

$$\phi^{\wedge}(g) = \left(\frac{1}{2\pi}\right)^q \int_{Re(H)=Y} E(g, \Phi(H), H)\,|dH|$$

Since we have still to complete the proof of Lemma 4.4 we take $P^{(i)}$, $i = 1, 2$ to be cuspidal subgroups, $V^{(i)}$ to be an admissible subspace of $\mathcal{L}_0(\Theta^{(i)}\backslash M^{(i)})$, and $W^{(i)}$ to be an admissible subspace of the space of functions on K. Suppose ϕ belongs to $\mathcal{J}(V^{(1)}, W^{(1)})$ and Ψ belongs to $\mathcal{E}(V^{(2)}, W^{(2)})$; then it has to be shown that if the rank of $P^{(2)}$ is less than the rank of $P^{(1)}$ the integral

(4.j) $$\int_{T^{(1)}\backslash G} \phi(g) \left\{ \int_{\Delta^{(1)}\backslash T^{(1)}} \overline{E}(tg, \Psi, \overline{H}^{(2)}) dt \right\} dg$$

vanishes for all $H^{(2)}$ in $\mathfrak{U}^{(2)}$. As usual we write this as a sum over the double cosets $\Delta^{(2)}\backslash\Gamma/\Delta^{(1)}$ of

$$\int \phi(g) \left\{ \int \exp\left\langle H^{(2)}(\gamma t g), H^{(2)} + H_\rho \right\rangle \overline{\Psi}(\gamma t g) dt \right\} dg$$

The outer integral is over $T^{(1)}\backslash G$; the inner over $\Delta^{(1)} \cap \gamma^{-1}\Delta^{(2)}\gamma\backslash T^{(1)}$. We shall show that each term vanishes. A typical term equals

(4.k) $$\int_{\Delta^{(1)}\cap\gamma^{-1}\Delta^{(2)}\gamma\backslash G} \exp\left\langle H^{(2)}(\gamma g), H^{(2)} + H_\rho \right\rangle \phi(g)\overline{\Psi}(\gamma g) dg$$

which equals

$$\int_{\Delta^{(2)}\cap\gamma\Delta^{(1)}\gamma^{-1}\backslash G} \exp\left\langle H^{(2)}(g), H^{(2)} + H_\rho \right\rangle \phi(\gamma^{-1}g)\overline{\Psi}(g) dg$$

If

$$\phi(g) = \left(\frac{1}{2\pi}\right)^q \int_{\mathrm{Re}(H^{(1)})=Y} \exp\left\langle H^{(1)}(g), H^{(1)} + H_\rho \right\rangle \Phi(H^{(1)}, g) \, |dH^{(1)}|$$

with Y in $\mathfrak{U}^{(1)}$ and $\xi(H^{(1)}, H^{(2)})$ is obtained by integrating

$$\exp\left\langle H^{(1)}(\gamma^{-1} t \exp H^{(2)} m k^{-1}), H^{(1)} + H_\rho \right\rangle \Phi(H^{(1)}, \gamma^{-1} t \exp H^{(2)} m k^{-1}) \overline{\Psi}(m k^{-1})$$

first over $\Delta^{(2)} \cap \gamma^{-1}\Delta^{(1)}\gamma\backslash T^{(2)}$ with respect to dt and afterwards over $\Theta^{(2)}\backslash M^{(2)} \times K$ with respect to $dm\,dk$, then (4.k) equals

(4.ℓ) $$\int_{\mathfrak{U}^{(2)}} \exp\left\langle H, H^{(2)} - H_\rho \right\rangle \left\{ \int_{\mathrm{Re}\,H^{(1)}=Y} \xi(H^{(1)}, H) \, |dH^{(1)}| \right\} dH$$

Since $\xi(H^{(1)}, H)$ vanishes when rank $P^{(1)}$ is greater than rank $P^{(2)}$ so does (4.k). Suppose now that $P^{(1)}$ and $P^{(2)}$ are associate, that $V^{(1)}$ and $V^{(2)}$ are associate, and that $W^{(1)} = W^{(2)}$. Then $\xi(H^{(1)}, H^{(2)})$ equals

$$\exp(\langle H^{(2)}, sH^{(1)}\rangle + \rho(H^{(2)}))(N_{\gamma^{-1}}(s, H^{(1)})\Phi(H^{(1)}), \Psi)$$

where s is some element of $\Omega(\mathfrak{n}^{(1)}, \mathfrak{n}^{(2)})$ determined by γ. Substitute in (4.ℓ) to obtain

$$\int_{\mathfrak{n}^{(1)}} \exp(-\langle H, -s^{-1}H^{(2)}\rangle)\left\{\int_{\operatorname{Re} H^{(1)}=Y} \exp\langle H, H^{(1)}\rangle (N_{\gamma^{-1}}(s, H^{(1)})\Phi(H^{(1)}), \Psi)\,|dH^{(1)}|\right\} dH$$

The outer integral and the corresponding integral for

$$\text{(4.m)} \quad \int_{\mathfrak{n}^{(1)}} \exp(-\langle H, -s^{-1}H^{(2)}\rangle)\left\{\int_{\operatorname{Re} H^{(1)}=Y} \exp\langle H, H^{(1)}\rangle (N(s, H^{(1)})\Phi(H^{(1)}), \Psi)\,|dH^{(1)}|\right\} dH$$

which is obtained by summing over double cosets, are absolutely convergent. On the other hand (4.k) equals

$$\left(\frac{1}{2\pi}\right)^q \int_{\mathfrak{n}^{(1)}} \exp(-\langle H, -s^{-1}H^{(2)}\rangle \int_{\operatorname{Re}(H^{(1)})=Y} \exp\langle H, H^{(1)}\rangle (\Phi(H^{(1)}), N_\gamma(s^{-1}, \overline{H}^{(2)})\Psi)$$

The sum over double cosets equals

$$\text{(4.n)} \quad \left(\frac{1}{2\pi}\right)^q \int_{\mathfrak{n}^{(1)}} \exp(-\langle H, -s^{-1}H^{(2)}\rangle \int_{\operatorname{Re}(H^{(1)})=Y} \exp\langle H, H^{(1)}\rangle (\Phi(H^{(1)}), N(s^{-1}, \overline{H}^{(2)})\Psi)$$

Thus (4.m) and (4.n) are equal. From the Fourier inversion formula (4.n) equals

$$(\Phi(-s^{-1}H^{(2)}),\ N(s^{-1}, \overline{H}^{(2)})\Psi)$$

On the other hand the inner integral in (4.m) is the Fourier transform of a function analytic on $\mathfrak{U}^{(1)}$ and uniformly integrable along vertical "lines." Thus its product with $\exp -\langle H, H_0^{(1)}\rangle$ is absolutely integrable if $H_0^{(1)}$ is in $\mathfrak{U}^{(1)}$. Referring to (4.m) we see that this product is also integrable if $H_0^{(1)}$ is in $-s^{-1}(\mathfrak{U}^{(2)})$. From Hölder's inequality the product is integrable if $H_0^{(1)}$ is in the convex hull of these two sets and then the integral must give us the analytic continuation of

$$(N(s, H^{(1)})\Phi(H^{(1)}, \Psi)$$

to this region. Consequently

$$(N^*(s^{-1}, \bar{H}^2)\Phi(-s^{-1}H^{(2)}), \Psi) = (N(s, -s^{-1}H^{(2)})\Phi(-s^{-1}H^{(2)}), \Psi)$$

which proves (iii).

Finally we prove (i). We start from the observation made at the end of Lemma 4.1 that if C is a compact subset of $\Gamma\backslash G$ or of G then

$$\Sigma_{\Delta\backslash\Gamma} |\exp\langle H^{(1)}(\gamma g), H^{(1)} + H_\rho\rangle| |\Phi(\gamma g)| \leq c\|\Phi\| \exp\langle \mathrm{Re}\, H^{(1)}, H_0^{(1)}\rangle$$

for g in C. If $\omega \subseteq N^{(1)}$ and

$$N^{(1)} = (\Gamma \cap N^{(1)})\omega$$

and if $\omega g \subseteq C$ then

$$\Sigma_{\Delta^{(1)}\backslash\Gamma/\Delta^{(2)}} \int_{\Delta^{(2)}\cap\gamma^{-1}\Delta^{(1)}\gamma\backslash T^{(2)}} |\exp\langle H^{(1)}(\gamma t g), H^{(1)} + H_\rho\rangle| |\Phi(\gamma t g)| dt$$

is at most

$$c\|\Phi\| \exp\langle \mathrm{Re}\, H^{(1)}, H_0^{(1)}\rangle$$

This remains true if for each s in $\Omega(\mathfrak{a}^{(1)}, \mathfrak{a}^{(2)})$ we sum only over those γ such that $N(s, H^{(1)})$ is not identically zero. Then

$$|N(s, H^{(1)})\Phi(g)| \leq c\|\Phi\| \exp\langle H^{(1)}, H_0^{(1)}\rangle \exp - \langle H^{(2)}(g), \mathrm{Re}(s(H^{(1)})\rangle$$

which proves the assertion since the linear functionals on $\mathcal{E}(V^{(2)}, W)$ obtained from evaluating a function at a point span the space of linear functionals on $\mathcal{E}(V^{(2)}, W)$.

The relation of being associate breaks up the cuspidal subgroups into equivalence classes. A set of representatives $\{P\}$ for the conjugacy classes under Γ in one of these equivalence classes will be called a complete family of associate cuspidal subgroups. If $P_0 \in \{P\}$ and V_0 is a simple admissible subspace of $\mathcal{L}_0(\Theta_0\backslash M_0)$ then for each P in $\{P\}$ there is a simple admissible subspace associate to V_0. The family $\{V\}$ so obtained will be called a complete family of associate admissible subspaces. Let W be a simple admissible subspace of the space of functions on K. If P belongs to $\{P\}$ and V, which is a subspace of $\mathcal{L}(\Theta\backslash M)$, belongs to $\{V\}$, and if ϕ belongs to $\mathcal{J}(V, W)$ then $\phi^\wedge(g)$ belongs to $\mathcal{L}(\Gamma\backslash G)$. Let the closed space spanned by the functions $\phi^\wedge$ as P varies over $\{P\}$ be denoted by $\mathcal{L}(\{P\}, \{V\}, W)$. Whenever we have $\{P\}$, $\{V\}$, and W as above we will denote by $\mathfrak{a}^{(1)}, \ldots, \mathfrak{a}^{(r)}$ the distinct split components of the elements of $\{P\}$, by $P^{(i,1)}, \ldots, P^{(i,m_i)}$ those elements of $\{P\}$ with $\mathfrak{a}^{(i)}$ as split component, and by $\mathcal{E}^{(i)}$ the direct sum

$$\oplus_{k=1}^{m_i} \mathcal{E}(V^{(i,k)}, W)$$

Moreover if $H^{(i)}$ belongs to $\mathfrak{a}^{(i)}$ and s belongs to $\Omega(\mathfrak{a}^{(i)}, \mathfrak{a}^{(j)})$ we will denote the linear transformation from $\mathcal{E}^{(i)}$ to $\mathcal{E}^{(j)}$ which takes Φ in $\mathcal{E}(V^{(i,k)}, W)$ to that element of $\mathcal{E}^{(j)}$ whose component in $\mathcal{E}(V^{(j,\ell)}, W)$ is $N(H^{(i)}, s)\Phi$ by $M(H^{(i)}, s)$. Of course $N(H^{(i)}, s)$ depends on $P^{(i,k)}$ and $P^{(j,\ell)}$ and is not everywhere defined. Finally if

$$\Phi = \oplus_{k=1}^{m_i} \Phi_k$$

belongs to $\mathcal{E}^{(i)}$ we let

$$E(g, \Phi, H^{(i)}) = \Sigma_{k=1}^{m_i} E(g, \Phi_k, H^{(i)})$$

LEMMA 4.6. (i) <u>Suppose</u> $\{P\}_i$, $\{V\}_i$, W_i, $i = 1, 2$, <u>are respectively a complete family of associate cuspidal subgroups, a complete family of associate admissible subspaces, and a simple admissible subspace of the space of functions on</u> K; <u>then</u> $\mathcal{L}(\{P\}_1, \{V\}_1, W_1)$ **<u>is orthogonal to</u>** $\mathcal{L}(\{P\}_2, \{V\}_2, W_2)$ <u>unless</u> $\{P\}_1$ <u>and</u> $\{P\}_2$

are representatives of the same equivalence class, the elements of $\{V\}_1$ and $\{V\}_2$ are associate, and $W_1 = W_2$. Moreover $\mathcal{L}(\Gamma\backslash G)$ is the direct sum of all the spaces $\mathcal{L}(\{P\}, \{V\}, W)$ and, for a fixed $\{P\}$ and $\{V\}$, $\oplus_W \mathcal{L}(\{P\}, \{V\}, W)$ is invariant under G.

(ii) If $\{P\}$, $\{V\}$, and W are given and if, for $1 \le i \le r$ and $1 \le k \le m_i$, $\phi_{i,k}$ and $\psi_{i,k}$, which belong to $\mathcal{H}(V^{(i,k)}, W)$, are the Fourier transforms of $\Phi_{i,k}(H^{(i)})$ and $\Psi_{i,k}(H^{(i)})$ respectively let

$$\Phi_i(H^{(i)}) = \oplus_{k=1}^{m_i} \Phi_{i,k}(H^{(i)})$$

and

$$\Psi_i(H^{(i)}) = \oplus_{k=1}^{m_i} \Psi_{i,k}(H^{(i)})$$

Then

$$\int_{\Gamma\backslash G} \Sigma_{i=1}^{r} \Sigma_{j=1}^{r} \Sigma_{k=1}^{m_i} \Sigma_{\ell=1}^{m_j} \hat{\phi}_{i,k}(g) \bar{\hat{\psi}}_{j,\ell}(g)\, dg \tag{4.o}$$

is equal to

$$\left(\frac{1}{2\pi}\right)^q \int_{\mathrm{Re}(H^{(i)})=Y^{(i)}} (M(s, H^{(i)})\Phi(H^{(i)}), \Psi_j(-s\bar{H}^{(i)}))\, |dH^{(i)}| \tag{4.p}$$

summed over s in $\Omega(\mathfrak{a}^{(i)}, \mathfrak{a}^{(j)})$ and $1 \le i, j \le r$. Here q is the rank of the elements of $\{P\}$ and $Y^{(i)}$ is a real point in $\mathfrak{A}^{(i)}$.

Suppose $P^{(i)}$, $i = 1, 2$, are cuspidal subgroups, suppose $V^{(i)}$ is an admissible subspace of $\mathcal{L}_0(\Theta^{(i)}\backslash M^{(i)})$, and $W^{(i)}$ is an admissible subspace of the space of functions on K. If ϕ belongs to $\mathcal{H}(V^{(1)}, W^{(1)})$ and ψ belongs to $\mathcal{H}(V^{(2)}, W^{(2)})$ let

$$\phi(g) = \left(\frac{1}{2\pi}\right)^q \int_{\mathrm{Re}(H^{(1)})=Y^{(1)}} \exp\langle H^{(1)}(g), H^{(1)} + H_\rho\rangle \Phi(H^{(1)}, g)\, |dH^{(1)}|$$

$$\psi(g) = \left(\frac{1}{2\pi}\right)^q \int_{\mathrm{Re}\, H^{(2)}=Y^{(2)}} \exp\langle H^{(2)}(g), H^{(2)} + H_\rho\rangle \Psi(H^{(2)}, g)\, |dH^{(2)}|$$

Then

$$\int_{\Gamma\backslash G} \phi^\wedge(g)\bar{\psi}^\wedge(g)$$

is equal to

$$\left(\frac{1}{2\pi}\right)^q \int_{\operatorname{Re} H^{(1)}=Y^{(1)}} \left\{ \int_{\Delta^{(2)}\backslash G} \bar{\psi}(g)E(g,\ \Phi(H^{(1)}),\ H^{(1)})dg \right\} |dH^{(1)}| \tag{4.q}$$

if $Y^{(1)}$ belongs to $U^{(1)}$. The inner integral is of the same form as (4.j) and as we know vanishes unless $P^{(1)}$ and $P^{(2)}$ are associate. If $P^{(1)}$ and $P^{(2)}$ are associate and $V^{(i)}$ and $W^{(i)}$ are simple admissible spaces for $i = 1, 2$ then it is zero unless $V^{(1)}$ and $V^{(2)}$ are associate and $W^{(1)} = W^{(2)}$. Finally if $P^{(1)}$ and $P^{(2)}$ and $V^{(1)}$ and $V^{(2)}$ are associate and $W^{(1)} = W^{(2)} = W$ the inner integral is readily seen to equal

$$\left(\frac{1}{2\pi}\right)^q \sum_{s\in\Omega(\mathfrak{a}^{(1)},\ \mathfrak{a}^{(2)})} (N(H^{(1)},\ s)\Phi(H^{(1)}),\ \Psi(-s\bar{H}^{(1)}))$$

This proves part (ii) of the lemma and the first assertion of part (i). The second assertion follows readily from the second corollary to Lemma 3.7.

To complete the proof of part (i) it is enough to show that

$$\oplus_W \mathfrak{L}(\{P\},\ \{V\},\ W)$$

is invariant under $\lambda(f)$ when f is continuous with compact support. If W_1 and W_2 are simple admissible subspaces of the space of functions on K define $\mathcal{C}(W_1, W_2)$ to be the set of all continuous functions on G with compact support such that $f(k^{-1}g)$ belongs to W_1 for each g in G and $f(gk^{-1})$ belongs to W_2 for each g in G. It is enough to show that for any W_1 and W_2 the space

$$\oplus_W \mathfrak{L}(\{P\},\ \{V\},\ W)$$

is invariant under $\lambda(f)$ for all f in $\mathcal{C}(W_1, W_2)$. Suppose $\phi(g)$ belongs to $\mathfrak{J}(V, W)$ for some V in $\{V\}$ and some W and

$$\phi(g) = \left(\frac{1}{2\pi}\right)^q \int_{\operatorname{Re} H = Y} \exp(\langle H(g), H\rangle + \rho(H(g)))\Phi(H, g)\,|dH|$$

If f belongs to $\mathcal{C}(W_1, W_2)$ then

$$\lambda(f)\phi(g) = \int_G \phi(gh)f(h)dh$$

equals 0 unless $W_2 = W$. If $W_2 = W$ it is readily seen that $\lambda(f)\phi$ belongs to $\mathcal{V}(V, W_1)$; since

$$\lambda(f)\phi^\wedge = (\lambda(f)\phi)^\wedge$$

the third assertion of part (i) is proved. Moreover

$$\text{(4.r)} \qquad \lambda(f)\phi(g) = \left(\frac{1}{2\pi}\right)^q \int_{\operatorname{Re} H = Y} \exp(\langle H(g), H\rangle + \rho(H(g))\Phi'(H, g)dH$$

if $\Phi'(H) = \pi(f, H)\Phi(H)$.

Let us now introduce some notation which will be useful later. Suppose $\{P\}$, $\{V\}$, and W are given. Suppose that, for $1 \leq i \leq r$, $\Phi_i(H^{(i)})$ is a function defined on some subset of $\mathfrak{a}_c^{(i)}$ with values in $\mathcal{E}^{(i)}$. We shall use the notation $\Phi(H)$ for the r-tuple $(\Phi_1(H^{(1)}), \ldots, \Phi_r(H^{(r)}))$ of functions and occasionally talk of Φ as though it were a function. If $\Phi_1(H^{(1)}), \ldots, \Phi_r(H^{(r)})$ arise as in part (ii) of the lemma let us denote

$$\Sigma_{i=1}^r \Sigma_{k=1}^{m_i} \phi_{i,k}^\wedge$$

by $\phi^\wedge$. If $R^2 > \langle \rho, \rho\rangle$ the map $\Phi(\cdot) \longrightarrow \phi^\wedge$ can be extended to the space of all functions $\Phi(H) = (\Phi_1(H^{(1)}), \ldots, \Phi_r(H^{(r)}))$ which are such that $\Phi_i(H^{(i)})$ is analytic on

$$\{H^{(i)} \in \mathfrak{a}_c^{(i)} \mid \|\operatorname{Re}(H^{(i)})\| < R\}$$

and dominated on this set by a square integrable function of $\operatorname{Im}(H^{(i)})$. The formula of part (ii) of the lemma will still be valid. In particular the map can be extended to the set $\mathcal{H}$ of all functions $\Phi(H)$ such that $\Phi_i(H^{(i)})$ is analytic on the above set and $\| p(\mathrm{Im}(H^{(i)}))\Phi(H^{(i)}) \|$ is bounded on the above set if p is any polynomial. $\mathcal{H}$ is invariant under multiplication by polynomials.

5. Miscellaneous lemmas.

In order to avoid interruptions later we collect together in this section a number of lemmas necessary in the proof of the functional equations of the Eisenstein series.

LEMMA 5.1. *Let* ϕ *be a continuous function on* $\Gamma\backslash G$ *and suppose that there is a constant* r *so that if* $\mathfrak{S}'$ *is a Siegel domain associated to a percuspidal subgroup* P' *there is a constant* c' *such that* $|\phi(g)| \leq c'\eta^{r}(a'(g))$ *if* g *belongs to* $\mathfrak{S}'$. *Suppose that there is an integer* q *so that if* ${}^{*}P$ *is any cuspidal subgroup then the cuspidal component of*

$$ {}^{*}\phi^{\wedge}(g) = \int_{\Gamma \cap {}^{*}N \backslash {}^{*}N} \phi(ng)\,dn $$

is zero unless the rank of ${}^{*}P$ *equals* q. *Let* $\{P_1, \ldots, P_s\}$ *be a set of representatives for the conjugacy classes of cuspidal subgroups of rank* q *and for each* i *let* V_i *be an admissible subspace of* $\mathcal{L}_0(\Theta_i\backslash M_i)$; *let* W *be an admissible space of functions on* K. *Suppose there is an integer* N *so that if* $\{p_i^{(k)} \mid 1 \leq k \leq t\}$ *is a basis for the polynomials on* $\mathfrak{a}_i$ *of degree at most* N *then*

$$ (5.a) \qquad \int_{\Gamma\cap N_i\backslash N_i} \phi(ng)\,dn = \Sigma_{j=1}^{s_i} \exp\left\langle H_i(g),\, H_i^{(j)}\right\rangle \Sigma_{k=1}^{t}\, p_i^{(k)}(H_i(g))\,\Phi_i^{(j,k)}(g) $$

with $\Phi_i^{(j,k)}$ *in* $\mathcal{E}(V_i, W)$. *Let* $\{p_i \mid 1 \leq i \leq u\}$ *be a basis for the polynomials on* $\mathfrak{h}$ *of degree at most* N; *then given any percuspidal subgroup* P *and any Siegel domain* $\mathfrak{S}$ *associated to* P *there is a constant* c *so that on* $\mathfrak{S}$

$$ (5.b) \qquad |\phi(g)| \leq c\left\{\Sigma_{i=1}^{s}\,\Sigma_{j=1}^{s_i} \exp\left\langle H(g),\, \operatorname{Re} H_i^{(j)}\right\rangle\right\}\left\{\Sigma_{k=1}^{u}\, |p_k(H(g))|\right\} $$

Suppose f is an infinitely differentiable function on G with compact support such that $f(kgk^{-1}) = f(g)$ for all g and k. Let $\phi_1 = \lambda(f)\phi$. If ${}^{*}P$ is any cuspidal subgroup, ${}^*V$ an admissible subspace of $\mathcal{L}_0({}^*\Theta\backslash{}^*M)$, ${}^*W$ an admissible space of functions on K, and ψ an element of $\mathfrak{I}({}^*V, {}^*W)$ we have

$$\int_{{}^*T\backslash G} \psi(g)\,{}^*\phi_1^\wedge(g)dg = \int_{{}^*T\backslash G} \lambda(f^*)\psi(g)\,{}^*\phi^\wedge(g)dg$$

If ψ belongs to $\mathfrak{I}({}^*V, {}^*W)$ so does $\lambda(f^*)\psi$ so that both integrals are zero if the rank of ${}^*P$ is not q. On the other hand if H_i belongs to the complexification of the split component of P_i and $\Phi_i^{(k)}$, $1 \leq k \leq t$, belongs to $\mathcal{E}(V_i, W)$ then the result of applying $\lambda(f)$ to the function

$$\exp\langle H_i(g), H_i\rangle\left\{\Sigma_{k=1}^t p_i^{(k)}(H_i(g))\Phi_i^{(k)}(g)\right\}$$

is the function

$$\exp\langle H_i(g), H_i\rangle\left\{\Sigma_{k=1}^t p_i^{(k)}(H_i(g))\left(\Sigma_{\ell=1}^t \pi^{(k,\ell)}(f, H_i)\Phi_i^{(\ell)}\right)(g)\right\}$$

where $\pi^{(k,\ell)}(f, H_i)$ is a linear transformation on $\mathcal{E}(V^{(i)}, W)$. The matrix $(\pi^{(k,\ell)}(f, H_i))$ defines a linear transformation on

$$\oplus_{k=1}^t \mathcal{E}(V^{(i)}, W)$$

which we will denote by $\pi(f, H_i)$ even though $\pi(f, H_i)$ usually has another meaning. Given the finite sets of points $H_i^{(1)}, \ldots, H_i^{(s_i)}$ we readily see that we can choose f so that $\pi(f, H_i^{(j)})$ is the identity for $1 \leq i \leq s$, $1 \leq j \leq s_i$.

$$\int_{\Gamma\cap N_i\backslash N_i} \phi(ng)dn = \int_{\Gamma\cap N_i\backslash N_i} \phi_1(ng)dn$$

for $1 \leq i \leq s$. It follows from Lemma 3.7 that $\lambda(f)\phi = \phi$. Arguing the same way as in the proof of Lemma 4.1 we see that if X is in the centre of the universal enveloping algebra then the result of applying $\lambda(X)$ to the function

$$\exp\langle H_i(g), H_i\rangle\left\{\Sigma_{k=1}^t p_i^{(k)}(H_i(g))\Phi_i^{(k)}(g)\right\}$$

is the function

$$\exp\left\langle H_i(g),\ H_i\right\rangle\left\{\Sigma_{k=1}^{t} p_i^{(k)}(H_i(g))\left(\Sigma_{\ell=1}^{t} \pi^{(k,\ell)}(X,\ H_i)\Phi_i^{(\ell)}\right)(g)\right\}$$

where $\pi^{(k,\ell)}(X, H_i)$ is a linear transformation on $\mathcal{L}(V^{(i)}, W)$. It then follows readily that there are points $Z_1, \ldots, Z_m$ in $\mathfrak{j}_c$, irreducible representations $\sigma_1, \ldots, \sigma_n$ of K, and an integer ℓ_0 so that ϕ belongs to

$$\mathcal{J}(Z_1, \ldots, Z_m; \sigma_1, \ldots, \sigma_n; \ell_0)$$

If $q = 0$ the inequality (5.b) merely asserts that $\phi(g)$ is bounded on any Siegel domain. That this is so follows of course from Lemma 3.5 and the corollary to Lemma 3.4. The lemma will be proved for a general value of q by induction. Suppose q is positive. If $\{\alpha_1, \ldots, \alpha_p\}$ is the set of simple roots of $\mathfrak{f}$ let ${}^*P_i$ be the cuspidal subgroup belonging to P determined by $\{\alpha_j, \,|j \neq i\}$. It follows from Lemma 4.2 that

$$\int_{\Gamma\cap {}^*N_i\backslash {}^*N_i} \phi(ng)dn = \Sigma_{j=1}^{j_i} \exp\left\langle {}^*H_i(g),\ {}^*H_i^{(j)}\right\rangle \Sigma_{k=1}^{k_i} q_i^{(k)}({}^*H_i(g))\phi_i^{(j,k)}(g)$$

where $\phi_i^{(j,k)}$ is a function on ${}^*A_i\, {}^*T_i\backslash G$, the elements ${}^*H_i^{(j)}$, $1 \leq j \leq j_i$, are distinct, and the set of homogeneous polynomials $q_i^{(1)}, \ldots, q_i^{(k_i)}$ is linearly independent. Let us consider $\phi_i^{(j,k)}$ as a function on ${}^*\Theta_i \times \{1\} \backslash {}^*M_i \times K$ and show that it satisfies the conditions of the lemma. Since the functions

$$\exp\left\langle {}^*H_i,\ {}^*H_i^{(j)}\right\rangle q_i^{(k)}({}^*H_i),\ 1 \leq j \leq j_i,\ 1 \leq k \leq k_i$$

are linearly independent, $\phi_i^{(j,k)}(m, k)$ is a linear combination of functions of the form

$$\int_{\Gamma\cap {}^*N_i\backslash {}^*N_i} \phi(namk^{-1})dn$$

with a in ${}^*A_i$. Any condition of the lemma which is satisfied by the latter functions will also be satisfied by each of the functions $\phi_i^{(j,k)}$. Lemma 3.3 shows that the condition on the rate of growth on Siegel domains is satisfied. The proof of Lemma 3.7 shows that if ${}^{\Psi}P \times K$ is a cuspidal subgroup of ${}^{*}M_i \times K$ then the cuspidal component of

$$\int_{{}^{*}\Theta_i \cap {}^{\Psi}N \backslash {}^{\Psi}N} \left\{ \int_{\Gamma \cap {}^{*}N_i \backslash {}^{*}N_i} \phi(nan_1 mk^{-1}) dn \right\} dn_1$$

is zero unless the rank of ${}^{\Psi}P$, or equivalently ${}^{\Psi}P \times K$, is $q-1$. Finally we must find the analogue of the form (5.a).

In order to free the indices i, j, and k for other use we set $i = i_0$, $j = j_0$, and $k = k_0$. If P' is a cuspidal subgroup of rank q to which ${}^{*}P_{i_0}$ belongs suppose for simplicity that $P' = P_i$ for some i. If F is the subset of $\{1, \ldots, s_i\}$ consisting of those j such that the projection of $H_i^{(j)}$ on the complexification of ${}^{*}\mathfrak{a}_{i_0}$ equals ${}^{*}H_{i_0}^{(j_0)}$ and if $r^{(1)}, \ldots, r^{(t_i)}$ is a basis for the polynomials on the orthogonal complement ${}^{\Psi}\mathfrak{a}_i$ of ${}^{*}\mathfrak{a}_{i_0}$ in $\mathfrak{a}_i$ of degree at most $N-M$, with M equal to the degree of $q_{i_0}^{(k_0)}$, then

$$\int_{{}^{*}\Theta_{i_0} \cap {}^{\Psi}N_i \backslash {}^{\Psi}N_i} \phi_{i_0}^{(j_0,k_0)}(nm, k)dn$$

is equal to

$$(5.c) \qquad \Sigma_{j \in F} \exp \langle {}^{\Psi}H_i(m), {}^{\Psi}H_i^{(j)} \rangle \Sigma_{k=1}^{t_i} r^{(k)}({}^{\Psi}H_i(m)) \Psi_i^{(j,k)}(mk^{-1})$$

Here

$${}^{\Psi}P_i = {}^{*}N_{i_0} \backslash P_i \cap {}^{*}S_{i_0}, \quad H_i^{(j)} = {}^{*}H_i^{(j)} + {}^{\Psi}H_i^{(j)}$$

with ${}^{*}H_i^{(j)}$ in the complexification of $\mathfrak{a}_{i_0}$ and ${}^{\Psi}H_i^{(j)}$ in the complexification of ${}^{\Psi}\mathfrak{a}_i$. The functions $\Psi_i^{(j,k)}$ are linear combinations of the functions $\Phi_i^{(j,k)}$. Considered as functions on ${}^{*}M_{i_0} \times K$ they belong to $\mathcal{E}(V_i \times W, W^{*})$ as we saw when proving Lemma 3.5.

Applying the induction assumption to each of the functions $\phi_i^{(j,k)}$ we see that if ${}^{\Psi}\mathfrak{S}_i$ is a Siegel domain associated to a percuspidal subgroup of ${}^*M_i$ there is a constant c_i so that if $g = n_i a_i m_i k_i$ and m_i belongs to ${}^{\Psi}\mathfrak{S}_i$ then

$$\left|\int_{\Gamma\cap {}^*N_i\backslash {}^*N_i} \phi(ng)dn\right| \leq c_i\left\{\Sigma_{i=1}^{s}\Sigma_{j=1}^{s_i}\exp\left\langle H(g),\ \operatorname{Re}H_i^{(j)}\right\rangle\right\}\left\{\Sigma_{k=1}^{u}|p_k(H(g))|\right\}$$

Suppose $\mathfrak{S}$ is a Siegel domain associated to P. It is enough to establish the inequality (5.b) on each

$$\mathfrak{S}_i = \{g \in \mathfrak{S} \,|\, \xi_{\alpha_{i,}}(a(g)) \geq \xi_{\alpha_{j,}}(a(g)),\ 1 \leq j \leq p\}$$

It is not difficult to see that there is a Siegel domain ${}^{\Psi}\mathfrak{S}_i$ associated to a percuspidal subgroup of ${}^*M_i$ so that $\mathfrak{S}$ is contained in ${}^*N_i\,{}^*A_i\,{}^{\Psi}\mathfrak{S}_i K$; the simple calculations necessary for a complete verification are carried out later in this section. Since $\lambda(f)\phi = \phi$ we see from Lemma 3.4 that if b is any real number there is a constant c_i' so that

$$\left|\phi(g) - \int_{\Gamma\cap {}^*N_i\backslash {}^*N_i} \phi(ng)dn\right| \leq c_i'\eta^b(a(g))$$

on $\mathfrak{S}_i$. For b sufficiently small $\eta^b(a(g))$ is bounded on $\mathfrak{S}$ by a constant times the expression in brackets on the right side of (5.b) so the lemma is proved.

COROLLARY. <u>If, for each</u> i <u>and</u> j,

$$\operatorname{Re}(\alpha_{,k}(H_i^{(j)})) < \langle \alpha_{,k},\ \rho\rangle,\ 1 \leq k \leq p$$

<u>then</u> ϕ <u>is square integrable on</u> $\Gamma\backslash G$.

It has only to be verified that the right side of (5.b) is square integrable on any Siegel domain. This is a routine calculation.

LEMMA 5.2. *Let* $\{\phi_n\}$ *be a sequence of functions on* $\Gamma\backslash G$ *and suppose that for each* n *there is a constant* $r(n)$ *so that if* $\mathfrak{S}'$ *is a Siegel domain associated to a percuspidal subgroup there is a constant* $c'(n)$ *so that*

$$|\phi(g)| \le c'(n)\eta^{r(n)}(a'(g))$$

if g *belongs to* $\mathfrak{S}'$. *Suppose that there is an integer* q *so that if* ${}^*P$ *is any cuspidal subgroup then the cuspidal component of*

$$\int_{\Gamma\cap {}^*N\backslash {}^*N} \phi_n(ng)dn$$

is zero unless the rank of ${}^*P$ *is* q. *Let* $\{P_1, \dots, P_s\}$ *be a set of representatives for the conjugacy classes of cuspidal subgroups of rank* q *and for each* i *let* V_i *be an admissible subspace of* $\mathcal{L}_0(\Theta_i\backslash M_i)$; *let* W *be an admissible space of functions on* K. *Suppose there is an integer* N *so that if* $\{p_i^{(k)} | 1 \le k \le t\}$ *is a basis for the polynomials on* $\mathfrak{a}_i$ *of degree at most* N *then*

$$\int_{\Gamma\cap N_i\backslash N_i} \phi_n(ng)dn = \Sigma_{j=1}^{s_i} \exp\langle H_i(g), H_{n,i}^{(j)}\rangle \Sigma_{k=1}^{t} p_i^{(k)}(H_i(g))\Phi_{n,i}^{(j,k)}(g)$$

with $H_{n,i}^{(j)}$ *in the complexification of* $\mathfrak{a}_i$ *and* $\Phi_{n,i}^{(j,k)}$ *in* $\mathcal{E}(V_i, W)$. *Finally suppose that*

$$\lim_{n\to\infty} H_{n,i}^{(j)} = H_i^{(j)}$$

and

$$\lim_{n\to\infty} \Phi_{n,i}^{(j,k)} = \Phi_i^{(j,k)}$$

exist for all i, j, *and* k. *Then there is a function* ϕ *on* $\Gamma\backslash G$ *so that*

$$\lim_{n\to\infty} \phi_n(g) = \phi(g)$$

uniformly on compact sets. Moreover if $\mathfrak{S}$ is any Siegel domain associated to a percuspidal subgroup there is a constant c so that $|\phi_n(g)|$ is less than or equal to

$$(5.d)\quad c\left\{\Sigma_{i=1}^{s}\Sigma_{j=1}^{s_i}\Sigma_{k=1}^{t}\|\Phi_{n,i}^{(j,k)}\|\right\}\left\{\Sigma_{i=1}^{s}\Sigma_{j=1}^{s_i}\exp\left\langle H(g),\ \operatorname{Re} H_{n,i}^{(j)}\right\rangle\right\}\left\{\Sigma_{k=1}^{u}|p_k(H(g))|\right\}$$

The polynomials p_k are the same as in the previous lemma. If f is an infinitely differentiable function on G with compact support such that

$$f(kgk^{-1}) = f(g)$$

for all g and k then define $\pi(f, H_i)$ as in the proof of the previous lemma. Choose f so that $\pi(f, H_i^{(j)})$ is the identity for $1 \leq i \leq s$, $1 \leq j \leq s_i$. If we take the direct sum

$$\oplus_{i=1}^{s}\oplus_{j=1}^{s_i}\oplus\left(\Sigma_{k=1}^{t}\ \mathcal{E}(V^{(i)},\ W)\right)$$

then we can define the operator

$$\oplus_{i=1}^{s}\oplus_{j=1}^{s_i}\oplus\ \pi(f,\ H_{n,i}^{(j)}) = \pi_n$$

on this space. For n sufficiently large the determinant of π_n will be at least $\frac{1}{2}$. Thus for n sufficiently large there is a polynomial p_n of a certain fixed degree with no constant term and with uniformly bounded coefficients so that $p_n(\pi_n)$ is the identity. Because of Lemmas 3.7 and 5.1 we can ignore any finite set of terms in the sequence so we suppose that p_n is defined for all n. The function $f_n = p_n(f)$ is defined in the group algebra and the argument used in the proof of Lemma 5.1 shows that $\lambda(f_n)\phi_n = \phi_n$. There is a fixed compact set which contains the support of all of the functions f_n and if X belongs to $\mathfrak{g}$ there is a constant μ so that

$$|\lambda(X)f_n(g)| \leq \mu$$

for all n and all g.

In the statement of the lemma the limit as n approaches infinity of $\Phi_{n,i}^{(j,k)}$ is to be taken in the norm on $\mathcal{L}(V_i, W)$ that has been introduced earlier. This, as we know, implies uniform convergence. Thus if q equals zero the first assertion of the lemma is immediate. The second is also; so we suppose that q is positive and proceed by induction. Let

$$\nu(n) = \Sigma_{i=1}^{s} \Sigma_{j=1}^{s_i} \Sigma_{k=1}^{t} \| \Phi_{n,i}^{(j,k)} \|$$

If $\lim_{n\to\infty} \nu_n = 0$ we have only to establish the inequality (5.d) because we can then take ϕ to be zero. Since ϕ_n is zero when $\nu(n)$ is, the lemma will be valid for the given sequence if it is valid for the sequence which results when all terms with $\nu(n)$ equal to zero are removed. We thus suppose that $\nu(n)$ is different from zero for all n. If the lemma were false for a given sequence with $\lim_{n\to\infty} \nu(n) = 0$ then from this sequence we could select a subsequence for which the lemma is false and for which

$$\lim_{n\to\infty} \nu^{-1}(n) \Phi_{n,i}^{(j,k)}$$

exists for all i, j, and k; replacing the elements of this subsequence by $\nu^{-1}(n)\phi_n$ we obtain a sequence for which the lemma is false and for which $\lim_{n\to\infty} \nu(n) = 1$. We now prove the lemma in the case that $\lim_{n\to\infty} \nu(n)$ is not zero.

Let $\mathfrak{S}^{(1)}, \ldots, \mathfrak{S}^{(v)}$ be a set of Siegel domains, associated to the percuspidal subgroups $P^{(1)}, \ldots, P^{(v)}$ respectively, which cover $\Gamma\backslash G$. If $\{\phi_n\}$ is any sequence satisfying the conditions of the lemma it follows from Lemmas 3.7 and 5.1 that for $1 \leq n < \infty$ there is a constant $c_1(n)$ so that

$$(5.e) \qquad |\phi_n(g)| \leq c_1(n) \left\{ \Sigma_{i=1}^{s} \Sigma_{j=1}^{s_i} \exp \left\langle H^{(x)}(g), \operatorname{Re} H_{n,i}^{(j)} \right\rangle \right\} \left\{ \Sigma_{k=1}^{u} | p_k(H^{(x)}(g)) | \right\}$$

if g belongs to $\mathfrak{S}^{(x)}$. It may be supposed that $c_1(n)$ is the smallest number for which (5.e) is valid. Since we can always take $\mathfrak{S}$ to be one of $\mathfrak{S}^{(1)}, \ldots, \mathfrak{S}^{(v)}$

the inequality (5.d) will be proved, at least when $\lim_{n\to\infty} \nu(n)$ is not zero, if it is shown that the sequence $\{c_1(n)\}$ is bounded. At the moment however there are still two possibilities: either the sequence is bounded or it is not. In the second case replace ϕ_n by $c_1^{-1}(n)\phi_n$ and, for the present at least, assume that $\{c_1(n)\}$ is bounded. It follows from Ascoli's lemma and the relation $\lambda(f_n)\phi_n = \phi_n$ that we can choose a subsequence $\{\phi'_n\}$ so that

$$\lim_{n\to\infty} \phi'_n(g) = \phi(g)$$

exists for each g and the convergence is uniform on compact sets. Lemma 3.3 and the dominated convergence theorem imply that if ${}^*P$ is a cuspidal subgroup of rank different from q then the cuspidal component of

$$\int_{\Gamma\cap {}^*N\backslash {}^*N} \phi(ng)dn$$

is zero. Moreover

$$\int_{\Gamma\cap N_i\backslash N_i} \phi(ng)dn = \Sigma_{j=1}^{s_i} \exp\left\langle H_i(g), \lim H_{n,i}^{(j)}\right\rangle \Sigma_{k=1}^{t} p_i^{(k)}(H_i(g)) \lim \Phi_i^{(j,k)}(g)$$

If $\{\phi_n\}$ did not converge to ϕ uniformly on all compact sets we could choose another subsequence which converged to ϕ' which is not equal to ϕ; but the cuspidal component of

$$\int_{\Gamma\cap {}^*N\backslash {}^*N} \phi(ng) - \phi'(ng)dn$$

would be zero for any cuspidal subgroup. According to Lemma 3.7 this is impossible. For the same reasons, if

$$\lim_{n\to\infty} \Phi_{n,i}^{(j,k)} = 0$$

for all i, j, and k then ϕ is zero. In order to exclude the second possibility for (5.e) it has to be shown that if (5.e) is satisfied with a bounded sequence $\{c_1(n)\}$

and

$$\lim_{n\to\infty} \Phi^{(j,k)}_{n,i} = 0$$

for all i, j, and k then

$$\lim_{n\to\infty} c_1(n) = 0$$

Once the second possibility is excluded the lemma will be proved.

We will suppose that $\lim_{n\to\infty} c_1(n)$ is not zero and derive a contradiction. Passing to a subsequence if necessary it may be supposed that there is a definite Siegel domain, which we again call $\mathfrak{S}$, among $\mathfrak{S}^{(1)}, \ldots, \mathfrak{S}^{(v)}$ so that

$$(5.f) \quad \sup_{g\in\mathfrak{S}} |\phi_n(g)| \left\{\Sigma_{i=1}^{s}\Sigma_{j=1}^{s_i} \exp\langle H(g), \operatorname{Re} H^{(j)}_{n,i}\rangle\right\}^{-1} \left\{\Sigma_{k=1}^{u} |p_k(H(g))|\right\}^{-1} = c_1(n)$$

is greater than or equal to $\varepsilon > 0$ for all n; $\mathfrak{S}$ is of course associated to the percuspidal subgroup P. Let ${}^*P_i$, $1 \le i \le p$, be the cuspidal subgroup belonging to P determined by the set $\{\alpha_{j,} \mid j \neq i\}$ and let

$$\mathfrak{S}_i = \{g \in \mathfrak{S} \mid \xi_{\alpha_{i,}}(a(g)) \ge \xi_{\alpha_{j,}}(a(g)),\ 1 \le j \le p\}$$

Suppose it could be shown that there is a sequence $\{c'_1(n)\}$ of numbers converging to zero so that

$$\left|\int_{\Gamma\cap {}^*N_\ell \backslash {}^*N_\ell} \phi_n(ng)\,dn\right|$$

is less than or equal to

$$(5.g) \qquad c'_1(n)\left\{\Sigma_{i=1}^{s}\Sigma_{j=1}^{s_i} \exp\langle H(g), \operatorname{Re} H^{(j)}_{n,i}\rangle\right\} \left\{\Sigma_{k=1}^{u} |p_k(H(g))|\right\}$$

if g belongs to $\mathfrak{S}_\ell$. Then it would follow from Lemma 3.4 that there was a constant c' independent of n so that, for g in $\mathfrak{S}$, $|\phi(g)|$ is at most

$$(c'_1(n) + c'c_1(n)\eta^{-1}(a(g)))\left\{\Sigma_{i=1}^{s}\Sigma_{j=1}^{s_i} \exp\langle H(g), \operatorname{Re} H^{(j)}_{n,i}\rangle\right\} \left\{\Sigma_{k=1}^{u} |p_k(H(g))|\right\}$$

There is a conditionally compact subset C of $\mathcal{J}$ so that $c'\eta^{-1}(a(g)) \leq \frac{1}{2}$ if g is not in C. If in the left side of (5.d) g is allowed to vary only over the complement of C the result would be at most $c_1'(n) + \frac{1}{2}c_1(n)$. Thus if n were so large that $c_1'(n) < \frac{1}{2}\varepsilon$

$$\sup_{g\in C} |\phi_n(g)| \left\{\Sigma_{i=1}^{s} \Sigma_{j=1}^{s_i} \exp\left\langle H(g),\ H_{n,i}^{(j)}\right\rangle\right\}^{-1} \left\{\Sigma_{k=1}^{u} |p_k(H(g))|\right\}^{-1} \geq \varepsilon$$

This is however impossible since $\phi_n(g)$ converges to zero uniformly on compact sets.

The induction assumption will be used to establish (5.g). As in the proof of Lemma 5.1 let

$$\int_{\Gamma\cap {}^*N_i \backslash {}^*N_i} \phi_n(ng)dn = \Sigma_{j=1}^{j_i(n)} \exp\left\langle {}^*H_i(g),\ {}^*H_{n,i}^{(j)}\right\rangle \Sigma_{k=1}^{k_i} q_i^{(k)}({}^*H_i(g))\phi_{n,i}^{(j,k)}(g)$$

where $\phi_i^{(j,k)}$ is a function on ${}^*A_i\,{}^*T_i\backslash G$, the elements

$${}^*H_{n,i}^{(j)},\ 1 \leq j \leq j_i(n)$$

are distinct, and the set of homogeneous polynomials $q_i^{(1)}, \ldots, q_i^{(k_i)}$ is linearly independent. We have already seen in the proof of Lemma 5.1 that if the $\phi_{n,i}^{(j,k)}$ are considered as functions on ${}^*\Theta_i \times \{1\} \backslash {}^*M_i \times K$ then the sequences $\{\phi_{n,i}^{(j,k)}\}$ satisfy all conditions of the lemma, with q replaced by q-1, except perhaps the last. We again replace i by i_0, j by j_0, and k by k_0 in order to free the indices i, j, and k. For each n and each i define a partition $P_i(n)$ of $\{1, \ldots, s_i\}$ by demanding that two integers j_1 and j_2 belong to the same class of the partition if and only if $H_{n,i}^{(j_1)}$ and $H_{n,i}^{(j_2)}$ have the same projection on ${}^*\mathfrak{a}_{i_0}$. Breaking the sequence into a number of subsequences we can suppose that $j_i(n) = j_i$ and $P_i(n) = P_i$ are independent of n. With this assumption we can verify the last condition of the lemma for the sequence $\phi_{n,i_0}^{(j_0,k_0)}$. If P is a cuspidal subgroup of rank q to which ${}^*P_{i_0}$ belongs suppose for simplicity that $P = P_i$ for some i.

Let M be the degree of $q_{i_0}^{(k_0)}$. If F is the subset of $\{1, \ldots, s_i\}$ consisting of those j such that the projection of $H_{n,i}^{(j)}$ on ${}^*\mathfrak{n}_{i_0}$ equals ${}^*H_{n,i_0}^{(j_0)}$ and if $r^{(1)}, \ldots, r^{(t_i)}$ is a basis for the polynomials on the orthogonal complement ${}^\Psi\mathfrak{n}_i$ of ${}^*\mathfrak{n}_{i_0}$ in $\mathfrak{n}_i$ of degree at most $N-M$ then

$$\int_{{}^*\Theta_{i_0} \cap {}^\Psi N_i \backslash {}^\Psi N_i} \phi_{i_0}^{(j_0, k_0)}(nm, k)dn$$

is equal to

$$\Sigma_{j\epsilon F} \exp\left\langle {}^\Psi H_i(m), {}^\Psi H_{n,i}^{(j)}\right\rangle \Sigma_{k=1}^{t_i} r^{(k)}({}^\Psi H_i(m))\Psi_{n,i}^{(j,k)}(mk^{-1})$$

Here

$${}^\Psi P_i = {}^*N_{i_0} \backslash P_i \cap {}^*S_{i_0}$$

$$H_{n,i}^{(j)} = {}^*H_{n,i}^{(j)} + {}^\Psi H_{n,i}^{(j)}$$

with ${}^*H_{n,i}^{(j)}$ in the complexification of $\mathfrak{n}_{i_0}$ and ${}^\Psi H_{n,i}^{(j)}$ in the complexification of ${}^\Psi\mathfrak{n}_i$. It is clear that $\lim_{n\to\infty} {}^\Psi H_{n,i}^{(j)}$ exists for each j. The functions $\Psi_{n,i}^{(j,k)}$ are linear combinations of the functions $\Phi_{n,i}^{(j,k)}$ with coefficients which do not depend on n; consequently

$$\lim_{n\to\infty} \Psi_{n,i}^{(j,k)} = 0$$

for each i, j, and k. The inequality (5.g) follows immediately from the induction assumption.

In the next section it will be necessary to investigate the integral over $\Gamma\backslash G$ of various expressions involving the terms of a sequence $\{\phi_n\}$ which satisfies the conditions of the lemma with $q = 1$. In order to do this we must be able to estimate the integral of $|\phi(g)|^2$ over certain subsets of G and $\Gamma\backslash G$. For example if C is a compact subset of G then

$$\int_C |\phi_n(g)|^2 dg = O(\nu^2(n))$$

if $\nu(n)$ has the same meaning as in the proof of the lemma. Suppose that $\mathfrak{S}$ is a Siegel domain associated to the percuspidal subgroup P. If $\alpha_1, \ldots, \alpha_p$, are the simple roots of $\mathfrak{f}$ let ${}^*P_i$ be the cuspidal subgroup of rank one determined by $\{\alpha_j, | j \neq i\}$ and let

$$\mathfrak{S}_i = \{g \in \mathfrak{S} \,|\, \xi_{\alpha_i,}(a(g)) \geq \xi_{\alpha_j,}(a(g)),\ 1 \leq j \leq p\}$$

It follows from Lemmas 3.4 and 5.2 that if

$${}^*\hat{\phi}_{n,i}(g) = \int_{\Gamma \cap {}^*N_i \backslash {}^*N_i} \phi_n(ng)dn$$

and r is any real number then

$$|\phi_n(g) - \hat{\phi}_{n,i}(g)| = O(\nu(n))\eta^r(a(g))$$

for all g in $\mathfrak{S}_i$. Since $\eta^r(a(g))$ is square integrable on $\mathfrak{S}_i$ for $r \leq 0$

$$\int_{\mathfrak{S}_i} |\phi(g) - {}^*\hat{\phi}_i(g)|^2 dg = O(\nu^2(n))$$

If $1 \geq b > 0$ let

$$\mathfrak{S}_i(b) = \{g \in \mathfrak{S}_i \,|\, \xi_{\alpha_j,}(a(g)) \geq \xi^b_{\alpha_i,}(a(g)) \text{ for some } j \neq i\}$$

We shall show that

(5.h) $$\int_{\mathfrak{S}_i(b)} |{}^*\hat{\phi}_{n,i}(g)|^2 dg = O(\nu^2(n))$$

and hence that

(5.i) $$\int_{\mathfrak{S}_i(b)} |\phi_n(g)|^2 dg = O(\nu^2(n))$$

It will be better to prove a slightly stronger assertion than (5.g). Suppose that $\mathfrak{S} = \mathfrak{S}(c, \omega)$. If g is in G let $g = namk$ with n in ${}^*N_i$, a in ${}^*A_i$, m in ${}^*M_i$, and k in K. If ${}^\Psi A_i = A \cap {}^*M_i$ then ${}^\Psi A_i$ is the split component of the cuspidal subgroup ${}^\Psi P_i = {}^*N_i \backslash P \cap {}^*S_i$ of ${}^*M_i$. If g belongs to S and $j \neq i$ then

$$\xi_{\alpha_{j,}}({}^\Psi a_i(m)) = \xi_{\alpha_{j,}}(a(g)) \geq c$$

It follows readily from Lemma 2.6 that

$$\xi^{-1}_{\alpha_{i,}}({}^\Psi a_i(m)) = \prod_{j \neq i} \xi^{\delta_j}_{\alpha_{j,}}({}^\Psi a_i(m))$$

with $\delta_j \geq 0$; consequently

$$\xi_{\alpha_{i,}}(a) \geq c\xi^{-1}_{\alpha_{i,}}({}^\Psi a_i(m)) \geq c_1$$

with some constant c_1. If g belongs to $\mathfrak{S}_i(b)$ then, for some $j \neq i$,

$$\xi_{\alpha_{j,}}({}^\Psi a_i(m)) \prod_{k \neq i} \xi^{b\delta_k}_{\alpha_{k,}}({}^\Psi a_i(m)) \geq \xi^{b}_{\alpha_{i,}}(a)$$

Consequently there is a constant $b_1 > 0$ so that, for some other j,

$$\xi_{\alpha_{j,}}({}^\Psi a_i(m)) \geq \xi^{b_1}_{\alpha_{i,}}(a)$$

Suppose ω_1 and ω_2 are compact subsets of ${}^*N_i$ and ${}^\Psi S_i$ respectively so that ω is contained in $\omega_1\omega_2$; then we can choose n in ω_1 and m in ${}^\Psi\mathfrak{S}_i(c, \omega_2)$. For each a in ${}^*A_i$ let

$$U(a) = \{m \in {}^\Psi\mathfrak{S}_i(c, \omega_2) \,|\, \eta({}^\Psi a_i(m)) \geq \eta^{b_1}(a)\}$$

The integral of (5.h) is at most a constant, which does not depend on n, times

(5.j) $$\int_{{}^*A_i^+(c_1, \infty)} \omega^2(a) \left\{ \int_{U(a) \times K} |{}^*\hat{\phi}_{n,i}(amk)|^2 \, dm\, dk \right\} da$$

To estimate (5.j) we can replace ${}^*P_i$ by any cuspidal subgroup conjugate to it. In particular we can suppose that ${}^*P_i$ is one of the groups $P_1, \ldots, P_s$. If ${}^*P_i$ equals P_{i_0} the above integral equals

$$\int_{{}^*A^+_{i_0}(c_1,\infty)} \omega^2(a) \left\{ \int_{U(a)\times K} \left| \Sigma_{j=1}^{s_{i_0}} \exp\left\langle H, H^{(j)}_{n,i_0} \right\rangle \Sigma_{k=1}^{t} p^{(k)}_{i_0}(H)\Phi^{(j,k)}_{n,i_0}(mk^{-1}) \right|^2 dmdk \right\} da$$

if $a = \exp H$. Given any real number r there is a constant $c(r)$ so that

$$|\Phi^{(j,k)}_{n,i_0}(m, k^{-1})| \leq c(r) \|\Phi^{(j,k)}_{n,i_0}\| \eta^r({}^\Psi a_i(m))$$

if m belongs to ${}^\Psi \mathfrak{S}_i$. Thus if r is less than or equal to zero the above integral is

$$O(\nu^2(n)) \int_{{}^*A^+_{i_0}(c_1,\infty)} \omega^2(a)\eta^{2rb_1}(a) \Sigma_{j=1}^{s_{i_0}} \Sigma_{k=1}^{t} \left| \exp\left\langle H, H^{(j)}_{n,i_0} \right\rangle p^{(k)}_{i_0}(H) \right|^2 da$$

which is $O(\nu^2(n))$ for r sufficiently small.

For each i let $P_i^{(1)}, \ldots, P_i^{(n_i)}$ be a set of percuspidal subgroups to which P_i belongs which are such that there are Siegel domains ${}^\Psi\mathfrak{S}_i^{(j)}$, $1 \leq j \leq n_i$, associated to

$${}^\Psi P_i^{(j)} = N_i \backslash P_i^{(j)} \cap S_i$$

whose union covers $\Theta_i \backslash M_i$. It may be supposed that $\{P_i^{(j)} | 1 \leq j \leq n_i\}$ contains a complete set of representatives for the conjugacy classes of percuspidal subgroups to which P_i belongs and hence that $\{P_i^{(j)} | 1 \leq i \leq s, \ 1 \leq j \leq n_i\}$ contains a complete set of representatives for the conjugacy classes of percuspidal subgroups. It should perhaps be recalled that we have seen in Section 2 that if two percuspidal subgroups to which P_i belongs are conjugate then the conjugation can be effected by an element of Δ_i. Let t be a positive number and for each i and j let $\omega_i^{(j)}$ be a compact subset of $S_i^{(j)}$; let $\mathfrak{S}_i^{(j)}$ be the set of all g in the Siegel domain

$\mathfrak{S}_i^{(j)}(t, \omega_i^{(j)})$ such that

$$\xi_\alpha(a_i^{(j)}(g)) \geq \xi_\beta(a_i^{(j)}(g))$$

if β is any simple root of $\mathfrak{f}$ and α is the unique simple root which does not vanish on $\mathfrak{n}_i$. Let us now verify that $\bigcup_{i=1}^{s} \bigcup_{j=1}^{n_i} \mathfrak{S}_i^{(j)}$ covers $\Gamma\backslash G$ if t is sufficiently small and the sets $\omega_i^{(j)}$ are sufficiently large. Since $\Gamma\backslash G$ is covered by a finite number of Siegel domains it is enough to show that if t and the sets $\omega_i^{(j)}$ are suitably chosen the projection of the above set on $\Gamma\backslash G$ contains the projection on $\Gamma\backslash G$ of any given Siegel domain $\mathfrak{S}$. Suppose $\mathfrak{S}$ is associated to the percuspidal subgroup P and ${}^*P_k$ is the cuspidal subgroup belonging to P determined by $\{\alpha_{\ell,} \mid \ell \neq k\}$. It is enough to show that the projection of the above set contains the projection on $\Gamma\backslash G$ of

$$\mathfrak{S}_k = \{g \in \mathfrak{S} \mid \xi_{\alpha_{k,}}(a(g)) \geq \xi_{\alpha_{\ell,}}(a(g)),\ 1 \leq \ell \leq p\}$$

for each k. Given k there is an i and a j and a γ in Γ so that $\gamma\, {}^*P_k \gamma^{-1} = P_i$ and $\gamma P \gamma^{-1} = P_i^{(j)}$. Let $\mathfrak{S} = \mathfrak{S}(c, \omega)$. The projection of $\mathfrak{S}_k$ on $\Gamma\backslash G$ is the same as the projection on $\Gamma\backslash G$ of $\gamma\mathfrak{S}_k$. The set $\mathfrak{S}_k$ is contained in

$$\gamma\omega\gamma^{-1} N_i^{(j)} A_i^{(j)+}(c, \infty)\gamma K$$

since $\Delta_i^{(j)} \backslash S_i^{(j)}$ is compact there is a Siegel domain $\mathfrak{S}_i^{(j)}(t, \omega_i^{(j)})$ so that $\gamma\mathfrak{S}_k$ is contained in $\Delta_i^{(j)} \mathfrak{S}_i^{(j)}(t, \omega_i^{(j)})$. The set $\gamma\mathfrak{S}_k$ will then be contained in $\Delta_i^{(j)}\mathfrak{S}_i^{(j)}$ because

$$\xi_\alpha(a_i^{(j)}(\gamma g \gamma^{-1})) = \xi_{\alpha_{k,}}(a(g))$$

if α is the unique simple root which does not vanish on $\mathfrak{n}_i$.

If $1 \geq b > 0$ and $u > 0$ let $\mathfrak{S}_i^{(j)}(b, u)$ be the set of all g in $\mathfrak{S}_i^{(j)}$ such that

$$\xi_\beta(a_i^{(j)}(g)) < \xi_\alpha^b(a_i^{(j)}(g))$$

for all simple roots β of $\mathfrak{f}$ different from α and such that $\xi_\alpha(a_i(g)) > u$. Let F be the projection on $\Gamma\backslash G$ of $\bigcup_{i=1}^{s}\bigcup_{j=1}^{n_i}\mathfrak{S}_i^{(j)}(b, u)$. We now know that

(5.k) $$\int_{\Gamma\backslash G-F} |\phi_n(g)|^2 dg = O(\gamma^2(n))$$

Let F_i be the projection on $\Delta_i\backslash G$ of $\bigcup_{j=1}^{n_i}\mathfrak{S}_i^{(j)}(b, u)$. It follows from Lemma 2.12 that if u is sufficiently large and b is sufficiently small the projections on $\Gamma\backslash G$ of F_i and F_j are disjoint unless $i = j$ and that the projection of F_i into $\Gamma\backslash G$ is injective. Thus if $\psi(g)$ is any function on $\Gamma\backslash G$ for which

$$\int_{\Gamma\backslash G} \psi(g)dg$$

is defined the integral is equal to

(5.ℓ) $$\Sigma_{i=1}^{s}\int_{F_i}\psi(g)dg + \int_{\Gamma\backslash G-F}\psi(g)dg$$

We also know that

(5.m) $$\int_{F_i} |\phi_n(g) - \hat{\phi}_{n,i}(g)|^2 dg = O(\nu^2(n))$$

if

$$\hat{\phi}_{n,i}(g) = \int_{\Gamma\cap N_i\backslash N_i}\phi_n(ng)dn$$

There is one more lemma which should be established before we go on to the proof of the functional equations.

LEMMA 5.3. Let U be an open subset of the n-dimensional complex coordinate space. Suppose that to each point z in U there is associated a continuous function $E(g, z)$ on $\Gamma\backslash G$. Suppose that for each z in U there is a constant r so that if $\mathfrak{S}$ is any Siegel domain, associated to a percuspidal subgroup P, there is a constant c, which may also depend on z, so that

$$|E(g, z)| \leq c\eta^r(a(g))$$

if g belongs to γ. Suppose that there is an integer q so that if ${}^*P$ is any cuspidal subgroup then the cuspidal component of

$$\int_{\Gamma \cap {}^*N \backslash {}^*N} E(ng, z)dn$$

is zero for all z unless the rank of ${}^*P$ equals q. Let $\{P_1, \ldots, P_s\}$ be a set of representatives for the conjugacy classes of cuspidal subgroups of rank q and for each i let V_i be an admissible subspace of $\mathcal{L}_0(\Theta_i \backslash M_i)$; let W be an admissible space of functions on K. Suppose there is an integer N so that if $\{p_i^{(k)} | 1 \leq k \leq t\}$ is a basis for the polynomials on $\mathfrak{a}_i$ of degree at most N then

$$\int_{\Gamma \cap N_i \backslash N_i} E(ng, z)dn = \Sigma_{j=1}^{s_i} \exp\left\langle H_i(g), H_i^{(j)}(z)\right\rangle \Sigma_{k=1}^{t} p_i^{(k)}(H_i(g))\Phi_i^{(j,k)}(g, z)$$

For each z and each i and j the point $H_i^{(j)}(z)$ belongs to the complexification of $\mathfrak{a}_i$; $\Phi_i^{(j,k)}(g, z)$ is the value at g of an element $\Phi_i^{(j,k)}(z)$ of $\mathcal{E}(V_i, W)$. If $H_i^{(j)}(z)$ and $\Phi_i^{(j,k)}(z)$ are holomorphic functions on U, with values in the complexification of $\mathfrak{a}_i$ and $\mathcal{E}(V_i, W)$ respectively, for all i, j, and k then $E(g, z)$ is a continuous function on $\Gamma\backslash G \times U$ which is holomorphic in z for each fixed g.

It follows immediately from Lemma 5.2 that $E(g, z)$ is a continuous function on $\Gamma\backslash G \times U$. Let $z^o = (z_1^o, \ldots, z_n^o)$ be a point in U and let

$$B = \{z = (z_1, \ldots, z_n) | |z_i - z_i^o| < \varepsilon\}$$

be a polycylinder whose closure is contained in U. It is enough to show that $E(g, z)$ is analytic in B for each g. To do this we show that if C_i is the contour consisting of the circle of radius ε about z_i^o transversed in the positive direction then

$$E(g, z) = \left(\frac{1}{2\pi i}\right)^n \int_{C_1} d\zeta_1 \ldots \int_{C_n} d\zeta_n\, E(g, \zeta)\prod_{\ell=1}^{n}(\zeta_i - z_i)^{-1}$$

when z is in B. Denote the right hand side by $E_1(g, z)$. It follows from Lemma 5.2 that if $\mathfrak{S}$ is any Siegel domain there are constants c and r so that $|E(g, z)| \le c\eta^r(a(g))$ for all g in $\mathfrak{S}$ and all z in the closure of B. Consequently for all z in B the function $E(g, z) - E_1(g, z)$ satisfies the first condition of Lemma 3.7. If ${}^*P$ is a cuspidal subgroup then

$$\int_{\Gamma\cap {}^*N\backslash {}^*N} E_1(ng, z)dn$$

is equal to

$$\left(\frac{1}{2\pi i}\right)^n \int_{C_1} d\zeta_1 \ldots \int_{C_n} d\zeta_n \left\{\int_{\Gamma\cap {}^*N\backslash {}^*N} E(ng, \zeta)dn\right\}\prod_{\ell=1}^{n}(\zeta_i - z_i)^{-1}$$

It follows from Fubini's theorem that the cuspidal component of

$$\int_{\Gamma\cap {}^*N\backslash {}^*N} E_1(ng, z)dn$$

is zero if the rank of ${}^*P$ is not q. However

$$\int_{\Gamma\cap N_i\backslash N_i} E_1(ng, z)dn$$

is equal to the sum over j, $1 \le j \le s_i$, and k, $1 \le k \le t$ of

$$\left(\frac{1}{2\pi i}\right)^n \int_{C_1} d\zeta_1 \ldots \int_{C_n} d\zeta_n \left\{\exp\left\langle H_i(g), H_i^{(j)}(\zeta)\right\rangle p_i^{(k)}(H_i(g))\Phi_i^{(j,k)}(g, \zeta)\right\}\prod_{\ell=1}^{n}(\zeta - z_i)^{-1}$$

Since the expression in brackets is a holomorphic function of ζ this equals

$$\Sigma_{j=1}^{s_i}\Sigma_{k=1}^{t} \exp\left\langle H_i(g), H_i^{(j)}(z)\right\rangle p_i^{(k)}(H_i(g))\Phi_i^{(j,k)}(g, z)$$

and the lemma follows from Lemma 3.7.

6. Some functional equations.

We are now ready to prove the functional equations for the Eisenstein series associated to cusp forms. Let $\{P\}$ be a complete family of associate cuspidal subgroups; let $\{V\}$ be a complete family of associate admissible subspaces; and let W be a simple admissible subspace of the space of functions on K. If $\mathfrak{a}^{(1)}, \ldots, \mathfrak{a}^{(r)}$ are the distinct subspaces of $\mathfrak{h}$ occurring among the split components of the elements of $\{P\}$ then for each transformation s in $\Omega(\mathfrak{a}^{(i)}, \mathfrak{a}^{(j)})$ we have defined a holomorphic function $M(s, H^{(i)})$ on $\mathfrak{U}^{(i)}$ with values in the space of linear transformations from $\mathcal{E}^{(i)}$ to $\mathcal{E}^{(j)}$ and for each point Φ in $\mathcal{E}^{(i)}$ we have defined a continuous function $E(g, \Phi, H^{(i)})$ on $\Gamma\backslash G \times \mathfrak{U}^{(i)}$ which is holomorphic in $H^{(i)}$ for each fixed g. In order to avoid some unpleasant verbosity later we introduce some conventions now. As usual $M(s, H^{(i)})$ is said to be holomorphic or meromorphic on some open set V containing $\mathfrak{U}^{(i)}$ if there is a holomorphic or meromorphic function, which is still denoted by $M(s, H^{(i)})$, on V whose restriction to $\mathfrak{U}^{(i)}$ is $M(s, H^{(i)})$. The function $E(\cdot, \Phi, H^{(i)})$ is said to be holomorphic on V if there is a continuous function on $\Gamma\backslash G \times V$ which is holomorphic in $H^{(i)}$ for each fixed g and equals $E(g, \Phi, H^{(i)})$ on $\Gamma\backslash G \times U^{(i)}$. Of course this function on $\Gamma\backslash G \times V$ is still denoted by $E(g, \Phi, H^{(i)})$. $E(\cdot, \Phi, H^{(i)})$ is said to be meromorphic on V if it is holomorphic on an open dense subset V' of V and if for each point $H_0^{(i)}$ in V there is a non-zero holomorphic function $f(H^{(i)})$ defined in a neighbourhood U of $H_0^{(i)}$ so that $f(H^{(i)})E(g, \Phi, H^{(i)})$ is the restriction to $\Gamma\backslash G \times (U \cap V')$ of a continuous function on $\Gamma\backslash G \times (U \cap V)$ which is holomorphic on $U \cap V$ for each fixed g. If V' is the complement of the intersection of V with a set of hyperplanes and if $f(H^{(i)})$ can always be taken as a product of linear functions we will say that the singularities of $E(\cdot, \Phi, H^{(i)})$ in V lie along hyperplanes. A similar convention applies to the functions $M(s, H^{(i)})$.

LEMMA 6.1. For each i and each j and each transformation s in $\Omega(\mathfrak{a}^{(i)}, \mathfrak{a}^{(j)})$ the function $M(s, H^{(i)})$ is meromorphic on $\mathfrak{a}^{(i)}$ and its singularities lie along hyperplanes. For each i and each Φ in $E^{(i)}$ the function $E(\cdot, \Phi, H^{(i)})$ is meromorphic on $\mathfrak{a}^{(i)}$ and its singularities lie along hyperplanes. If s belongs to $\Omega(\mathfrak{a}^{(i)}, \mathfrak{a}^{(j)})$, t belongs to $\Omega(\mathfrak{a}^{(j)}, \mathfrak{a}^{(k)})$, and Φ belongs to $\mathcal{E}^{(i)}$ then

$$M(ts, H^{(i)}) = M(t, sH^{(i)})M(s, H^{(i)})$$

and

$$E(g, M(s, H^{(i)})\Phi, sH^{(i)}) = E(g, \Phi, H^{(i)})$$

There are a number of other properties of the functions $E(\cdot, \Phi, H^{(i)})$ which it is important to remark.

LEMMA 6.2. Fix i and fix $H_0^{(i)}$ in $\mathfrak{a}^{(i)}$. Suppose that for every j and every s in $\Omega(\mathfrak{a}^{(i)}, \mathfrak{a}^{(j)})$ the function $M(s, H^{(i)})$ is analytic at $H_0^{(i)}$. Then for every Φ in $\mathcal{E}^{(i)}$ the function $E(\cdot, \Phi, H^{(i)})$ is analytic at $H_0^{(i)}$ and if $\mathfrak{S}$ is a Siegel domain, associated to a percuspidal subgroup P, there are constants c and r so that, for g in S,

$$|E(g, \Phi, H_0^{(i)})| \leq c\eta^r(a(g))$$

Moreover if ${}^*P$ is a cuspidal subgroup the cuspidal component of

$$\int_{\Gamma\cap {}^*N\backslash {}^*N} E(ng, \Phi, H_0^{(i)})dn$$

is zero unless ${}^*P$ belongs to $\{P\}$ but

$$\int_{\Gamma\cap N^{(j,\ell)}\backslash N^{(j,\ell)}} E(ng, \Phi, H_0^{(i)})dn$$

is equal to

$$\sum_{s\in\Omega(\mathfrak{a}^{(i)},\mathfrak{a}^{(j)})} \exp(\langle H^{(j,\ell)}(g), sH_0^{(i)}\rangle + \rho(H^{(j,\ell)}(g)))(E^{(j,\ell)}M(s, H^{(i)})\Phi)(g)$$

if $E^{(j,\ell)}$ *is the projection of* $\mathcal{E}^{(j)}$ *on* $\mathcal{E}(V^{(j,\ell)}, W)$.

It should be observed immediately that this lemma is true if $H_0^{(i)}$ belongs to $u^{(i)}$. Let us begin the proof of these two lemmas with some remarks of a general nature. We recall that if $\Phi(\cdot)$ and $\Psi(\cdot)$ belong to the space $\mathfrak{H}$ introduced in Section 4 then

$$\int_{\Gamma\backslash G} \phi^\wedge(g)\bar{\psi}^\wedge(g)dg$$

is equal to

$$(6.a)\ \Sigma_{i=1}^{r}\Sigma_{j=1}^{r}\Sigma_{s\in\Omega(\mathfrak{a}^{(i)},\mathfrak{a}^{(j)})}\left(\frac{1}{2\pi}\right)^q \int_{\operatorname{Re} H^{(i)}=Y^{(i)}} (M(s, H^{(i)})\Phi_i(H^{(i)}), \Psi_j(-s\bar{H}^{(i)}))\,|dH^{(i)}|$$

If, for $1 \le i \le r$, $f_i(H)$ is a bounded analytic function on

$$D_i = \{H^{(i)} \in \mathfrak{a}_c^{(i)} \mid \|\operatorname{Re} H^{(i)}\| < R\}$$

and if $\Phi(H)$ is in $\mathfrak{H}$ then

$$f(H)\Phi(H) = (f_1(H^{(1)})\Phi_1(H^{(1)}), \ldots, f_r(H^{(r)})\Phi_r(H^{(r)}))$$

is in $\mathfrak{H}$. Suppose that, for all s in $\Omega(\mathfrak{a}^{(i)}, \mathfrak{a}^{(j)})$, $f_j(sH^{(i)}) = f_i(H^{(i)})$ and let $f_i^*(H) = \bar{f}_i(-\bar{H})$. If (6.a) is denoted by $(\Phi(\cdot), \Psi(\cdot))$ it is readily verified that

$$(f(\cdot)\Phi(\cdot), \Psi(\cdot)) = (\Phi(\cdot), f^*(\cdot)\Psi(\cdot))$$

In particular $(f^*(\cdot)f(\cdot)\Phi(\cdot), \Psi(\cdot))$ is a positive definite hermitian symmetric form on $\mathfrak{H}$. Suppose k is a positive number and, for each i and all H in D_i, $|f_i(H)| < k$ then

$$(k^2 - f_i^*(H)f_i(H))^{\frac{1}{2}} = g_i(H)$$

is defined, analytic, and bounded on D_i and $g_i^*(H) = g_i(H)$. If the square root is properly chosen then $g_j(sH) = g_i(H)$ for all s in $\Omega(\mathfrak{a}^{(i)}, \mathfrak{a}^{(j)})$. Since

$$k^2 - f_i^*(H)f_i(H) = g_i^*(H)g_i(H)$$

we see that

$$(f(\cdot)\Phi(\cdot),\ f(\cdot)\Phi(\cdot)) \le k^2(\Phi(\cdot),\ \Phi(\cdot))$$

Consequently f defines a bounded linear operator $\lambda(f)$ on $\mathcal{L}(\{P\}, \{V\}, W)$. If $s_i(f)$ is the closure of the range of $f_i(H)$ for H in D_i then the spectrum of $\lambda(f)$ is contained in $\bigcup_{i=1}^r s_i(f)$. It is clear that $\lambda^*(f) = \lambda(f^*)$ so that if $f = f^*$ then $\lambda(f)$ is self-adjoint. If H belongs to D_i, let $H = H_1 + iH_2$ with H_1 and H_2 in $\mathfrak{a}^{(i)}$; then

$$\langle H, H\rangle = \langle H_1, H_1\rangle - \langle H_2, H_2\rangle + 2i\langle H_1, H_2\rangle$$

so that $\mathrm{Re}\langle H, H\rangle < R^2$. If $\mathrm{Re}\,\mu > R^2$ let $f_i^{\mu}(H) = (\mu - \langle H, H\rangle)^{-1}$; then $\lambda(f^{\mu})$ is a bounded operator on $\mathcal{L}(\{P\}, \{V\}, W)$. Since the map $\Phi(\cdot) \longrightarrow f^{\mu}(\cdot)\Phi(\cdot)$ is one-to-one map of $\mathfrak{H}$ onto itself the range of $\lambda(f^{\mu})$ is dense. Consequently if $f_i(H) = \langle H, H\rangle$ the map

$$\Phi(\cdot) \longrightarrow (f_1(\cdot)\Phi_1(\cdot),\ \ldots,\ f_r(\cdot)\Phi_r(\cdot))$$

defines a closed, self-adjoint, linear operator A on $\mathcal{L}(\{P\}, \{V\}, W)$ and

$$\lambda(f^{\mu}) = (\mu - A)^{-1} = R(\mu, A)$$

$R(\mu, A)$ is an analytic function of μ off the infinite interval $(-\infty, R^2]$.

Suppose $\Phi_{i,k}$ belongs to $\mathcal{E}(V^{(i,k)}, W)$ and $H^{(i)}$ belongs to $\mathfrak{U}^{(i)}$; consider

$$\sum_{\Delta^{(i,k)}\backslash\Gamma} \exp(\langle H^{(i,k)}(\gamma g),\ H^{(i)}\rangle + \rho(H^{(i,k)}(\gamma g)))\Phi_{i,k}(\gamma g)$$

Let $\mathfrak{S}$ be a Siegel domain, associated to a percuspidal subgroup P, and let C

be a fixed compact set. For each i let $\alpha^{(i)}_{1,}, \ldots, \alpha^{(i)}_{p,}$ be the simple roots of $\mathfrak{f}$ so numbered that $\alpha^{(i)}_{q+1,}, \ldots, \alpha^{(i)}_{p,}$ vanish on $\mathfrak{n}^{(i)}$; we will also denote the restriction of $\alpha^{(i)}_{j,}$ to $\mathfrak{n}^{(i)}$ by $\alpha^{(i)}_{j,}$ if $1 \le j \le q$. The methods used to prove Lemma 2.11 can be used to show that there is a constant x so that if g belongs to $\mathfrak{r}$ and h belongs to C then

$$\alpha^{(i)}_{,j}(H^{(i,k)}(\gamma gh)) \le x + \alpha^{(i)}_{,j}(H(g))$$

for $1 \le j \le q$. Let $F'(h, \Phi_{i,k}, H^{(i)})$ equal

$$\exp(\langle H^{(i,k)}(h), H^{(i)}\rangle + \rho(H^{(i,k)}(h))\Phi_{i,k}(h)$$

if, for all j,

$$\alpha^{(i)}_{,j}(H^{(i,k)}(h)) \le x + \alpha^{(i)}_{,j}(H(g))$$

and let it equal zero otherwise; then set

$$E'(h, \Phi_{i,k}, H^{(i)}) = \Sigma_{\Delta^{(i,k)}\backslash\Gamma} F'(\gamma h, \Phi_{i,k}, H^{(i)})$$

The functions $E(h, \Phi_{i,k}, H^{(i)})$ and $E'(h, \Phi_{i,k}, H^{(i)})$ are equal on gC. The Fourier transform of $F'(h, \Phi_{i,k}, H^{(i)}_1)$ is

$$a\left\{\prod_{j=1}^{q}\alpha^{(i)}_{j,}(H^{(i)}_1 - H^{(i)})\right\}^{-1}\exp(\langle X, H^{(i)}_1 - H^{(i)}\rangle + \langle H(g), H^{(i)}_1 - H^{(i)}\rangle)\Phi_{i,k}$$

if X in $\mathfrak{h}$ is such that $\alpha_{,j}(X) = x$, $1 \le j \le p$, and a is the volume of

$$\{H \in \mathfrak{n}^{(i)} \mid 0 \le \alpha^{(i)}_{,j}(H) \le 1,\ 1 \le j \le q\}$$

If

$$\Phi_i = \oplus_{k=1}^{m_i} \Phi_{i,k}$$

and

$$E'(h, \Phi_i, H^{(i)}) = \Sigma_{k=1}^{m_i} E'(g, \Phi_{i,k}, H^{(i)})$$

then Lemma 4.6 together with some simple approximation arguments shows that $E'(\cdot, \Phi_i, H^{(i)})$ is an analytic function on $\mathfrak{U}^{(i)}$ with values in $\mathfrak{L}(\{P\}, \{V\}, W)$ and that

$$\int_{\Gamma\backslash G} E'(h, \Phi_i, H_1^{(i)})\overline{E}'(h, \Psi_j, H_2^{(j)})dh$$

is equal to

$$(6.b) \qquad \Sigma_{s\in\Omega(\mathfrak{U}^{(i)}, \mathfrak{U}^{(j)})} \frac{a^2}{(2\pi)^q} \int_{\operatorname{Re} H^{(i)}=Y^{(i)}} (M(s, H^{(i)})\Phi_i, \Psi_j)\xi(H^{(i)})\,|dH^{(i)}|$$

with

$$\xi(H^{(i)}) = \exp|\langle X(g), H_1^{(i)} - H^{(i)}\rangle + \langle X(g), \overline{H}_1^{(j)} + sH^{(j)}\rangle| \left\{\prod_{k=1}^{q} \alpha_{k,}^{(i)}(H_1^{(i)} - H^{(i)})\alpha_{k,}^{(j)}(\overline{H}_2^{(j)} + sH^{(i)})\right\}^{-1}$$

if $Y^{(i)}$ is suitably chosen and $X(g) = X + H(g)$.

Suppose that for any choice of x and g and all Φ_i the function $E'(g, \Phi_i, H^{(i)})$ is analytic in a region V containing $\mathfrak{U}^{(i)}$. If f is a continuous function on G choose C so that it contains the support of f; then

$$(6.c) \qquad \lambda(f)E(g, \Phi_i, H^{(i)}) = \int_G E'(gh, \Phi_i, H^{(i)})f(h)dh$$

is a continuous function on $\Gamma\backslash G \times V$ which is an analytic function of $H^{(i)}$ for each fixed g. In particular if $f(kgk^{-1}) = f(g)$ for all g in G and all k in K then $E(g, \pi(f, H^{(i)})\Phi_i, H^{(i)})$ is analytic on V for each g. Of course

$$\pi(f, H^{(i)})\Phi_i = \Sigma_{k=1}^{m_i} \pi(f, H^{(i)})\Phi_{i,k}$$

But f can be so chosen that $\pi(f, H^{(i)})$ is non-singular in the neighbourhood of any given point $H_0^{(i)}$. Consequently $E(g, \Phi_i, H^{(i)})$ is, for each g and each Φ_i, analytic on V. In the course of proving the lemmas for the Eisenstein series in more than one variable we will meet a slightly different situation. There will be a function f_0 such that $f_0(kgk^{-1}) = f_0(g)$ for all g and k, the determinant of the linear transformation $\pi(f_0, H^{(i)})$ on $\mathcal{E}^{(i)}$ does not vanish identically, and $\lambda(f_0)E'(\cdot, \Phi_i, H^{(i)})$ is analytic on V for all Φ_i, all g, and all x. Arguing as above we see that $E(\cdot, \pi(f_0, H^{(i)})\Phi_i, H^{(i)})$ is analytic on V and hence that $E(\cdot, \Phi_i, H^{(i)})$ is meromorphic on V.

If $\mathfrak{S}$ is a Siegel domain and C a compact subset of G let us choose x as above. Suppose that given any compact subset U of V there are constants c and r so that

(6.d) $$\|E'(\cdot, \Phi_i, H^{(i)})\| \leq c\eta^r(a(g))\|\Phi_i\|$$

if $H^{(i)}$ belongs to U and g belongs to $\mathfrak{S}$. If we refer to the formula (6.c) and the proof of the corollary to Lemma 3.7 we see that there are constants c' and r' so that

$$|E(g, \Phi_i, H^{(i)})| \leq c'\eta^{r'}(a(g))\|\Phi_i\|$$

if g is in $\mathfrak{S}$ and $H^{(i)}$ is in U. If all the functions $M(s, H^{(i)})$ are analytic on V we see by combining the dominated convergence theorem and the estimates of Section 3 with the principle of permanence of functional relations that Lemma 6.2 is valid for any point of V. On the other hand suppose only that $\lambda(f_0)E'(\cdot, \Phi_i, H^{(i)})$ is analytic on V for all Φ_i but that for any $\mathfrak{S}$ and any C and any compact subset U of V there are constants c and r so that

$$\|\lambda(f_0)E'(\cdot, \Phi_i, H^{(i)})\| \leq c\eta^r(a(g))\|\Phi_i\|$$

if g is in $\mathfrak{S}$ and $H^{(i)}$ is in U. If all the functions $M(s, H^{(i)})$ are meromorphic on V we see just as above that Lemma 6.2 is valid at those points where the determinant of $\pi(f_0, H^{(i)})$ is not zero. It is a little more difficult to obtain Lemma 6.2 for a point $H_0^{(i)}$ at which the determinant of $\pi(f_0, H^{(i)})$ vanishes. If the assumption of the lemma is satisfied we can apply Lemma 5.2 to define $E(\cdot, \Phi, H^{(i)})$ in a neighbourhood of $H_0^{(i)}$ by continuity. That every assertion of the lemma except the first is valid for each point in a neighbourhood of $H_0^{(i)}$ follows immediately from the earlier lemma. Once we are assured of this we can immediately deduce the first assertion from Lemma 5. 3.

The prefatory remarks over we will now prove the lemmas for the case that the elements of $\{P\}$ have rank one. The case of rank greater than one will then be proved by induction. If the elements of $\{P\}$ have rank one then, as follows from Lemma 2. 13, r is either 1 or 2 and if r is 2 then $\Omega(\mathfrak{a}^{(i)}, \mathfrak{a}^{(i)})$, $i = 1, 2$, contains only the identity transformation. If z is a complex number let $H^{(i)}(z)$ be that element of $\mathfrak{a}_c^{(i)}$ such that

$$\alpha^{(i)}(H^{(i)}(z)) = z\langle \alpha^{(i)}, \alpha^{(i)}\rangle^{\frac{1}{2}}$$

if $\alpha^{(i)}$ is the one simple root of $\mathfrak{a}^{(i)}$. Let $\mathcal{E}$ be $\mathcal{E}^{(1)}$ or $\mathcal{E}^{(1)} \oplus \mathcal{E}^{(2)}$ according as r is 1 or 2. If r is 1 and there is an s in $\Omega(\mathfrak{a}^{(1)}, \mathfrak{a}^{(1)})$ different from the identity then $sH = -H$ for all H in $\mathfrak{a}^{(1)}$ so that s is uniquely determined; in this case let $M(z) = M(s, H^{(1)}(z))$. If there is no such s let $M(z)$ be 0; as we shall see this possibility cannot occur. If r is 2 and s belongs to $\Omega(\mathfrak{a}^{(1)}, \mathfrak{a}^{(2)})$ then $s(H^{(1)}(z)) = -H^{(2)}(z)$ for all z so that s is again uniquely determined. In this case let

$$M(z) = \begin{pmatrix} 0 & M(s^{-1}, H^{(2)}(z)) \\ M(s, H^{(1)}(z)) & 0 \end{pmatrix}$$

If r is 1 and Φ belongs to $\mathcal{E}$ we set

$$E(g, \Phi, z) = E(g, \Phi, H^{(1)}(z))$$

and if r is 2 and $\Phi = \Phi_1 \oplus \Phi_2$ belongs to $\mathcal{E}$ we set

$$E(g, \Phi, z) = E(g, \Phi_1, H^{(1)}(z)) + E(g, \Phi_2, H^{(2)}(z))$$

Lemma 6.1 can be reformulated as follows.

LEMMA 6.3. The function $M(z)$ is meromorphic on the complex plane and for each Φ in $\mathcal{E}$ the function $E(\cdot, \Phi, z)$ is meromorphic on the complex plane. Moreover $M(z)M(-z) = I$ and, for all Φ,

$$E(g, M(z)\Phi, -z) = E(g, \Phi, z)$$

There is no value in reformulating Lemma 6.2. As we observed in the introduction this lemma will be proved by the method of [19]. The space $\mathfrak{H}$ can be considered as a space of functions defined in a region of the complex plane with values in $\mathcal{E}$. If $\Phi(\cdot)$ is in $\mathfrak{H}$ we denote $\Phi_1(H^{(1)}(z))$ or $\Phi_1(H^{(1)}(z)) \oplus \Phi_2(H^{(2)}(z))$ if r is 2 by $\Phi(z)$. If

$$(\phi^\wedge, \psi^\wedge) = \int_{\Gamma\backslash G} \phi^\wedge(g)\overline{\psi}^\wedge(g)dg$$

then

$$(\phi^\wedge, \psi^\wedge) = \frac{1}{2\pi i}\int_{c-i\infty}^{c+i\infty} (\Phi(z), \Psi(-\overline{z})) + (M(z)\Phi(z), \Psi(\overline{z}))dz \tag{6.e}$$

if c is greater than but sufficiently close to

$$\langle a^{(i)}, a^{(i)}\rangle^{-\frac{1}{2}}\langle a^{(i)}, \rho\rangle = \langle \rho, \rho\rangle$$

If $c_1 > \mathrm{Re}\,\lambda > c$ then

$$(R(\lambda^2, A)\phi^\wedge, \psi^\wedge) = \frac{1}{2\pi i}\int_{c-i\infty}^{c+i\infty} (\lambda^2 - z^2)^{-1}\{(\Phi(z), \Psi(-\overline{z})) + (M(z)\Phi(z), \Psi(\overline{z}))\}dz$$

and the latter integral is the sum of

$$(2\lambda)^{-1}\{(\Phi(\lambda), \Psi(-\overline{\lambda})) + (M(\lambda)\Phi(\lambda), \Psi(\overline{\lambda}))\} \tag{6.f}$$

and

$$\frac{1}{2\pi i}\int_{c_1-i\infty}^{c_1+i\infty} (\lambda^2 - z^2)^{-1}\{(\Phi(z), \Psi(-\overline{z})) + (M(z)\Phi(z), \Psi(\overline{z}))\}dz \tag{6.g}$$

$(R(\lambda^2, A)\phi^\wedge, \psi^\wedge)$ is analytic if λ^2 does not belong to $(-\infty, R^2)$, that is, λ is not imaginary and not in the interval $[-\langle\rho, \rho\rangle^{\frac{1}{2}}, \langle\rho, \rho\rangle^{\frac{1}{2}}]$. If $\Phi(z) = e^{z^2}\Phi$ and $\Psi(z) = e^{z^2}\Psi$ with constant Φ and Ψ then (6.g) is an entire function of λ and (6.f) equals

$$(2\lambda)^{-1}e^{2\lambda^2}\{(\Phi, \Psi) + (M(\lambda)\Phi, \Psi)\}$$

Consequently $M(\lambda)$ is analytic for $\operatorname{Re}\lambda > 0$, $\lambda \notin (0, \langle\rho, \rho\rangle^{\frac{1}{2}}]$.

We next show that $E(\cdot, \Phi, \lambda)$ is holomorphic for $\operatorname{Re}\lambda > 0$, $\lambda \notin (0, \langle\rho, \rho\rangle^{\frac{1}{2}}]$. If x is given and $\Phi_{i,k}$ belongs to $\mathcal{E}(V^{(i,k)}, W)$ let $F'(g, \Phi_{i,k}, H^{(i)})$ equal

$$\exp(\langle H^{(i,k)}(g), H^{(i)}\rangle + \rho(H^{(i,k)}(g)))\Phi_{i,k}(g)$$

if

$$\alpha^{(i)}(H^{(i,k)}(g)) \leq x \langle\alpha^{(i)}, \alpha^{(i)}\rangle$$

and let it equal zero otherwise. Let

$$E'(g, \Phi_{i,k}, H^{(i)}) = \Sigma_{\Delta^{(i,k)}\backslash\Gamma} F'(\gamma g, \Phi_{i,k}, H^{(i)})$$

and if

$$\Phi = \oplus_{i=1}^{r} \Sigma_{k=1}^{m_i} \Phi_{i,k}$$

belongs to $\mathcal{E}$ let

$$E'(g, \Phi, z) = \Sigma_{i=1}^{r}\Sigma_{k=1}^{m_i} E'(g, \Phi_{i,k}, H^{(i)}(z))$$

It follows from (6.b) that

$$\int_{\Gamma\backslash G} E'(g, \Phi, \lambda)\overline{E}'(g, \Psi, \mu)dg$$

is equal to

$$\frac{1}{2\pi i} e^{ax(\lambda+\bar{\mu})}\left\{\int_{c-i\infty}^{c+i\infty} (\Phi, \Psi)\{(\lambda-z)(\bar{\mu}+z)\}^{-1} + e^{-2axz}(M(z)\Phi, \Psi)\{(\lambda-z)(\bar{\mu}-z)^{-1}\}dz\right\}$$

if c is as in (6.e). If x is sufficiently large one sees readily, making use of Lemma 4.5(i), that the above integral equals

$$(6.h)\ e^{ax(\lambda+\bar{\mu})}(\lambda+\bar{\mu})^{-1}(\Phi, \Psi) + e^{ax(\bar{\mu}-\lambda)}(\bar{\mu}-\lambda)^{-1}(M(\lambda)\Phi, \Psi) + e^{ax(\lambda-\bar{\mu})}(\lambda-\bar{\mu})^{-1}(\Phi, M(\mu)\Psi)$$

In general we obtain

$$\left(\frac{\partial^n E'}{\partial\lambda^n}(\cdot, \Phi, \lambda), \frac{\partial^n E'}{\partial\mu^n}(\cdot, \Psi, \mu)\right)$$

by differentiating (6.h) n times with respect to λ and $\bar{\mu}$. Thus

$$\Sigma_{n=0}^{\infty} \frac{1}{n!} |\lambda-\lambda_0|^n \left\|\frac{\partial^n E'}{\partial\lambda^n}(\cdot, \Phi, \lambda)\right\|$$

is seen to converge in the largest circle about λ_0 which does not meet the imaginary axis or the real axis. Since the above formulae persist in any subset of

$$\{\lambda | \mathrm{Re}\,\lambda > 0,\ \lambda \notin (0, \langle\rho, \rho\rangle^{\frac{1}{2}}]\}$$

in which $E'(\cdot, \Phi, \lambda)$ is defined we conclude that $E'(\cdot, \Phi, \lambda)$ is an analytic function in this region. Since the analogue of (6.d) is readily deduced from (6.h) we also see that Lemma 6.2 is valid if $H_0^{(i)} = H^{(i)}(z)$ and z is in this region.

The next step in the proof is to show that there are a finite number of points $z_1, \ldots, z_n$ in the interval $(0, \langle\rho, \rho\rangle^{\frac{1}{2}}]$ so that $M(z)$ and $E(\cdot, \Phi, z)$ are analytic in the region $\mathrm{Re}\, z > 0$ except perhaps at $z_1, \ldots, z_n$. It is enough to establish this for the function $M(z)$ because we can then apply Lemmas 5.2 and 5.3 to obtain the assertion for $E(\cdot, \Phi, z)$. Suppose that either there is a sequence $\{z_n\}$ converging to a point z_0 of the positive real axis and a sequence $\{\Phi_n\}$ in $\mathcal{E}$ with $\|\Phi_n\| = 1$ so that

$$\{\|M(z_n)\Phi_n\|\} = \{\nu_n\}$$

is unbounded or there are two sequences $\{z_n\}$ and $\{z'_n\}$ approaching z_0 and an element Φ of $\mathcal{E}$ so that

$$\lim_{n\to\infty} M(z_n)\Phi \neq \lim_{n\to\infty} M(z'_n)\Phi$$

In the first case select a subsequence so that $\lim_{n\to\infty} \nu_n = \infty$ and

$$\lim_{n\to\infty} \nu_n^{-1} M(z_n)\Phi_n = \Phi_0$$

exists; then $\{E(\cdot, \nu_n^{-1}\Phi_n, z_n)\}$ satisfies the conditions of Lemma 5.2. In the second case $\{E(\cdot, \Phi, z_n) - E(\cdot, \Phi, z'_n)\}$ does; let

$$\lim_{n\to\infty} M(z_n)\Phi - M(z'_n)\Phi = \Phi_0$$

In either case let the limit function be ϕ_0. If P is a cuspidal subgroup not in $\{P\}$ the cuspidal component of

$$\int_{\Gamma\cap N\backslash N} \phi_0(ng)dn$$

is zero. However

$$\int_{\Gamma\cap N^{(i,k)}\backslash N^{(i,k)}} \phi_0(ng)dn = \exp(-\langle H^{(i,k)}(g), H^{(i)}(z_0)\rangle + \rho(H^{(i,k)}(g)))(E^{(i,k)}\Phi_0)(g)$$

if $E^{(i,k)}$ is the projection of $\mathcal{E}$ on $\mathcal{E}(V^{(i,k)}, W)$. By the corollary to Lemma 5.1 the function ϕ_0 belongs to $\mathcal{L}(\Gamma\backslash G)$. It is clear that it belongs to $\mathcal{L}(\{P\}, \{V\}, W)$. For each z in $(0, \langle \rho, \rho\rangle^{\frac{1}{2}}]$ let $\mathcal{L}(z)$ be the set of all functions ψ in $\mathcal{L}(\{P\}, \{V\}, W)$ such that

$$\int_{\Gamma\cap N^{(i,k)}\backslash N^{(i,k)}} \psi(ng)dn = \exp(-\langle H^{(i,k)}(g), H^{(i)}(z)\rangle + \rho(H^{(i,k)}(g)))E^{(i,k)}\Psi(g)$$

for some Ψ in $\mathcal{E}$. Since $\Psi = 0$ implies $\psi = 0$ the space $\mathcal{L}(z)$ is finite-dimensional. If $\Phi(z)$ is in $\mathcal{H}$ then

$$\int_{\Gamma\backslash G} \phi^{\wedge}(g)\bar{\psi}(g)dg = (\Phi(z), \Psi)$$

from which we conclude that ψ is in the domain of A and $A\psi = z^2\psi$. In particular $L(z_1)$ and $L(z_2)$ are orthogonal if z_1 and z_2 are different. It is clear that there is a constant c which is independent of z so that $\|\Psi\| \leq c\|\psi\|$ for any z in $(0, \langle\rho, \rho\rangle^{\frac{1}{2}}]$ and all ψ in $\mathcal{L}(z)$. If there was a sequence $\{z_n\}$ in $(0, \langle\rho, \rho\rangle^{\frac{1}{2}}]$, converging to a point in $(0, \langle\rho, \rho\rangle^{\frac{1}{2}}]$, such that $\mathcal{L}(z_n) \neq \{0\}$ for all n it is clear that we could construct a sequence $\{\psi_n\}$ with ψ_n in $\mathcal{L}(z_n)$ and $\|\psi_n\| = 1$ which satisfied the hypotheses of Lemma 5.2. It would follow from the dominated convergence theorem, applied as in the corollary to Lemma 5.1, that $\lim_{n\to\infty} \psi_n$ exists in $\mathcal{L}(\Gamma\backslash G)$. This is impossible for an orthonormal sequence. Thus the set of points for which $\mathcal{L}(z) \neq \{0\}$ is discrete in $(0, \langle\rho, \rho\rangle^{\frac{1}{2}}]$. If z is not in this set then $M(w)$ is bounded on the complement of the real axis in a neighbourhood of z and $\lim_{w\to z} M(w)$ exists. It follows from the reflection principle that $M(z)$ is analytic in the right half plane except at this set of points.

We have still to exclude the possibility that the above set of points has 0 as a limit point. If it does let $\{z_n\}$ be a monotone decreasing sequence of points converging to 0 with $\mathcal{L}(z_n) \neq \{0\}$ for all n. Let $\{\psi_n\}$ be a sequence of functions so that ψ_n belongs to $\mathcal{L}(z_n)$ and $\|\psi_n\| = 1$. Let Ψ_n be that element of E such that

$$\int_{\Gamma\cap N^{(i,k)}\backslash N^{(i,k)}} \psi_n(ng)dn = \exp(-\langle H^{(i,k)}(g), H^{(i)}(z_n)\rangle + \rho(H^{(i,k)}(g)))(E^{(i,k)}\Psi_n)(g)$$

for all i and k. If $\Psi'_n = \|\Psi_n\|^{-1}\Psi_n$ it may be supposed that $\lim_{n\to\infty} \Psi'_n$ exists. To obtain a contradiction we make use of the formulae (5.k), (5.ℓ), and (5.m). The first and second show us that

$$\int_{\Gamma\backslash G} \psi_m(g)\overline{\psi}_n(g)dg = \Sigma_{i=1}^{s} \int_{F_i} \psi_m(g)\overline{\psi}_n(g) + O(\|\Psi_m\|)$$

The third shows us that

$$\int_{F_i} \psi_m(g)\overline{\psi}_n(g)dg = \int_{F_i} \hat{\psi}_{m,i}(g)\overline{\psi}_n(g)dg + O(\|\Psi_m\|)$$

The integral on the right is equal to

$$\int_{F_i'} \psi^\wedge_{m,i} \bar{\psi}^\wedge_{n,i}(g)dg$$

if F_i' is the projection of F_i on $T_i\backslash G$ for we can suppose that the inverse image in $\Delta_i\backslash G$ of F_i' is F_i. If we then apply the estimate obtained for (5.j) we see that

$$\int_{\Gamma\backslash G} \psi_m(g)\bar{\psi}_n(g)dg$$

is equal to

$$\Sigma_{i=1}^{s} \int_{A_i^+(u,\infty)} \omega^2(a) \left\{ \int_{\Theta_i\backslash M_i \times K} \psi^\wedge_{m,i}(amk)\bar{\psi}^\wedge_{n,i}(amk)dmdk \right\} da + O(\|\Psi_m\|)$$

The only integrals on the right which are different from zero are those for which P_i belongs to $\{P\}$. If however P_i is conjugate to $P^{(j,\ell)}$ and we suppose, for simplicity, that $\{P_1, \ldots, P_s\}$ contains $\{P^{(k,m)} \mid 1 \leq k \leq r,\ 1 \leq m \leq m_i\}$ the corresponding integral equals

$$(z_m + z_n)^{-1}\exp(-a^{-1}(z_m + z_n)\log u)(E^{(j,\ell)}\Psi_m,\ E^{(j,\ell)}\Psi_n)$$

The number a has been introduced in the expression (6.b). Summing we obtain

$$\delta_{m,n} = (z_m + z_n)^{-1}\exp(-a^{-1}(z_m + z_n)\log u)(\Psi_m,\ \Psi_n) + O(\|\Psi_m\|$$

Set $m = n$ to see that $\lim_{m\to\infty} \|\Psi_m\| = 0$ and

$$1 = (2z_m)^{-1}\exp(-2a^{-1}z_m\log u)\|\Psi_m\|^2 + O(\|\Psi_m\|)$$

Hence $\|\Psi_m\| = O(z_m^{\frac{1}{2}})$; consequently if $m \neq n$

$$0 = 2(z_n z_m)^{\frac{1}{2}}(z_n + z_m)^{-1}(\Psi_m',\ \Psi_n') + O(z_m^{\frac{1}{2}})$$

If we divide by $z_m^{\frac{1}{2}}$ and recall that $\lim_{m,n\to\infty} (\Psi_m',\ \Psi_n') = 1$ we conclude that $z_n^{\frac{1}{2}}(z_m + z_n)^{-1}$ is bounded for all m and n. But that is clearly impossible.

Let

$$\Phi = \oplus_{i=1}^{r} \oplus_{k=1}^{m_i} \Phi_{i,k}$$

belong to $\mathcal{E}$ and let

$$M(z)\Phi = \oplus_{i=1}^{r} \oplus_{k=1}^{m_i} \Phi_{i,k}(z)$$

If x is given and $M(z)$ is defined let $F''(g, \Phi_{i,k}, H^{(i)}(z))$ equal $F(g, \Phi_{i,k}, H^{(i)}(z))$ if

$$a^{(i)}(H^{(i,k)}(g)) \leq x\langle a^{(i)}, a^{(i)}\rangle$$

and let it equal $-F(g, \Phi_{i,k}(z), -H^{(i)}(z))$ otherwise. Observe that the notation is deceptive. The Fourier transform of $F''(g, \Phi_{i,k}, H^{(i)}(\lambda))$ evaluated at $H^{(i)}(z)$ is equal to

$$(\lambda - z)^{-1}\exp(ax(\lambda - z))\Phi_{i,k} - (\lambda + z)^{-1}\exp(-ax(\lambda + z))\Psi_{i,k}(\lambda)$$

It follows from Lemma 4.1 that the series

$$\Sigma_{\Delta^{(i,k)}\backslash\Gamma} F''(\gamma g, \Phi_{i,k}, H^{(i)}(z))$$

converges for $\operatorname{Re} z > \langle\rho, \rho\rangle^{\frac{1}{2}}$; denote its sum by $E''(g, \Phi_{i,k}, H^{(i)}(z))$. If

$$E''(g, \Phi, z) = \Sigma_{i=1}^{r}\Sigma_{k=1}^{m_i} E''(g, \Phi_{i,k}, H^{(i)}(z))$$

then Lemma 4.6, together with a simple approximation argument, shows that $E''(g, \Phi, z)$ is a square integrable on $\Gamma\backslash G$ for $\operatorname{Re} z > \langle\rho, \rho\rangle^{\frac{1}{2}}$. We need an explicit formula for

$$(E''(g, \Phi, \lambda), E''(g, \Psi, \mu))$$

If we use formula (4.p) we see that this inner product is equal to the sum of eight integrals which we list below.

(i) $$\frac{1}{2\pi i}\int_{\mathrm{Re}\,z=c} (\lambda-z)^{-1}(\bar{\mu}+z)^{-1}\exp(ax(\lambda+\bar{\mu}))(\Phi,\ \Psi)dz$$

(ii) $$\frac{-1}{2\pi i}\int_{\mathrm{Re}\,z=c} (\lambda-z)^{-1}(\bar{\mu}-z)^{-1}\exp(ax(\lambda-\bar{\mu}))(\Phi,\ M(\mu)\Psi)dz$$

(iii) $$\frac{-1}{2\pi i}\int_{\mathrm{Re}\,z=c} (\lambda+z)^{-1}(\bar{\mu}+z)^{-1}\exp(ax(\bar{\mu}-\lambda))(M(\lambda)\Phi,\ \Psi)dz$$

(iv) $$\frac{1}{2\pi i}\int_{\mathrm{Re}\,z=c} (\lambda+z)^{-1}(\bar{\mu}-z)^{-1}\exp(-ax(\lambda+\bar{\mu}))(M(\lambda)\Phi,\ M(\mu)\Psi)dz$$

(v) $$\frac{1}{2\pi i}\int_{\mathrm{Re}\,z=c} (\lambda-z)^{-1}(\bar{\mu}-z)^{-1}\exp(ax(\lambda+\bar{\mu}-2z))(M(z)\Phi,\ \Psi)dz$$

(vi) $$\frac{-1}{2\pi i}\int_{\mathrm{Re}\,z=c} (\lambda-z)^{-1}(\bar{\mu}+z)^{-1}\exp(ax(\lambda-\bar{\mu}-2z))(M(z)\Phi,\ M(\mu)\Psi)dz$$

(vii) $$\frac{-1}{2\pi i}\int_{\mathrm{Re}\,z=c} (\lambda+z)^{-1}(\bar{\mu}-z)^{-1}\exp(ax(\bar{\mu}-\lambda-2z))(M(z)M(\lambda)\Phi,\ \Psi)dz$$

(viii) $$\frac{1}{2\pi i}\int_{\mathrm{Re}\,z=c} (\lambda+z)^{-1}(\bar{\mu}+z)^{-1}\exp(-ax(\lambda+\bar{\mu}+2z))(M(z)M(\lambda)\Phi,\ M(\mu)\Psi)dz$$

If we then make use of Lemma 4.5(i) these integrals can be evaluated when x is sufficiently large by using the residue theorem. The result when $\lambda+\bar{\mu} \neq 0$ and $\lambda-\bar{\mu} \neq 0$ follows

(i) $$(\lambda+\bar{\mu})^{-1}\exp(ax(\lambda+\bar{\mu}))(\Phi,\ \Psi)$$

(ii) $$0$$

(iii) $$0$$

(iv) $$(\lambda+\bar{\mu})^{-1}\exp(-ax(\lambda+\bar{\mu}))(M(\lambda)\Phi,\ M(\mu)\Psi)$$

(v) $$(\bar{\mu}-\lambda)^{-1}\exp(ax(\bar{\mu}-\lambda))(M(\lambda)\Phi,\ \Psi) + (\lambda-\bar{\mu})^{-1}\exp(ax(\lambda-\bar{\mu}))(\Phi,\ M(\mu)\Psi)$$

(vi) $$-(\lambda+\bar{\mu})^{-1}\exp(-ax(\lambda+\bar{\mu}))(M(\lambda)\Phi,\ M(\mu)\Psi)$$

(vii) $$-(\lambda+\bar{\mu})^{-1}\exp(-ax(\lambda+\bar{\mu}))(M(\lambda)\Phi,\ M(\mu)\Psi)$$

(viii) $$0$$

Adding up these eight terms we see that

$$(E''(g,\ \Phi,\ \lambda),\ E''(g,\ \Psi,\ \mu))$$

is equal to the sum of

$$(\lambda+\bar{\mu})^{-1}\{\exp(ax(\lambda+\bar{\mu}))(\Phi,\ \Psi) - \exp(-ax(\lambda+\bar{\mu}))(M(\lambda)\Phi,\ M(\mu)\Psi)\}$$

and

$$(\lambda-\bar{\mu})^{-1}\{\exp(ax(\lambda-\bar{\mu}))(\Phi,\ M(\mu)\Psi) - \exp(ax(\bar{\mu}-\lambda))(M(\lambda)\Phi,\ \Psi)\}$$

It is known that $M(z)$ is analytic in the right half-plane except at a finite number of points; it can be shown in a number of ways and, in particular, will follow from the discussion below that this is also true of $E''(\cdot,\ \Phi,\ z)$ considered as a function with values in $\mathcal{L}(\Gamma\backslash G)$. The formula for $(E''(g,\ \Phi,\ \lambda),\ E''(g,\ \Psi,\ \mu)$ is valid in this larger region. If $\lambda = \sigma + i\tau$ and $\mu = \lambda$ the above formula reduces to the sum of

$$(2\sigma)^{-1}\{\exp(2ax\sigma)(\Phi,\ \Psi) - \exp(-2ax\sigma)(M(\lambda)\Phi,\ M(\lambda)\Psi)\}$$

and

$$(2i\tau)^{-1}\{\exp(2iax\tau)(\Phi,\ M(\lambda)\Psi) - \exp(-2iax\tau)(M(\lambda)\Phi,\ \Psi)\}$$

The sum will be labelled (6.i). If we choose Φ so that $\|\Phi\| = 1$ and $\|M(\lambda)\Phi\| = \|M(\lambda)\|$ and then take $\Phi = \Psi$ we can conclude that

$$(2\sigma)^{-1}\{\exp(2ax\sigma) - \exp(-2ax\sigma)\}\|M(\lambda)\|^2 + |\tau|^{-1}\|M(\lambda)\| \geq 0$$

As a consequence

$$\|M(\lambda)\| \leq \max\{2\exp 4ax\sigma,\ 4\sigma/|\tau|\exp 2ax\sigma\}$$

We conclude first of all that $\|M(\lambda)\|$ is bounded in the neighbourhood of any point different from zero on the imaginary axis. Let us show next that $\|E''(\cdot, \Phi, \lambda)\|$ is bounded in the neighbourhood of any such point.

To be more precise we will show that $E''(\cdot, \Phi, \lambda)$ is holomorphic in any region U in which both $M(\lambda)$ and $E(\cdot, \Phi, \lambda)$ are holomorphic and in which $E(\cdot, \Phi, \lambda)$ satisfies the analogue of Lemma 6.2 and that if B is a bounded set of this region on which $\|M(\lambda)\|$ is bounded then $\|E''(\cdot, \Phi, \lambda)\|$ is bounded on B. As above if Φ belongs to $\mathcal{E}$ let

$$M(z)\Phi = \oplus_{i=1}^{r} \oplus_{k=1}^{m_i} \Phi_{i,k}(z)$$

If x is given and $M(z)$ is defined let $F'''(g, \Phi_{i,k}, H^{(i)}(z))$ equal

$$F(g, \Phi_{i,k}, H^{(i)}(z)) + F(g, \Phi_{i,k}(z), -H^{(i)}(z))$$

if $\alpha^{(i)}(H^{(i,k)}(g)) > x\langle\alpha^{(i)}, \alpha^{(i)}\rangle$ and let it equal zero otherwise. Let

$$E'''(g, \Phi_{i,k}, H^{(i)}(z)) = \Sigma_{\Delta^{(i,k)}\backslash\Gamma} F'''(g, \Phi_{i,k}, H^{(i)}(z))$$

the series converges whenever it is defined. As usual let

$$E'''(g, \Phi, z) = \Sigma_{i=1}^{r} \Sigma_{k=1}^{m_i} E'''(g, \Phi_{i,k}, H^{(i)}(z))$$

then

$$E''(g, \Phi, z) = E(g, \Phi, z) - E'''(g, \Phi, z)$$

Consequently the function $E''(\cdot, \Phi, z)$ can be defined, although it may not be square integrable, whenever $M(z)$ and $E(\cdot, \Phi, z)$ are both defined. In particular it can be defined on U. We will show that if z_0 is any complex number and if $\|M(z)\|$ is bounded on the intersection of B with some neighbourhood of z_0 then there is another neighbourhood of z_0 so that $\|E''(\cdot, \Phi, z)\|$ is finite and bounded on the intersection of this neighbourhood with B. This will establish the second part of the assertion. To see that the first part will also follow we observe that the above statement implies that $\|E''(\cdot, \Phi, z)\|$ is bounded on any compact subset of U; thus we have only to prove that

$$\int_{\Gamma\backslash G} E''(g, \Phi, \lambda)\bar{\psi}(g)dg$$

is holomorphic on U if ψ is a continuous function on $\Gamma\backslash G$ with compact support. However, this follows from the fact that if C is a compact subset of G the set $\{E''(g, \Phi, \cdot) \mid g \in C\}$ of functions on U is equicontinuous. We have to show that if $\{z_n\}$ is any sequence of points in B converging to z_0 then the sequence $\{\|E''(g, \Phi, z_n)\|\}$ is bounded. Let the sets F and F_i, $1 \leq i \leq s$, be the same as in (5.k), (5.ℓ), and (5.m). We suppose again that F_i is the inverse image of its projection F_i' on $T_i\backslash G$. The set $\{P_1, \ldots, P_s\}$ can be so chosen that it contains the set $\{P\}$; then for each j and ℓ there is a unique i so that $P^{(j,\ell)} = P_i$. Let $F_1'''(g, \Phi_{j,\ell}, H^{(j)}(z))$ equal $F'''(g, \Phi_{j,\ell}, H^{(j)}(z))$ if g belongs to F_i and let it equal zero otherwise; let $F_2'''(g, \Phi_{j,\ell}, H^{(j)}(z))$ equal

$$F'''(g, \Phi_{j,\ell}, H^{(j)}(z)) - F_1'''(g, \Phi_{j,\ell}, H^{(j)}(z))$$

If the sets $\omega_i^{(k)}$ used to define the sets F_i have been appropriately chosen, as we assume, the functions $F_2'''(g, \Phi_{j,\ell}, H^{(j)}(z_n))$ satisfy, uniformly in n, the conditions of the corollary to Lemma 3.6. Thus if

$$E_2'''(g, \Phi_{j,\ell}, H^{(j)}(z)) = \Sigma_{\Delta^{(j,\ell)}\backslash\Gamma} F_2'''(g, \Phi_{j,\ell}, H^{(j)}(z))$$

we know that the sequence $\{\|E_2'''(\cdot,\ \Phi_{j,\ell},\ H^{(j)}(z_n))\|\}$ is bounded. Let

$$E_1'''(g,\ \Phi_{j,\ell},\ H^{(j)}(z)) = E'''(g,\ \Phi_{j,\ell},\ H^{(j)}(z)) - E_2'''(g,\ \Phi_{j,\ell},\ H^{(j)}(z))$$

The function $E_1'''(g,\ \Phi_{j,\ell},\ H^{(j)}(z))$ is zero on $\Gamma\backslash G\text{-}F$. Thus

$$\int_{\Gamma\backslash G-F} |E''(g,\ \Phi,\ z_n)|^2 dg = \int_{\Gamma\backslash G-F} |E(g,\ \Phi,\ z_n) - \Sigma_{j=1}^{r}\Sigma_{\ell=1}^{m_j} E_2'''(g,\ \Phi_{j,\ell},\ H^{(j)}(z_n)|^2 dg$$

It follows from (5.j) that the latter integrals are uniformly bounded. Moreover the integrals

$$\int_{F_i} |E''(g,\ \Phi,\ z_n)|^2 dg$$

are uniformly bounded if and only if the integrals

$$\int_{F_i} |E(g,\ \Phi,\ z_n) - \Sigma_{j=1}^{r}\Sigma_{\ell=1}^{m_j} E_1'''(g,\ \Phi_{j,\ell},\ H^{(j)}(z_n)|^2 dt$$

are. But it follows from the definition of the sets F_i that on F_i the sum

$$\Sigma_{j=1}^{r}\Sigma_{\ell=1}^{m_j} E_1'''(g,\ \Phi_{j,\ell},\ H^{(j)}(z_n))$$

is zero if P_i does not belong to $\{P\}$ and is $F'''(g,\ \Phi_{j,\ell},\ H^{(j)}(z))$ if $P_i = P^{(j,\ell)}$. If the number u used in the definition of the sets F_i is sufficiently large, as we suppose, then in all cases the sum equals

$$\int_{\Gamma\cap N_i\backslash N_i} E(ng,\ \Phi,\ z_n)dn$$

We can complete our argument by appealing to the estimate (5.m).

It now follows from (6.i) that

$$(2\sigma)^{-1}\{\exp(2ax\sigma)(\Phi,\ \Psi) - \exp(-2ax\sigma)(M(\lambda)\Phi,\ M(\lambda)\Psi)\}$$

is bounded in the neighbourhood of any point λ_0 on the imaginary axis different from zero. Hence

$$\lim_{\lambda \to \lambda_0} M^*(\lambda)M(\lambda) = I$$

or

$$\lim_{\lambda \to \lambda_0} M(\bar{\lambda})M(\lambda) = I$$

since $M^*(\lambda) = M(\bar{\lambda})$. Moreover if the interval $[a, b]$ does not contain zero there is an $\varepsilon > 0$ so that $\|M^{-1}(\lambda)\|$ is bounded for $0 < \sigma \leq \varepsilon$ and $a \leq \tau \leq b$; consequently

$$\lim_{\sigma \searrow 0} \|M^{-1}(\sigma - i\tau) - M(\sigma + i\tau)\| = 0$$

Define $M(\lambda)$ for $\operatorname{Re} \lambda < 0$ by $M(\lambda) = M^{-1}(-\lambda)$. Let C be the contour consisting of the lines joining $ia-\varepsilon$, $ia+\varepsilon$, $ib+\varepsilon$, $ib-\varepsilon$, and then $ia-\varepsilon$ again. It is clear that, for $0 < |\sigma| < \varepsilon$, $a < \tau < b$,

$$M(\lambda) = \frac{1}{2\pi i} \int_C (z-\lambda)^{-1} M(z)\,dz + \lim_{\delta \to 0} \frac{1}{2\pi} \int_a^b \{M(-\delta+it) - M(\delta+it)\}(it-\lambda)^{-1}\,dt$$

The final integral equals

$$\int_a^b \{M^{-1}(\delta - it) - M(\delta+it)\}\,dt$$

so the limit is zero. This shows that the function $M(\lambda)$ defined in the left half-plane is the analytic continuation of the function $M(\lambda)$ defined in the right half-plane. Thus $M(\lambda)$ is meromorphic except perhaps at a finite number of points $0, \pm z_1, \ldots, \pm z_n$ in the interval $[-\langle \rho, \rho \rangle^{\frac{1}{2}}, \langle \rho, \rho \rangle^{\frac{1}{2}}]$.

Let us verify that the same is true of $E(\cdot, \Phi, \lambda)$ for all Φ in $\mathcal{E}$. It follows from Lemma 5.2 that

$$\lim_{\sigma \searrow 0} E(g, \Phi, \sigma + i\tau) = E(g, \Phi, i\tau)$$

converges for all τ different from zero and all g and that the convergence is uniform on compact subsets of G for each τ. If we use this fact to define

$E(g, \Phi, z)$ for non-zero imaginary values of z all the assertions of Lemma 6.2, except perhaps the first, will be valid if $H_0^{(i)} = H^{(i)}(z)$ with z imaginary and different from zero. We define $E(g, \Phi, \lambda)$ when $\operatorname{Re}\lambda \leq 0$ by setting it equal to $E(g, M(\lambda)\Phi, -\lambda)$. With this definition all the assertions of Lemma 6.2, except perhaps the last, are valid if $H_0^{(i)} = H_0^{(i)}(z)$ with $\operatorname{Re} z < 0$ and z different from $-z_1, \ldots, -z_n$. Every assertion, except perhaps the first and last, is valid if $H_0^{(i)} = H^{(i)}(z)$ with z imaginary and different from zero. However

$$\int_{\Gamma\cap N^{(i,k)}\backslash N^{(i,k)}} E(ng, M(\lambda)\Phi, -\lambda)dn$$

is equal to the sum of

$$\exp(\langle H^{(i,k)}(g), H^{(i)}(-\lambda)\rangle + \rho(H^{(i,k)}(g)))(E^{(i,k)}M(\lambda)\Phi)(g)$$

and

$$\exp(\langle H^{(i,k)}(g), H^{(i)}(\lambda)\rangle + \rho(H^{(i,k)}(g)))(E^{(i,k)}M(-\lambda)M(\lambda)\Phi)(g)$$

which, since $M(-\lambda)M(\lambda) = I$, is equal to the sum of

$$\exp(\langle H^{(i,k)}(g), H^{(i)}(\lambda)\rangle + \rho(H^{(i,k)}(g)))(E^{(i,k)}\Phi)(g)$$

and

$$\exp(\langle H^{(i,k)}(g), H^{(i)}(-\lambda)\rangle + \rho(H^{(i,k)}(g)))(E^{(i,k)}M(\lambda)\Phi(g))$$

Consequently the last assertion is also valid. It follows from Lemma 3.7 that the two definitions of $E(g, \Phi, \lambda)$ agree when λ is imaginary and then from Lemma 5.3 that $E(\cdot, \Phi, \lambda)$ is analytic at the non-zero points on the imaginary axis.

It remains to examine the behaviour of $M(\lambda)$ and $E(\cdot, \Phi, \lambda)$ at the points $0, \pm z_1, \ldots, \pm z_n$. Since we readily see from Lemma 5.2 that the behaviour of $E(\cdot, \Phi, \lambda)$ is at least as good as that of $M(\lambda)$ we shall only study the latter. We shall show that $M(\lambda)$ is analytic at zero and has at most a simple pole at the points $z_1, \ldots, z_n$. If $\Phi(z)$ and $\Psi(z)$ belong to $\mathfrak{H}$ the formula (6.e) expresses the inner product $(\phi^\wedge, \psi^\wedge)$ as a contour integral. We shall replace the contour of

(6.e) by the sum of n+1 other contours $C, C_1, \ldots, C_n$. Let $\varepsilon > 0$ be so small that the closed discs of radius ε about $0, z_1, \ldots, z_n$ are disjoint. Let C_i, $1 \le i \le n$, be the circle of radius ε about z_i traversed in the positive direction; let C be the path running from $-i\infty$ to $-i\varepsilon$ along the imaginary axis, then in the positive direction on the circle of radius ε and centre zero to $i\varepsilon$, and then along the imaginary axis to $i\infty$. Our estimates of $\|M(\lambda)\|$ are good enough that we can replace the right side of (6.e) by the sum of

$$\frac{1}{2\pi i}\int_C (\Phi(z), \Psi(-\bar{z})) + (M(z)\Phi(z), \Psi(\bar{z}))dz$$

and

$$\Sigma_{i=1}^{n} \frac{1}{2\pi i}\int_{C_i} (M(z)\Phi(z), \Psi(\bar{z}))dz$$

This sum will be labelled (6.k). Suppose that $E(\cdot)$ is, in the terminology of [21], the resolution of the identity belonging to the linear transformation A. It is well known ([21], Theorem 5.10) that, if b is greater than a and c is positive,

$$(6.\ell) \qquad \tfrac{1}{2}\{(E(b)\phi^{\wedge}, \psi^{\wedge}) - (E(b-0)\phi^{\wedge}, \psi^{\wedge})\} - \tfrac{1}{2}\{(E(a)\phi^{\wedge}, \psi^{\wedge}) - (E(a-0)\phi^{\wedge}, \psi^{\wedge})\}$$

is given by

$$(6.m) \qquad \lim_{\delta \searrow 0} \frac{1}{2\pi i}\int_{C(a,b,c,\delta)} (R(\lambda, A)\phi, \psi)d\lambda$$

where the contour $C(a, b, c, \delta)$ consists of two polygonal paths whose vertices are in order $b+i\delta$, $b+ic$, $a+ic$, $a+i\delta$ and $a-i\delta$, $a-ic$, $b-ic$, $b-i\delta$ respectively. Since the spectrum of A is contained in $(-\infty, \langle\rho, \rho\rangle)$ we know that $E(\langle\rho, \rho\rangle) = I$. Choose a and b so that $b > a > 0$ and so that exactly one of the numbers $z_1^2, \ldots, z_n^2$, say z_i^2, belongs to the interval $[a, b]$. If we use the formula (6.k) to calculate (6.m) we find that (6.ℓ) is equal to

$$(6.n) \qquad \frac{1}{2\pi i}\int_{C_i} (M(z)\Phi(z), \Psi(\bar{z}))dz$$

Since this is true for any such a and b we conclude that (6.n) is equal to $E(b) - E(a)$. If we assume, as we may, that $M(z)$ is not analytic at any of the points $z_1, \ldots, z_n$ we see that $z_1^2, \ldots, z_n^2$ are isolated points in the spectrum of A. Consequently, for any ϕ and ψ in $\mathcal{L}(\{P\}, \{V\}, W)$,

$$(R(\lambda^2, A)\phi, \psi)$$

has only a simple pole at $z_1, \ldots, z_n$. Referring to the discussion following (6.f) and (6.g) we see that the same is true of $M(\lambda)$.

If we again use (6.k) to calculate (6.m) we find that $(E(x)\phi, \psi)$ is continuous except at $z_1^2, \ldots, z_n^2$ and, perhaps, zero and that, if ε is positive but sufficiently small,

$$(E(0)\phi^\wedge, \psi^\wedge) - (E(-\delta^2)\phi^\wedge, \psi^\wedge)$$

is equal to

$$\frac{1}{2\pi i}\int_{-i\varepsilon}^{i\varepsilon} (\Phi(z), \Psi(-\bar{z}))dz + \frac{1}{2\pi i}\int_{C(\varepsilon)} (M(z)\Phi(z), \Psi(\bar{z}))dz \tag{6.o}$$

if $C(\varepsilon)$ is the semi-circle of radius ε and centre zero traversed in the positive direction from $-i\varepsilon$ to $i\varepsilon$. Hence

$$(E(0)\phi^\wedge, \psi^\wedge) - (E(0-0)\phi^\wedge, \psi^\wedge) = \lim_{\varepsilon \searrow 0} \frac{1}{2\pi i}\int_{C(\varepsilon)} (M(z)\Phi(z), \Psi(\bar{z}))dz$$

The right side must be a positive definite hermitian symmetric form on $\mathfrak{H}_0$. However it is defined if $\Phi(z)$ and $\Psi(z)$ are merely defined and analytic in some neighbourhood of zero. A simple approximation argument shows that it remains positive definite on this larger space of functions. Consequently, if $\omega(z)$ is a scalar valued function, analytic in a neighbourhood of zero,

$$\lim_{\varepsilon \searrow 0} \frac{1}{2\pi i}\int_{C(\varepsilon)} (\bar{\omega}(\bar{z})\omega(z)M(z)\Phi(z), \Phi(\bar{z}))dz \geq 0 \tag{6.p}$$

If δ is positive we can take $\omega(z)$ to be either $(\delta+z)^{\frac{1}{2}}$ or $(\delta-z)^{\frac{1}{2}}$; then $\overline{\omega(\bar{z})}\omega(z)$ is $\delta+z$ or $\delta-z$. Substituting in the relation (6.p) we conclude that

$$\lim_{\varepsilon \searrow 0} \frac{1}{2\pi i} \int_{C(\varepsilon)} (zM(z)\Phi(z),\ \Phi(\bar{z}))dz = 0$$

Applying Schwarz's inequality to (6.p) we can conclude more generally that

$$\lim_{\varepsilon \searrow 0} \frac{1}{2\pi i} \int_{C(\varepsilon)} (zM(z)\Phi(z),\ \Psi(\bar{z}))dz = 0$$

Consequently

$$(6.q) \qquad (E(0)\phi^\wedge,\ \psi^\wedge) - (E(0-0)\phi^\wedge,\ \psi^\wedge) = \lim_{\varepsilon \searrow 0} \frac{1}{2\pi i} \int_{C(\varepsilon)} (M(z)\Phi(0),\ \Psi(0))dz$$

There is a linear transformation M on $\mathfrak{E}$ so that the right side of this equation equals $(M\Phi(0),\ \Psi(0))$.

We shall use the equation we have just found to show that $E(0) = E(0-0)$. It is enough to show that, for all functions $\Phi(z)$ in $\mathfrak{H}$, $E(0)\phi^\wedge = E(0-0)\phi^\wedge$. Suppose f is a continuous function on G with compact support such that $f(kgk^{-1}) = f(g)$ for all g in G and all k in K. For each $H^{(i)}$ in $\mathfrak{n}_c^{(i)}$ we have defined, in Section 3, a linear transformation $\pi(f,\ H^{(i)})$ on $\mathfrak{E}(V^{(i,k)},\ W)$. For each complex number z the direct sum of the linear transformations $\pi(f,\ H^{(i)}(z))$ is a linear transformation $\pi(f,\ z)$ on $\mathfrak{E}$. It follows from (4.r) that if $\Psi(z)$ belongs to $\mathfrak{H}$ and

$$\Psi_1(z) = \pi(f,\ z)\Psi(z)$$

then $\lambda(f)\psi^\wedge = \psi_1^\wedge$. As a consequence $\lambda(f)$ commutes with A and with $E(x)$ for all x. Choosing f so that $\pi(f,\ 0)$ is the identity we deduce from (6.q) that if $\phi' = E(0)\phi^\wedge - E(0-0)\phi^\wedge$ then $\lambda(f)\phi' = \phi'$. Hence ϕ' is continuous. Referring to Lemma 4.6(i) we see that if P is a cuspidal subgroup the cuspidal component of

$$\int_{\Gamma \cap N \backslash N} \phi'(ng)dn$$

is zero unless P is conjugate to an element of $\{P\}$. However it follows from (6.q) and the remark following the proof of Lemma 3.7 that

$$\int_{\Gamma \cap N^{(i,k)} \backslash N^{(i,k)}} \phi'(ng)dn = \exp \rho(H^{(i,k)}(g))(E^{(i,k)}M\Phi(0))(g)$$

If P is a percuspidal subgroup to which $P^{(i,k)}$ belongs and $\mathfrak{S}$ a Siegel domain associated to P then the left, and hence the right, side must be square integrable on $\mathfrak{S}$. A simple calculation shows that this is so only if $E^{(i,k)}M\Phi(0)$ is zero. Since i and k are arbitrary the function ϕ' is identically zero.

Now let C be the semi-circle of radius 1 and centre zero traversed in the positive direction from $-i$ to i. Suppose $0 < |\lambda| < 1$ and $\operatorname{Re}\lambda > 0$; since (6.q) vanishes and $M(z)$ is unitary for imaginary z the residue theorem implies

$$M(\lambda) = \frac{1}{2\pi i}\int_i^{-i} (z-\lambda)^{-1}M(z)dz + \frac{1}{2\pi i}\int_C (z-\lambda)^{-1}M(z)dz$$

Since the right side vanishes if λ is replaced by $-\bar{\lambda}$ we have

$$\text{(6.r)} \quad M(\lambda) = \frac{\sigma}{\pi}\int_{-1}^{1} (\sigma^2 + (y-\tau)^2)^{-1}M(iy)dy + \frac{1}{2\pi i}\int_C \{(z-\lambda)^{-1} + (z+\bar{\lambda})^{-1}\}M(z)dz$$

if $\lambda = \sigma + i\tau$. We shall use this equation to show that

$$\text{(6.s)} \quad \lim_{\sigma \searrow 0,\, \tau \to 0} M(\sigma + i\tau) = M(0)$$

exists. Since $M(0)$ must equal $\lim_{\tau \to 0} M(i\tau)$ which is unitary and, hence, invertible we shall conclude that there is an $\varepsilon > 0$ so that $M(\lambda)$ and $M^{-1}(\lambda)$ are uniformly bounded on

$$\{\lambda \mid 0 < |\lambda| < \varepsilon,\ \operatorname{Re}\lambda \geq 0\}$$

Consequently $M(\lambda)$ is bounded in a neighbourhood of zero and zero is a removable singularity.

Let

$$M'(z) = \frac{d}{dz} M(z)$$

It is a familiar, and easily proved, fact that (6.s) will follow from (6.r) if it is shown that $\lim_{y\to 0} M(iy) = M(0)$ exists and that, if $N > 0$, there are positive constants c' and r' so that $\|M'(iy)\| \le c'|y|^{r'-1}$ for $0 < |y| \le N$. We know that, for every Φ in $\mathcal{E}$, $\|E''(\cdot, \Phi, iy)\|$ is bounded on $\{y \mid 0 < |y| \le N\}$. If in (6.i) we replace τ by y and take the limit as σ approaches zero we find that $(E''(\cdot, \Phi, iy), E''(\cdot, \Psi, iy))$ is equal to

$$(M^{-1}(iy)M'(iy)\Phi, \Psi) - (2iy)^{-1}\{(M(iy)\Phi, \Psi) - (M^{-1}(iy)\Phi, \Psi)\}$$

if the number x is taken to be zero. Consequently the linear transformation

$$B(y) = M^{-1}(iy)M'(iy) - (2iy)^{-1}(M(iy) - M^{-1}(iy))$$

is positive definite for y different from zero and is bounded on $\{y \mid 0 < |y| \le N\}$. If we show that there is a $\delta > 0$ and positive constants c and r so that

$$\|M(iy) - M^{-1}(iy)\| \le 2cy^{r}$$

if $0 < y < \delta$ it will follow that, for some c' and r',

$$\|M'(iy)\| \le c'y^{r'-1}$$

if $0 < y \le N$. We shall conclude that

$$\lim_{y\searrow 0} M(iy) = M(0)$$

and

$$\lim_{y\nearrow 0} M(iy) = M^{-1}(0)$$

exist and that

$$\lim_{y\searrow 0} M(iy) = \lim_{y\searrow 0} M^{-1}(iy)$$

so that $M(0) = M^{-1}(0)$. Since

$$M'(-z) = M^{-1}(z)M'(z)M^{-1}(z)$$

we need only establish the above estimate on the interval $(0, N]$. Choose b so that $\|B(y)\| \leq b$ for $0 < y \leq N$. Suppose $0 < y$ and suppose $e^{i\theta}$ is an eigenvalue fo $M(iy)$ of multiplicity m. It is known that if y' is sufficiently close to y then $M(iy')$ has exactly m eigenvalues, counted with multiplicities, which are close to $e^{i\theta}$. If

$$8yb \leq |\sin\theta|$$

it is possible to obtain more precise information about the position of these m eigenvalues. Choose an orthonormal basis $\Phi_1, \ldots, \Phi_n$ for $\mathcal{C}$ consisting of eigenvectors of $M(iy)$ and let $e^{i\theta_1}, \ldots, e^{i\theta_n}$ be the corresponding eigenvalues. If

$$\Phi = \Sigma_{j=1}^{n} a_j \Phi_j$$

with

$$\Sigma_{j=1}^{n} |a_j|^2 = 1$$

is a unit vector then

$$(M(iy)\Phi, \Phi) = \Sigma_{j=1}^{n} e^{i\theta_j} |a_j|^2$$

and

$$(M'(iy)\Phi, \Phi) = \Sigma_{j=1}^{n} y^{-1} \sin\theta_j e^{i\theta_j} |a_j|^2 + (M(iy)B(y)\Phi, \Phi)$$

which is equal to

$$\Sigma_{j=1}^{n} (y^{-1} \sin\theta_j e^{i\theta_j} + \beta) |a_j|^2$$

if

$$\beta = (M(iy)B(y)\Phi,\ \Phi)$$

Certainly $|\beta| \leq b$. It follows from the first formula that $(M(iy)\Phi,\ \Phi)$ lies in the convex hull of the eigenvalues of $M(iy)$; a similar assertion is of course valid for any unitary transformation. For any positive y'

$$(M(iy')\Phi,\ \Phi) = \Sigma_{j=1}^{n} e^{i\theta_j} |a_j|^2 + i\int_{y}^{y'} (M'(is)\Phi,\ \Phi)ds$$

Let $t = y'-y$ and suppose $|t|$ is so small that $\|M'(is) - M'(iy)\| \leq b$ if $|s-y| \leq |t|$; then

$$(M(iy')\Phi,\ \Phi) = \Sigma_{j=1}^{n} e^{i\theta_j}(1 + ity^{-1}\sin\theta_j + \beta(t))|a_j|^2$$

with $|\beta(t)| \leq 2bt$. Set

$$v_j(t) = \mp ty^{-1}\sin\theta_j \pm i\beta(t)$$

and set

$$u_j(t) = e^{i\theta_j}(1 \mp iv_j(t))$$

The upper or the lower sign is taken according as $\sin\theta_j \geq 0$ or $\sin\theta_j < 0$. The number $v_j(t)$ equals

$$(\mp ty^{-1}\sin\theta_j + 8bt) + (-8bt \pm i\beta(t))$$

If $t < 0$ and $8yb \leq |\sin\theta_j|$ the second term lies in the sector $\{z \mid |\arg z| \leq \frac{\pi}{4}\}$ and the first term is positive.

Suppose $e^{i\theta}$ is an eigenvalue of $M(iy)$ of multiplicity m and $8yb \leq |\sin\theta|$; we shall show that if y' is less than but sufficiently close to y the m eigenvalues of $M(iy')$ which are close to $e^{i\theta}$ then lie in

$$X(t) = \{e^{i\theta}(1 \mp i(\mp ty^{-1}\sin\theta + 8tb + z)) \mid |\arg z| \leq \tfrac{\pi}{3}\}$$

Again the upper or the lower sign is taken according as $\sin\theta \geq 0$ or $\sin\theta < 0$. This will follow if it is shown that for some ε with $0 \leq \varepsilon < \frac{\pi}{12}$ these eigenvalues lie in

$$Y(t) = \{e^{i\theta}(1 \mp i(\mp ty^{-1}\sin\theta + 8tb + z)) \mid -\frac{\pi}{2} \pm \frac{\pi}{4} \mp \varepsilon \leq \arg z \leq \frac{\pi}{2} \pm \frac{\pi}{4} \mp \varepsilon\}$$

The set $e^{-i\theta}Y(t)$ is the shaded sector in the diagram below, and $e^{-i\theta}X(t)$ is the shaded half-plane

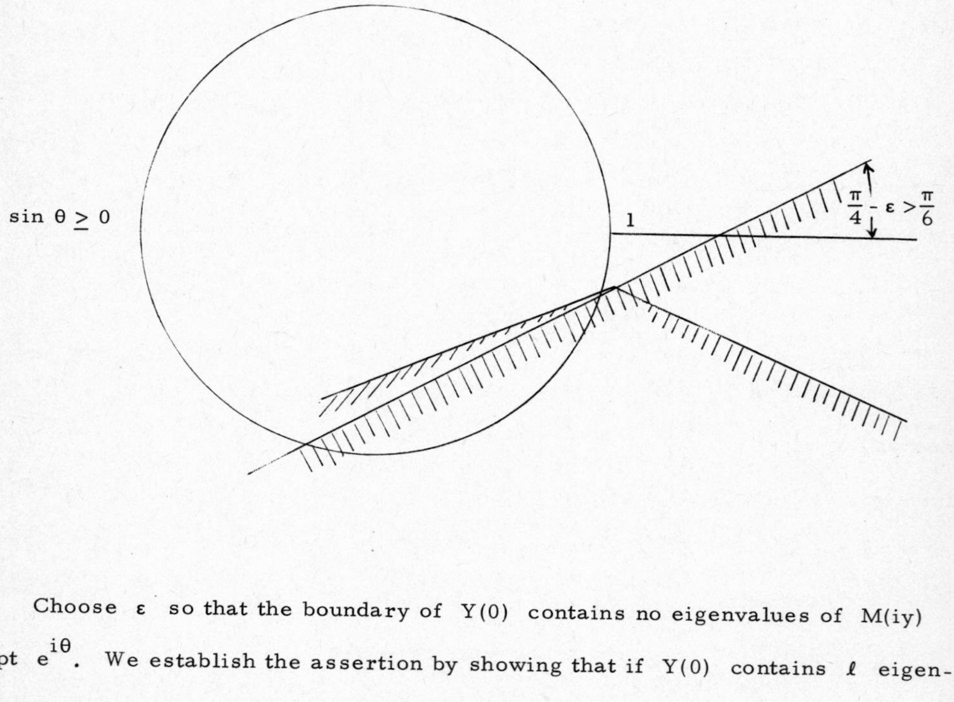

Choose ε so that the boundary of $Y(0)$ contains no eigenvalues of $M(iy)$ except $e^{i\theta}$. We establish the assertion by showing that if $Y(0)$ contains ℓ eigenvalues of $M(iy)$ then $Y(t)$ contains ℓ eigenvalues of $M(iy + it)$ when t is negative but sufficiently close to 0. Let $e^{i\theta_1}, \ldots, e^{i\theta_\ell}$ be the ℓ eigenvalues of $M(iy)$ which lie in $Y(0)$. If

$$\Phi = \Sigma_{j=1}^{\ell} a_j \Phi_j$$

is a unit vector then

$$(M(iy + it)\Phi, \Phi) = \Sigma_{j=1}^{\ell} u_j(t)|a_j|^2$$

If $1 \le j \le \ell$ and $e^{i\theta_j} \neq e^{i\theta}$ then, for $|t|$ sufficiently small, $u_j(t)$ lies in $Y(t)$ simply because it is close to $e^{i\theta_j}$. If $e^{i\theta_j} = e^{i\theta}$ the calculations above show that $u_j(t)$ lies in $Y(t)$. Since the set is convex $(M(iy + it)\Phi, \Phi)$ does also. If the assertion were false we could choose Φ to be a linear combination of eigenvectors of $M(iy + it)$ belonging to eigenvalues lying in the complement of $Y(t)$, and thereby force $(M(iy + it)\Phi, \Phi)$ to lie in the complement. This is a contradiction.

A glance at the diagram allows us to infer that if $e^{i\theta'}$ is an eigenvalue of $M(iy')$ lying close to $e^{i\theta}$ then

$$(6.t) \quad \pm(\theta-\theta') \ge \pm \sin(\theta-\theta') \ge (y-y')(y^{-1}|\sin\theta| - 8b)$$

provided of course that θ' is chosen near θ. We readily deduce that if $-1 < a < 1$ there is an $\varepsilon > 0$ so that the number of eigenvalues which lie on the arc

$$V(y) = \{e^{i\theta} \mid |\sin\theta| < 8yb, \ \cos\theta < 0\}$$

and the number of eigenvalues which lie on the arc $\{e^{i\theta} | \cos\theta < a\}$ are non-decreasing functions on $(0, \varepsilon)$. Indeed we can find ε and a so that these functions are equal and constant on $(0, \varepsilon)$. For example at a point y at which one of the eigenvalues $\theta = \theta(y)$ enters or leaves $V(y)$ we have

$$|\sin\theta| = 8yb$$

Hence (6.t) holds. Moreover if y' is close to y but less than it then

$$-8y'b + |\sin\theta'| = (8yb - |\sin\theta|) - (8y'b - |\sin\theta'|) = (y-y')\left\{8b \mp \frac{\theta-\theta'}{y-y'}\cos\theta''\right\}$$

with θ'' close to θ. Since $\cos\theta'' < 0$ the right hand side is greater than or equal to $8b(y-y')$. It follows that

$$8y'b - |\sin \theta'| < 0$$

so that V(y) has more elements than V(y').

We next observe that the eigenvalues of M(iy) which do not lie on V(y) must all approach 1. Suppose they did not. From all the eigenvalues $e^{i\theta}$ of M(iy) which lie outside of V(y) choose one $e^{i\theta(y)}$, with $0 \le |-\theta(y)| \le \pi$, for which $\cos\theta$ is a minimum and set $a(y) = \cos\theta(y)$; then $a(y) \ge a$. If $\liminf_{y \to 0} a(y) \neq 1$ then there is an $a' < 1$ so that $a(y) \le a'$ for all sufficiently small y. Consequently there is a constant c' so that

$$|y^{-1} \sin\theta(y)| - 8b \ge c'y^{-1}$$

for all sufficiently small y. It then follows from (6.t) that, for y' less than but sufficiently close to y,

$$|\theta(y)| - |\theta(y')| \ge c'y^{-1}(y-y')$$

Hence, for all $y' \le y$,

$$|\theta(y)| - |\theta(y')| \ge \tfrac{1}{2}c' \log y/y'$$

which is a patent impossibility. Choose $\delta > 0$ so that $|\sin\theta(\delta)| \le \frac{1}{2}$, $\cos\theta(\delta) \ge \frac{1}{2}$, and $32b\delta < 1$. Let $r = 1/5$ and choose c so that $c\delta^r = 1$. We shall show that if $0 < y \le \delta$ then $|\sin\theta(y)| \le cy^r$. If $\delta < \varepsilon < 1$, $c \ge 8b$, as we may suppose, we can combine this with our earlier assertion to see that $\|M(iy) - M^{-1}(iy)\| \le 2cy^r$ on the interval $(0, \delta]$. If the assertion is false for some number y' let y be the least upper bound of the numbers for which it is false. It is true for y and $|\sin\theta(y)| = cy^r$. If y' is less than, but sufficiently close to, y then

$$|\sin\theta(y)| - |\sin\theta(y')| \ge \tfrac{1}{2}(|\theta(y)| - |\theta(y')|) \ge \tfrac{1}{2}(y-y')(cy^{r-1} - 8b)$$

Since

$$cy^{r-1} = \tfrac{1}{2}\delta^{-r} y^{r-1} \geq \tfrac{1}{2}\delta^{-1} > 16b$$

we see that

$$|\sin \theta(y')| \leq cy^{r} - \tfrac{1}{4}cy^{r-1}(y-y')$$

However, for y' sufficiently close to y,

$$y^{r} - \tfrac{1}{4}y^{r-1}(y-y') \leq (y')^{r}$$

so that

$$|\sin \theta(y')| \leq c(y')^{r}$$

This is a contradiction.

We turn now to the proof of Lemmas 6.1 and 6.2 for families of cuspidal subgroups of rank greater than one. Let $\mathfrak{a}^{(i)}$ be one of $\mathfrak{a}^{(1)}, \ldots, \mathfrak{a}^{(r)}$. If $\alpha^{(i)}_{\ell,}$ is a simple root of $\mathfrak{a}^{(i)}$ let

$$\mathfrak{a}^{(i)}_{\ell} = \{H \in \mathfrak{a}^{(i)} \mid \alpha^{(i)}_{\ell}(H) = 0\}$$

If we fix i and ℓ then, as was remarked before stating Lemma 2.13, there is a unique j such that $\mathfrak{a}^{(j)}$ contains $\mathfrak{a}^{(i)}_{\ell}$ and such that $\Omega(\mathfrak{a}^{(i)}, \mathfrak{a}^{(j)})$ contains an element s so that $\alpha^{(i)}_{m,} \circ S^{-1}$ is a positive root of $\mathfrak{a}^{(j)}$ if and only if $m \neq \ell$. We first show that if Φ belongs to $\mathcal{E}^{(i)}$ then $E(g, \Phi, H)$ is meromorphic on the convex hull of $\mathfrak{a}^{(i)}$ and $s^{-1}\mathfrak{a}^{(j)}$ and, on this set,

$$E(g, M(s, H)\Phi, sH) = E(g, \Phi, H)$$

For each k there is a unique cuspidal subgroup ${}^{*}P^{(i,k)}$ belonging to $P^{(i,k)}$ which has the split component $\mathfrak{a}^{(i)}_{\ell}$. We define ${}^{*}P^{(j,k)}$ in the same manner. There is no harm in supposing that the elements of $\{P\}$ have been so chosen that if ${}^{*}P^{(i,k_1)}$ and ${}^{*}P^{(i,k_2)}$ or ${}^{*}P^{(j,k_2)}$ are conjugate they are equal. Choose ${}^{*}P$ in

$$\{{}^*P^{(i,k)} \mid 1 \le k \le m_i\} = \{{}^*P^{(j,k)} \mid 1 \le k \le m_j\}$$

and suppose ${}^*P = {}^*P^{(i,k)}$ for $1 \le k \le m_i'$ and ${}^*P = {}^*P^{(j,k)}$ for $1 \le k \le m_j'$. Let

$${}^{\Psi}\mathcal{E}^{(i)} = \oplus_{k=1}^{m_i'} \mathcal{E}(V^{(i,k)} \times W, W^*)$$

and let

$${}^{\Psi}\mathcal{E}^{(j)} = \oplus_{k=1}^{m_j'} \mathcal{E}(V^{(i,k)} \times W, W^*)$$

According to the remarks preceding Lemma 3.5 we can identify

$$\oplus_{k=1}^{m_i} \mathcal{E}(V^{(i,k)}, W)$$

or

$$\oplus_{k=1}^{m_j'} \mathcal{E}(V^{(j,k)}, W)$$

with the space of functions in ${}^{\Psi}\mathcal{E}^{(i)}$ or ${}^{\Psi}\mathcal{E}^{(j)}$ respectively which are invariant under right translations by elements of ${}^*K_0$. If H belongs to $\mathfrak{a}_c^{(i)}$ let $H = {}^*H + {}^{\Psi}H$ with ${}^*H$ in the complexification of $\mathfrak{a}_\ell^{(i)}$ and ${}^{\Psi}H$ orthogonal to $\mathfrak{a}_\ell^{(i)}$. The restriction of $M(s, H)$ to

$$\oplus_{k=1}^{m_i'} \mathcal{E}(V^{(i,k)}, W)$$

depends only on ${}^{\Psi}H$ and agrees with the restriction to this space of a linear transformation on ${}^{\Psi}E^{(i)}$ which, using a notation suggested by that of Lemma 4.5(ii), we call $M({}^{\Psi}s, {}^{\Psi}H)$. If Φ belongs to

$$\oplus_{k=1}^{m_i} \mathcal{E}(V^{(i,k)}, W)$$

then $M(s, H)\Phi$ belongs to

$$\oplus_{k=1}^{m_j'} \mathcal{E}(V^{(j,k)}, W)$$

It is enough to show that for each ${}^*P$ and each Φ in

$$\oplus_{k=1}^{m_i'} \mathcal{E}(V^{(i,k)}, W)$$

the function $E(g, \Phi, H)$ is meromorphic on the convex hull of $\mathfrak{U}^{(i)}$ and $s^{-1}\mathfrak{U}^{(j)}$ and

$$E(g, M({}^\Psi s, {}^\Psi H)\Phi, sH) = E(g, \Phi, H)$$

If

$$\Phi = \oplus_{k=1}^{m_i'} \Phi_k$$

then

$$E(g, \Phi, H) = \Sigma_{k=1}^{m_i'} E(g, \Phi_k, H)$$

and if H belongs to $\mathfrak{U}^{(i)}$ then

$$\text{(6.u)}\quad E(g, \Phi_k, H) = \Sigma_{{}^*\Delta\backslash\Gamma} \Sigma_{\Delta^{(i,k)}\backslash {}^*\Delta} \exp(\langle H^{(i,k)}(\delta\gamma g), H\rangle + \rho(H^{(i,k)}(\delta\gamma g))\Phi_k(\delta\gamma g)$$

If g is in G let $g = namk^{-1}$ with n in ${}^*N$, a in ${}^*A$, m in ${}^*M$, and k in K; then

$$\Sigma_{\Delta^{(i,k)}\backslash {}^*\Delta} \exp(\langle H^{(i,k)}(\delta g), H\rangle + \rho(H^{(i,k)}(\delta g)))\Phi_k(\delta g)$$

is equal to

$$\exp(\langle {}^*H(g), {}^*H\rangle + \rho({}^*H(g)))\Sigma_{{}^\Psi\Delta^{(i,k)}\backslash {}^*\Theta} \exp(\langle {}^\Psi H^{(i,k)}(\theta m), {}^\Psi H\rangle + \rho({}^\Psi H(\theta m)))\Phi_k(\theta m, k)$$

if

$${}^\Psi P^{(i,k)} = {}^*N\backslash P^{(i,k)} \cap {}^*S$$

and

$${}^\Psi\Delta^{(i,k)} = {}^*\Theta \cap {}^\Psi P^{(i,k)}$$

The sum on the right is, essentially, the Eisenstein series $E((m, k), \Phi_k, {}^{\Psi}H)$ associated to the function Φ_k, considered as an element of $\mathcal{E}(V^{(i,k)} \times W, W^*)$, and the cuspidal subgroup ${}^{\Psi}P^{(i,k)} \times K$. It is not quite this Eisenstein series because the Killing form on ${}^*\mathfrak{m}$ is not the restriction to ${}^*\mathfrak{m}$ of the Killing form on $\mathfrak{g}$. We ignore this difficulty. It is a function on ${}^*\Theta \times \{1\} \backslash {}^*M \times K$ which is invariant under right translations by elements of ${}^*K_0$ and can thus be considered a function on ${}^*T \backslash G$ which we write as $E(g, \Phi_k, {}^{\Psi}H)$. The right side of (6.u) equals

$$\Sigma_{{}^*\Delta\backslash\Gamma} \exp(\langle {}^*H(\gamma g), {}^*H \rangle + \rho({}^*H(\gamma g)))E(\gamma g, \Phi_k, {}^{\Psi}H)$$

Consequently

$$E(g, \Phi, H) = \Sigma_{{}^*\Delta\backslash\Gamma} \exp(\langle {}^*H(\gamma g), {}^*H \rangle + \rho({}^*H(\gamma g)))E(\gamma g, \Phi, {}^{\Psi}H) \tag{6.v}$$

if, for all g,

$$E(g, \Phi, {}^{\Psi}H) = \Sigma_{k=1}^{m_i'} E(g, \Phi_k, {}^{\Psi}H)$$

A similar result is valid if i is replaced by j. The cuspidal subgroups

$${}^{\Psi}P^{(i,k)} \times K, \quad 1 \leq k \leq m_i'$$

have a common split component ${}^{\Psi}\mathfrak{a}^{(i)}$ of dimension one. Since Lemmas 6.1 and 6.2 are valid for families of cuspidal subgroups of rank one $E(\cdot, \Phi, {}^{\Psi}H)$ is meromorphic on ${}^{\Psi}\mathfrak{a}^{(i)}$ and

$$E(\cdot, \Phi, {}^{\Psi}H) = E(\cdot, M({}^{\Psi}s, {}^{\Psi}H)\Phi, {}^{\Psi}H)$$

Let ${}^{\Psi}\mathfrak{S}$ be a Siegel domain associated to a percuspidal subgroup ${}^{\Psi}P$ of ${}^*M$. If U is a bounded subset of ${}^{\Psi}\mathfrak{a}^{(i)}$ let $p({}^{\Psi}H)$ be a polynomial so that $p({}^{\Psi}H)M({}^{\Psi}s, {}^{\Psi}H)$ is analytic on U. It follows readily from Lemmas 5.2 and 6.2 that there is a constant c so that, for all m in $\mathfrak{S}$, all k in K, and all ${}^{\Psi}H$ in U,

$$|p({}^{\Psi}H)E((m, k), \Phi, {}^{\Psi}H)|$$

is at most

$$c\{\exp(\langle {}^{\Psi}H(m), \operatorname{Re}{}^{\Psi}H\rangle) + \exp(\langle {}^{\Psi}H(m), {}^{\Psi}s(\operatorname{Re}{}^{\Psi}H)\rangle)\}\exp \rho({}^{\Psi}H^{(i)}(m))$$

${}^{\Psi}H(m)$ belongs, of course, to ${}^{\Psi}\mathfrak{f}$ the split component of ${}^{\Psi}P$ and its projection on ${}^{\Psi}\mathfrak{n}^{(i)}$ is ${}^{\Psi}H^{(i)}(m)$. The remarks following the proof of Lemma 4.1 imply that (6.v) converges absolutely if H is in the convex hull of $\mathfrak{U}^{(i)}$ and $s^{-1}\mathfrak{U}^{(j)}$ and $M({}^{\Psi}s, \cdot)$ is analytic at ${}^{\Psi}H$, that $E(\cdot, \Phi, H)$ is meromorphic on this set, and that every assertion of Lemma 6.2 except perhaps the last is true if $H_0^{(i)}$ belongs to this set. Since $sH = {}^{*}H + {}^{\Psi}s({}^{\Psi}H)$ the relation

$$E(g, \Phi, H) = E(g, M(s, H)\Phi, H)$$

is immediate. It is however the last assertion of Lemma 6.2 which is of importance to us.

Let Φ belong to $\mathcal{E}^{(i)}$ and let $P^{(h, \ell)}$ belong to $\{P\}$. Fix a split component, which we still call $\mathfrak{n}^{(h)}$, of $P^{(h, \ell)}$ and let X belong to $\mathfrak{n}^{(h)}$, m to $M^{(h, \ell)}$, and k to K. If H belongs to $\mathfrak{U}^{(i)}$ then

$$\int_{\Gamma\cap N^{(h, \ell)}\backslash N^{(h, \ell)}} E(n \exp X mk, \Phi, H)dn$$

is equal to

$$\Sigma_{t\in\Omega(\mathfrak{n}^{(i)}, \mathfrak{n}^{(j)})} \exp(\langle X, tH\rangle + \rho(X))(E^{(h, \ell)}M(t, H)\Phi)(mk)$$

if $E^{(h, \ell)}$ is the projection of $\mathcal{E}^{(h)}$ on $\mathcal{E}(V^{(h, \ell)}, W)$. If $t_1, \ldots, t_n$ are the elements of $\Omega(\mathfrak{n}^{(i)}, \mathfrak{n}^{(j)})$ there are elements $X_1, \ldots, X_n$ of $\mathfrak{n}^{(h)}$ so that $\det(\exp(\langle X_x, t_yH\rangle + \rho(X_x)))$ does not vanish identically. The inverse, $(a_{xy}(H))$, of the matrix $(\exp(\langle X_x, t_yH\rangle + \rho(X_x)))$ is a meromorphic function on $\mathfrak{n}^{(i)}$ and $(E^{(h, \ell)}M(t_x, H)\Phi)(m, k)$ is equal to

$$\Sigma_{y=1}^{n} a_{xy}(H) \int_{\Gamma\cap N^{(h,\ell)}\backslash N^{(h,\ell)}} E(n \exp X_y mk,\ \Phi,\ H)dn$$

which is meromorphic on the convex hull of $\mathfrak{U}^{(i)}$ and $s^{-1}\mathfrak{U}^{(j)}$. Since m and k are arbitrary this is also true of $E^{(h,\ell)}M(t, H)$ and hence of $M(t, H)$ for any t. Moreover

$$\int_{\Gamma\cap N^{(h,\ell)}\backslash N^{(h,\ell)}} E(ng,\ \Phi,\ H)dn$$

is equal to

$$\Sigma_{t\in\Omega(\mathfrak{v}^{(i)},\, \mathfrak{v}^{(h)})} \exp(\langle H^{(h,\ell)}(g),\ tH\rangle + \rho(H^{(h,\ell)}(g)))(E^{(h,\ell)}M(t,\ H)\Phi)(g)$$

at those points of the convex hull where both sides are defined. A similar result is of course valid if i is replaced by j. Use this together with the functional equation we have discovered to see that the left side of this equation also equals

$$\Sigma_{t\in\Omega(\mathfrak{v}^{(j)},\, \mathfrak{v}^{(h)})} \exp(\langle H^{(h,\ell)}(g),\ tsH\rangle + \rho(H^{(h,\ell)}(g)))(E^{(h,\ell)}M(t,\ sH)M(s,\ H)\Phi)(g)$$

This is so for every ℓ only if

$$M(t,\ sH)M(s,\ H) = M(ts,\ H)$$

If i and j are arbitrary and s is any element of $\Omega(\mathfrak{v}^{(i)},\ \mathfrak{v}^{(j)})$ then, according to the first corollary to Lemma 2.13, s can be written as a product of reflections, say $s = s_n \dots s_1$. Let us show by induction on n that $M(s, H^{(i)})$ is meromorphic on $\mathfrak{v}^{(i)}$ and that its singularities lie along hyperplanes. If $n = 1$ then the discussion above, together with the remarks following the proof of Lemma 4.5(ii), shows that $M(s, H^{(i)})$ depends, apart from an exponential factor, on only one variable and is a meromorphic function on $\mathfrak{v}^{(i)}$. On the set $\mathfrak{U}^{(i)}$

$$M(s,\ H^{(i)}) = M(s_n \dots s_2,\ s_1H^{(i)})M(s_1,\ H^{(i)})$$

The induction assumption implies that $M(s, H^{(i)})$ is meromorphic on all of $\mathfrak{a}^{(i)}$ and that its singularities lie along hyperplanes. It can also be shown by induction that if t belongs to $\Omega(\mathfrak{a}^{(j)}, \mathfrak{a}^{(k)})$ then

$$M(ts, H^{(i)}) = M(t, sH^{(i)})M(s, H^{(i)})$$

Indeed

$$M(ts, H^{(i)}) = M(ts_n \dots s_1, H^{(i)}) = M(ts_n \dots s_2, s_1H^{(i)})M(s_1, H^{(i)})$$

Apply the induction assumption to the first factor to see that $M(ts, H^{(i)})$ equals

$$M(t, sH^{(i)})M(s_n \dots s_2, s_1H^{(i)})M(s_1, H^{(i)}) = M(t, sH^{(i)})M(s, H^{(i)})$$

There is one more property of the functions $M(s, H^{(i)})$ which will be needed to complete the proof of Lemma 6.1. If, as above, s is in $\Omega(\mathfrak{a}^{(i)}, \mathfrak{a}^{(j)})$ choose $s_n, \dots, s_1$ so that $s = s_n \dots s_1$ and so that if $t_k = s_{k-1} \dots s_1$, $2 \le k \le n$, and s_k lies in $\Omega(\mathfrak{a}^{(i_k)}, \mathfrak{a}^{(j_k)})$ and belongs to the simple root $\alpha_{k,}$ then $t_k((\mathfrak{a}^{(i)})^+)$ is contained in

$$\{H \in \mathfrak{a}^{(i_k)} \mid \alpha_{k,}(H) > 0\}$$

Then

$$M(s, H^{(i)}) = M(s_n, t_nH^{(i)}) \dots M(s_2, t_1H^{(i)})M(s_1, H^{(i)}) \tag{6.w}$$

But there are only a finite number of singular hyperplanes of $M(s_k, H)$ which intersect the closure of

$$\{H \in \mathfrak{a}_c^{(i_k)} \mid \operatorname{Re} \alpha_{k,}(H) > 0\}$$

Consequently there are only a finite number of singular hyperplanes of $M(s, H^{(i)})$ which intersect the closure of the tube over $(\mathfrak{a}^{(i)})^+$.

For each i, $1 \le i \le r$, there are a finite number of points $Z_1^{(i)}, \dots, Z_{n_i}^{(i)}$

in the orthogonal complement of $\mathfrak{a}_c^{(i)}$ in $\mathfrak{j}_c$ so that for any X in $\mathfrak{Z}$, the centre of the universal enveloping algebra of $\mathfrak{g}$, for any $H^{(i)}$ in $\mathfrak{a}^{(i)}$, and for $1 \leq k \leq m_i$, the eigenvalues of $\pi(X, H^{(i)})$, the linear transformation on $\mathcal{E}(V^{(i,k)}, W)$ defined in Section 4, belong to the set

$$\{p_X(H^{(i)} + Z_1^{(i)}), \ldots, p_X(H^{(i)} + Z_{n_i}^{(i)})\}$$

There is certainly a polynomial p in $\mathfrak{Z}$ so that

$$p(H^{(i)} + Z_k^{(i)}) = 0, \; 1 \leq k \leq n_i$$

if, for some s, $H^{(i)}$ in $\mathfrak{a}_c^{(i)}$ belongs to a singular hyperplane of $M(s, \cdot)$ which intersects the closure of the tube over $(\mathfrak{a}^{(i)})^+$ but so that $p(H^{(i)} + Z_k^{(i)})$ does not vanish identically on $\mathfrak{a}_c^{(i)}$ for any choice of i and k. Thus there is an X in $\mathfrak{Z}$ so that for all i, all j, and all s in $\Omega(\mathfrak{a}^{(i)}, \mathfrak{a}^{(j)})$ the function

$$M(s, H^{(i)})\pi(X, H^{(i)})$$

is analytic on the closure of the tube over $(\mathfrak{a}^{(i)})^+$ but not identically zero. Let f be an infinitely differentiable function G so that $f(kgk^{-1}) = f(g)$ for all g and all k and so that the determinant of the linear transformation $\pi(f, H^{(i)})$ on $\mathcal{E}^{(i)}$ vanishes identically for no i. Set $f_0 = \lambda'(X)f$; then $\pi(f_0, H^{(i)}) = \pi(X, H^{(i)})\pi(f, H^{(i)})$ and its determinant does not vanish identically. If $\mathfrak{S}$ is a Siegel domain associated to a percuspidal subgroup then for each g in $\mathfrak{S}$ define $E'(h, \Phi_i, H^{(i)})$ as in the beginning of this section. According to (4.r) and (6.b) the inner product of $\lambda(f_0)E'(\cdot, \Phi_i, H_1^{(i)})$ and $\lambda(f_0)E'(\cdot, \Psi_j, H_2^{(j)})$ is equal to

$$\sum_{s \in \Omega(\mathfrak{a}^{(i)}, \mathfrak{a}^{(j)})} \frac{a^2}{(2\pi)^q} \int_{\operatorname{Re} H^{(i)} = Y^{(i)}} (M(s, H^{(i)})\pi(f_0, H^{(i)})\Phi_i, \pi(f_0, -s\overline{H}^{(i)})\Psi_j)\xi(s, H^{(i)})\,|dH^{(i)}|$$

with

$$\xi(s, H^{(i)}) = \exp\langle X(g), H_1^{(i)} + \overline{H}_2^{(j)} - H^{(i)} + sH^{(i)}\rangle \left\{\prod_{k=1}^{q} \alpha_{k,}^{(i)}(H_1^{(i)} - H^{(i)})\alpha_{k,}^{(j)}(\overline{H}_2^{(j)} + sH^{(i)})\right\}$$

If the relation (6.w) is combined with the estimates obtained for the function $M(z)$ of Lemma 6.3 when $\operatorname{Re} z \geq 0$ it is seen that in this integral $Y^{(i)}$ can be replaced by 0. Consequently the expression is an analytic function of $(H_1^{(i)}, H_2^{(j)})$ on the Cartesian product of the tubes over $(\mathfrak{a}^{(i)})^+$ and $(\mathfrak{a}^{(j)})^+$. Applying an argument similar to that used in the case of a single variable we see that $\lambda(f_0)E'(\cdot, \Phi_i, H^{(i)})$ is an analytic function on the tube over $(\mathfrak{a}^{(i)})^+$ with values in $\mathcal{L}(\Gamma\backslash G)$. The estimate of (6.d) is a manifest consequence of the above expression for the inner product. We conclude that $E(\cdot, \Phi_i, H^{(i)})$ is meromorphic on the tube over $(\mathfrak{a}^{(i)})^+$ and that Lemma 6.2 is true if $H_0^{(i)}$ is in this set. If $H^{(i)}$ lies on the boundary of this set and if, for every h and all t in $\Omega(\mathfrak{a}^{(i)}, \mathfrak{a}^{(h)})$, $M(t, \cdot)$ is analytic $H^{(i)}$ then, applying Lemma 5.2, we define $E(\cdot, \Phi, H^{(i)})$ by continuity. Suppose W is a Weyl chamber of $\mathfrak{a}^{(i)}$. Choose the unique j and the unique s in $\Omega(\mathfrak{a}^{(i)}, \mathfrak{a}^{(j)})$ such that $sW = (\mathfrak{a}^{(j)})^+$ and if $H^{(i)}$ is in the closure of the tube over W set

$$E(\cdot, \Phi_i, H^{(i)}) = E(\cdot, M(s, H^{(i)})\Phi_i, sH^{(i)})$$

when the right side is defined. Then

$$\int_{\Gamma\cap N^{(h,\ell)}\backslash N^{(h,\ell)}} E(ng, \Phi_i, H^{(i)})dn$$

is equal to

$$\sum_{t\in\Omega(\mathfrak{a}^{(j)}, \mathfrak{a}^{(h)})} \exp(\langle H^{(h,\ell)}(g), tsH^{(i)}\rangle + \rho(H^{(h,\ell)}(g)))(E^{(h,\ell)}M(t, sH^{(i)})M(s, H^{(i)})\Phi_i)(g)$$

which in turn equals

$$\sum_{t\in\Omega(\mathfrak{a}^{(i)}, \mathfrak{a}^{(h)})} \exp(\langle H^{(h,\ell)}(g), tH^{(i)}\rangle + \rho(H^{(h,\ell)}(g)))(E^{(h,\ell)}M(t, H^{(i)})\Phi_i)(g)$$

Since the cuspidal component of

$$\int_{\Gamma\cap N\backslash N} E(ng,\ \Phi_i,\ H^{(i)})dn$$

is zero if P is not conjugate to an element of $\{P\}$ and since $E(\cdot,\ \Phi_i,\ H^{(i)})$ has the proper rate of growth on Siegel domains it follows from Lemma 5.2 that $E(\cdot,\ \Phi_i,\ \cdot)$ can be defined at $H^{(i)}$ in the closure of W if, for all h and all t in $\Omega(\mathfrak{a}^{(i)},\ \mathfrak{a}^{(h)})$, $M(t,\ \cdot)$ is analytic at $H^{(i)}$. However a given point $H^{(i)}$ at which all the functions $M(t,\ \cdot)$ are analytic may lie in the closure of more than one Weyl chamber so that it is not clear that we have defined $E(\cdot,\ \Phi,\ H^{(i)})$ unambiguously; but to see that we have it is sufficient to refer to Lemma 3.7. Lemma 5.3 implies that $E(\cdot,\ \Phi_i,\ H^{(i)})$ is meromorphic on $\mathfrak{a}_c^{(i)}$ and that the first assertion of Lemma 6.2 is valid. It remains to verify the functional equations. Appealing again to Lemma 3.7 we see that it is enough to show that for all j, all s in $\Omega(\mathfrak{a}^{(i)},\ \mathfrak{a}^{(j)})$, and for $1 \le h \le r$, $1 \le \ell \le m_h$

$$\int_{\Gamma\cap N^{(h,\ell)}\backslash N^{(h,\ell)}} E(ng,\ \Phi_i,\ H^{(i)})dn = \int_{\Gamma\cap N^{(h,\ell)}\backslash N^{(h,\ell)}} E(ng,\ M(s,\ H^{(i)})\Phi_i,\ sH^{(i)})dn$$

The left side has just been calculated; the right side is

$$\sum_{t\in\Omega(\mathfrak{a}^{(j)},\,\mathfrak{a}^{(h)})} \exp(\langle H^{(h,\ell)}(g),\ tsH^{(i)}\rangle + \rho(H^{(h,\ell)}(g)))(E^{(h,\ell)}M(t,\ sH^{(i)})M(s,\ H^{(i)})\Phi_i)(g)$$

Since

$$M(t,\ sH^{(i)})M(s,\ H^{(i)}) = M(ts,\ H^{(i)})$$

they are equal.

7. The main theorem.

As was stressed in the introduction the central problem of this paper is to obtain a spectral decomposition for $\mathcal{L}(\Gamma\backslash G)$ with respect to the action of G. Referring to Lemma 4.6 we see that it is enough to obtain a spectral decomposition for each of the spaces $\mathcal{L}(\{P\}, \{V\}, W)$ with respect to the action of $\mathcal{C}(W, W)$. If q is the rank of the elements of $\{P\}$ it will be seen that $\mathcal{L}(\{P\}, \{V\}, W)$ is the direct sum of $q+1$ invariant and mutually orthogonal subspaces

$$\mathcal{L}_m(\{P\}, \{V\}, W), \ 0 \leq m \leq q$$

and that, in a sense which will become clear later, the spectrum $\mathcal{C}(W, W)$ in $\mathcal{L}_m(\{P\}, \{V\}, W)$ is of dimension m. The spectral decomposition of $\mathcal{L}_q(\{P\}, \{V\}, W)$ will be effected by means of the Eisenstein series discussed in Section 6, the Eisenstein series associated to cusp forms. The spectral decomposition of $\mathcal{L}_m(\{P\}, \{V\}, W)$, $m < q$, is effected by means of Eisenstein series in m-variables which are residues of the Eisenstein series in q variables associated to cusp forms. More precisely the series in m-variables are residues of the series in $m+1$ variables. In any case they are by definition meromorphic functions and it will be proved that they must satisfy functional equations similar to those of Lemma 6.1. It will also be shown that there are relations between the functions defined by Eisenstein series and certain other functions that arise in the process of taking residues but cannot be defined directly. It will be apparent, <u>a posteriori,</u> that the Eisenstein series described above are precisely those of Lemma 4.1.

It will be easy to define the space $\mathcal{L}_q(\{P\}, \{V\}, W)$; the other spaces $\mathcal{L}_m(\{P\}, \{V\}, W)$, $m < q$, will be defined by induction. Although the spaces $\mathcal{L}_m(\{P\}, \{V\}, W)$ can be shown, <u>a posteriori,</u> to be unique it is, unfortunately, necessary to define them by means of objects which are definitely not unique. Since the induction on m must be supplemented by an induction similar to that of the last section this lack of uniqueness will cause us trouble if we do not take the precaution of providing at each step the necessary material for the supplementary induction. To do this it is best to let $\{P\}$ denote a full class of associate cuspidal subgroups rather than a set of representatives for the conjugacy classes in an equivalence class. Then $\mathcal{L}(\{P\}, \{V\}, W)$ is just the closure of the space of functions spanned by the functions

$$\phi^\wedge(g) = \Sigma_{\Delta\backslash\Gamma}\phi(\gamma g)$$

where for some P in $\{P\}$, ϕ belongs to $\mathcal{J}(V, W)$. Suppose ${}^*P$ is a cuspidal subgroup belonging to some element P of $\{P\}$. The space $\mathcal{J}(V \otimes W, W^*)$ of functions on ${}^\Psi T \times \{1\} \backslash {}^*M \times K$ has been defined; it can be regarded as a space of functions on ${}^*AT \times \{1\} \backslash {}^*P \times K$. The subspace of $\mathcal{J}(V \otimes W, W^*)$ consisting of those functions ϕ such that $\phi(p_1, k_1) = \phi(p_2, k_2)$ when p_1 and p_2 belong to ${}^*P$, k_1 and k_2 belong to K, and $p_1k_1^{-1} = p_2k_2^{-1}$ can be regarded as a space of functions on ${}^*AT\backslash G$. It will be called ${}^*\mathcal{J}(V, W)$. ${}^*\mathcal{L}(\{P\}, \{V\}, W)$ will be the closure, in $\mathcal{L}({}^*\Theta \times \{1\} \backslash {}^*M \times K)$, of the space of functions on ${}^*A{}^*T\backslash G$ spanned by functions of the form

$$\phi^\wedge(g) = \Sigma_{\Delta\backslash{}^*\Delta}\phi(\gamma g)$$

where, for some P in ${}^*\{P\}$, the set of elements in $\{P\}$ to which ${}^*P$ belongs, ϕ belongs to ${}^*\mathcal{J}(V, W)$. If $\mathfrak{a}^{(1)}, \ldots, \mathfrak{a}^{(r)}$ are as before the distinct split components of the elements of $\{P\}$ we let ${}^*\{P\}^{(i)}$ be the set of elements in ${}^*\{P\}$ with the split component $\mathfrak{a}^{(i)}$. $\{P\}^{(i)}$ is defined in a similar fashion. Suppose P belongs to ${}^*\{P\}^{(i)}$ and ${}^\Psi\mathfrak{a}^{(i)}$ is the orthogonal complement of ${}^*\mathfrak{a}$ in $\mathfrak{a}^{(i)}$. Let ${}^*\mathcal{H}(V, W)$ be the set of all functions $\Phi(\cdot)$ with values in $\mathcal{E}(V, W)$ which are defined and analytic on

$$\{H \in {}^\Psi\mathfrak{a}_c^{(i)} \mid \|\mathrm{Re}\, H\| < R\}$$

and are such that, if p is any polynomial, $\|p(\operatorname{Im} H)\Phi(H)\|$ is bounded on this set. R is the number introduced at the end of Section 4 and $\|\operatorname{Re} H\|$ is the norm of Re H in $\mathfrak{a}^{(i)}$. If we are to use these new spaces effectively we have to realize that all of the facts proved earlier have analogues for these new types of spaces. Since the proof generally consists merely of regarding functions on ${}^*N\,{}^*A\backslash G$ as functions on ${}^*M \times K$ we will use the analogues without comment. In particular the analogue of the operator A on $\mathcal{L}(\{P\}, \{V\}, W)$ is defined on ${}^*\mathcal{L}(\{P\}, \{V\}, W)$; it will also be called A.

Since the entire discussion concerns one family $\{P\}$, one family $\{V\}$, and one space W we fix the three of them immediately and start by introducing some simple notions. Let $\mathfrak{a} = \mathfrak{a}^{(i)}$ with $1 \le i \le r$. If $\mathfrak{s}$ is a complex affine subspace of $\mathfrak{a}_c$ defined by equations of the form $\alpha(H) = \mu$ where α is a positive root of $\mathfrak{a}$ and μ is a complex number then $\mathfrak{s} = X(\mathfrak{s}) + \tilde{\mathfrak{s}}$ where $\tilde{\mathfrak{s}}$ is a complex subspace of $\mathfrak{a}_c$ defined by real linear equations which contains zero and $X(\mathfrak{s})$ is orthogonal to $\tilde{\mathfrak{s}}$. Let $S(\mathfrak{s})$ be the symmetric algebra over the orthogonal complement of $\tilde{\mathfrak{s}}$. Suppose ${}^*\mathfrak{a}$ is a distinguished subspace of $\mathfrak{a}$ and suppose $\tilde{\mathfrak{s}}$ contains ${}^*\mathfrak{s}$. If ${}^\psi\mathfrak{a}$ is the orthogonal complement of ${}^*\mathfrak{a}$ in $\mathfrak{a}$ there is a unique isomorphism $Z \longrightarrow D(Z)$ of $S(\mathfrak{s})$ with a subalgebra of the algebra of holomorphic differential operators on ${}^\psi\mathfrak{a}_c$ such that

$$D(Y)f(H) = \frac{df}{dt}(H + tY)\Big|_{t=0}$$

if Y belongs to the orthogonal complement of $\tilde{\mathfrak{s}}$. If $\mathfrak{E}$ is a finite dimensional unitary space and if $\Phi(\cdot)$ is a function with values in $\mathfrak{E}$ which is defined and analytic in a neighbourhood of the point H in ${}^\psi\mathfrak{a}_c$ let $d\Phi(H)$ be that element of $L(S(\mathfrak{s}), \mathfrak{E})$, the space of linear transformations from $S(\mathfrak{s})$ to $\mathfrak{E}$, defined by

$$d\Phi(H)(Z) = D(Z)\Phi(H)$$

$L(S(\mathfrak{s}), \mathfrak{E})$ can be identified with the space of formal power series over the orthogonal complement of $\tilde{\mathfrak{h}}$ with coefficients in $\mathcal{E}$ and we obtain $d\Phi(H)$ by expanding the function

$$\Phi_H(Y) = \Phi(H+Y)$$

about the origin. If $f(\cdot)$ is a function with values in the space of linear transformations from $\mathcal{E}$ to $\mathcal{E}'$ which is defined and analytic in a neighbourhood of H we can regard $df(H)$ as a power series; if F belongs to $L(S(\mathfrak{h}), \mathcal{E})$ the product $df(H)F$ is defined and belongs to $L(S(\mathfrak{h}), \mathcal{E}')$. There is a unique conjugate linear isomorphism $Z \longrightarrow Z^*$ of $S(\mathfrak{h})$ with itself such that $Y^* = -\overline{Y}$ if Y belongs to the orthogonal complement of $\tilde{\mathfrak{h}}$ and there is a unique function (T, F) on

$$S(\mathfrak{h}) \otimes \mathcal{E} \times L(S(\mathfrak{h}), \mathcal{E})$$

which is linear in the first variable and conjugate linear in the second so that

$$(Z \otimes \Phi, F) = (\Phi, F(Z^*))$$

if Z is in $S(\mathfrak{h})$, Φ is in $\mathcal{E}$, and F is in $L(S(\mathfrak{h}), \mathcal{E})$. It is easily seen that if Λ is any linear function on $S(\mathfrak{h}) \otimes \mathcal{E}$ there is an F in $L(S(\mathfrak{h}), \mathcal{E})$ so that $\Lambda(T) = (T, F)$ for all T in $S(\mathfrak{h}) \otimes \mathcal{E}$. If we define the order of F, denoted $O(F)$, to be the degree of the term of lowest degree which actually occurs in the power series expansion of F and if we say that a linear transformation N from $L(S(\mathfrak{h}), \mathcal{E})$ to some other vector space is of finite degree n if $NF = 0$ when $O(F)$ is greater than n and if $NF \neq 0$ for some F of order n then a linear function Λ on $L(S(\mathfrak{h}), \mathcal{E})$ is of finite degree if and only if there is a T in $S(\mathfrak{h}) \otimes \mathcal{E}$ so that $\Lambda(F)$ is the complex conjugate of (T, F) for all F. In particular if $\mathfrak{f}$ is a subspace of $\mathfrak{a}^{(j)}$ defined by linear equations of the form $\alpha(H) = \mu$ where α is a positive root of $\mathfrak{a}^{(j)}$ and μ is a complex number, if $\mathcal{E}'$ is another unitary space, and if N is a linear transformation from $L(S(\mathfrak{h}), \mathcal{E})$ to $S(\mathfrak{f}) \otimes E'$ which is of finite degree there is a unique linear transformation N^*

from $L(S(\mathfrak{f}), \mathcal{E}')$ to $S(\mathfrak{b}) \otimes \mathcal{E}$ such that (NF, F') is the complex conjugate of (N^*F', F) for all F and F' and N^* is of finite degree.

There is a unique isomorphism $Z \longrightarrow p_Z$ of $S(\mathfrak{b})$ with a subalgebra of the algebra of polynomials on ${}^\Psi\mathfrak{a}_c$ such that $p_Y(H) = \langle H, Y\rangle$ if Y belongs to the orthogonal complement of $\tilde{\mathfrak{b}}$. If P belongs to $\{P\}^{(i)}$, V is the corresponding element of $\{V\}$, and ${}^*P$ is the cuspidal subgroup with split component ${}^*\mathfrak{a}$ belonging to P and if ${}^*A$ is a split component of ${}^*P$ there is a unique map of $S(\mathfrak{b}) \otimes \mathcal{E}(V, W)$ into the space of functions on ${}^*AT\backslash G$ such that the image of $Z \otimes \Phi$ is $p_Z({}^\Psi H(g))\Phi(g)$ if ${}^\Psi H(g)$ is the projection of $H(g)$ on ${}^\Psi\mathfrak{a}$. We denote the image of T by $T(\cdot)$. If $\psi(g)$ belongs to ${}^*\mathfrak{D}(V, W)$ then we can represent $\psi(g)$ as a Fourier transform

$$\psi(g) = \frac{1}{(2\pi i)^p}\int_{\operatorname{Re} H=0} \exp(\langle {}^\Psi H(g), H\rangle + \rho({}^\Psi H(g)))\Psi(g, H)\,|dH|$$

where $\Psi(\cdot)$ is a holomorphic function on ${}^\Psi\mathfrak{a}_c$ with values in $\mathcal{E}(V, W)$ and $\Psi(g, H)$ is the value of $\Psi(H)$ at g, and p is the dimension of ${}^\Psi\mathfrak{a}$. We shall need the formula

$$(7.a)\qquad (T, d\Psi(-\overline{H})) = \int_{{}^\Psi T\backslash {}^*M\times K} \exp(\langle {}^\Psi H(m), H\rangle + \rho({}^\Psi H(m)))T(m)\overline{\psi}(m)\,dm\,dk$$

for H in ${}^\Psi\mathfrak{a}_c$ and T in $S(\mathfrak{b}) \otimes \mathcal{E}(V, W)$. We need only verify it for $T = Z \otimes \Phi$. If Y belongs to ${}^\Psi\mathfrak{a}$ then the function $\psi(\exp Y\,mk)$ on $M \times K$ belongs to $\mathcal{E}(V, W)$; call it $\Psi'(Y)$. Then

$$\Psi(H) = \int_{{}^\Psi\mathfrak{a}} \exp(-\langle Y, H\rangle - \rho(Y))\Psi'(Y)\,|dY|$$

Consequently

$$D(Z^*)\Psi(H) = \int_{{}^\Psi\mathfrak{a}} \exp(-\langle Y, H\rangle - \rho(Y))p_{Z^*}(-Y)\Psi'(Y)\,|dY|$$

Since the complex conjugate of $p_{Z^*}(-Y)$ is $p_Z(Y)$,

$$(\Phi,\ D(Z^*)\Psi(-\overline{H}))$$

is equal to

$$\int_{{}^{\Psi}\mathfrak{a}} \omega^2(a)da \int_{\Theta\backslash M} dm \int_K dk\,\{\exp(\langle Y,\ H\rangle + \rho(Y))p_Z(Y)\Phi(mk)\overline{\psi}(\exp Y\, mk)\}$$

or

$$\int_{{}^{\Psi}T\backslash {}^{*}M\times K} \exp(\langle {}^{\Psi}H(m),\ H\rangle + \rho({}^{\Psi}H(m)))(Z\otimes\Phi)(mk)\overline{\psi}(mk)dmdk$$

Suppose that $\mathfrak{t}$ is contained in $\mathfrak{s}$ and is also defined by equations of the form $\alpha(H) = \mu$ where α is a positive root and μ is a complex number. There are a number of simple relations between $S(\mathfrak{s})$ and $S(\mathfrak{t})$ which we state now although they are not needed till later. Let $S_0(\mathfrak{t})$ be the symmetric algebra over the orthogonal complement of $\mathfrak{t}$ in $\mathfrak{s}$; then $S(\mathfrak{t})$ is isomorphic in a natural manner to $S_0(\mathfrak{t})\otimes S(\mathfrak{s})$. If F belongs to $L(S(\mathfrak{t}),\ \mathfrak{E})$ and X_0 belongs to $S_0(\mathfrak{t})$ let $X_0 \vee F$ be that element of $L(S(\mathfrak{s}),\ \mathfrak{E})$ such that

$$(X_0 \vee F)(X) = F(X_0\otimes X)$$

It is clear that $S(\mathfrak{t})\otimes\mathfrak{E}$ is isomorphic to $S_0(\mathfrak{t})\otimes(S(\mathfrak{s})\otimes\mathfrak{E})$ and that if T belongs to $S(\mathfrak{s})\otimes\mathfrak{E}$ then

$$(T,\ X_0\vee F) = (X_0^*\otimes T,\ F)$$

If $F(\cdot)$ is a function defined in a neighbourhood of a point H in ${}^{\Psi}\mathfrak{a}$ with values in $L(S(\mathfrak{s}),\ \mathfrak{E})$ such that $F(\cdot)(X)$ is analytic at H for all X in $S(\mathfrak{s})$ we let $dF(H)$ be that element of $L(S(\mathfrak{t}),\ \mathfrak{E})$ such that

$$dF(H)(X_0\otimes X) = D(X_0)(F(H)(X))$$

It is clear that

$$d(d\Phi)(H) = d\Phi(H)$$

There is one more definition to make before we can begin to prove anything. Let $\mathfrak{S}$ be a subspace of $\mathfrak{a}^{(i)}$ as above and suppose that if ${}^*P$ is any cuspidal subgroup belonging to an element of $\{P\}$ whose split component ${}^*\mathfrak{a}$ is contained in $\tilde{\mathfrak{S}}$ and P is any element of ${}^*\{P\}^{(i)}$ there is given a function $E(g, F, H)$ on

$$ {}^*A\,{}^*T\backslash G \times L(S(\mathfrak{S}), \mathfrak{E}(V, W)) \times {}^{\Psi}\mathfrak{S} $$

${}^{\Psi}\mathfrak{S}$ is the projection of $\mathfrak{S}$ on the orthogonal complement of ${}^*\mathfrak{a}$. The space $\mathfrak{S}$ together with this collection of functions will be called an Eisenstein system belonging to $\mathfrak{S}$ if the functions $E(\cdot, \cdot, \cdot)$ do not all vanish identically and the following conditions are satisfied:

(i) Suppose ${}^*P$ and a P in ${}^*\{P\}^{(i)}$ are given. For each g in G and each F in $L(S(\mathfrak{S}), \mathfrak{E}(V, W))$ the function $E(g, F, H)$ on ${}^{\Psi}\mathfrak{S}$ is meromorphic. Moreover if H_0 is any point of ${}^{\Psi}\mathfrak{S}$ there is a polynomial $p(H)$ which is a product of linear polynomials $\alpha(H) - \mu$, where α is a positive root of ${}^{\Psi}\mathfrak{a}$ and μ is a complex number, and which does not vanish identically on ${}^{\Psi}\mathfrak{S}$ and a neighbourhood U of H_0 so that $p(H)E(g, F, H)$ is, for all F in $L(S(\mathfrak{S}), \mathfrak{E}(V, W))$, a continuous function on ${}^*AT\backslash G \times U$ which is analytic on U for each fixed g and so that if $\mathfrak{S}_0$ is a Siegel domain associated to a percuspidal subgroup P_0 of ${}^*M$ and F belongs to $L(S(\mathfrak{S}), \mathfrak{E}(V, W))$ there are constants c and b so that

$$ |p(H)E(mk, F, H)| \leq c\eta^{b}(a_0(m)) $$

for all m in $\mathfrak{S}_0$, k in K, and all H in U. The function $E(g, F, H)$ is for each g and H a linear function of F and there is an integer n so that $E(g, F, H)$ vanishes for all g and H if the order of F is greater than n.

(ii) If ${}^*\mathfrak{a}$ is a distinguished subspace of $\mathfrak{a}^{(i)}$ which is contained in $\tilde{\mathfrak{S}}$ and if $\mathfrak{a}^{(j)}$ contains ${}^*\mathfrak{a}$ let ${}^{\Psi}\Omega^{(j)}(\mathfrak{S})$ be the set of distinct linear transformations from $\mathfrak{S}$ into $\mathfrak{a}_c^{(j)}$ obtained by restricting the elements of ${}^{\Psi}\Omega(\mathfrak{a}^{(i)}, \mathfrak{a}^{(j)})$ to $\mathfrak{S}$. If s belongs to ${}^{\Psi}\Omega^{(j)}(\mathfrak{S})$ let

$$\mathfrak{S}_s = \{-(\overline{sH}) \mid H \in \mathfrak{S}\}$$

$\mathfrak{S}_s$ is a complex affine subspace of $\mathfrak{a}^{(j)}$. Suppose the cuspidal subgroup ${}^*P$ with split component ${}^*\mathfrak{a}$, the group P in ${}^*\{P\}^{(i)}$, and the group P' in ${}^*\{P\}^{(j)}$ are given. Then for every s in ${}^{\Psi}\Omega^{(j)}(\mathfrak{S})$ there is a function $N(s, H)$ on ${}^{\Psi}\mathfrak{S}$ with values in the space of linear transformations from $L(S(\mathfrak{S}), \mathcal{E}(V, W))$ to $S(\mathfrak{S}_s) \otimes \mathcal{E}(V', W)$ so that for all F in $L(S(\mathfrak{S}), \mathcal{E}(V, W))$ and all F' in $L(S(\mathfrak{S}_s), \mathcal{E}(V', W))$ the function $(N(s, H)F, F')$ is meromorphic on ${}^{\Psi}\mathfrak{S}$. If H_0 is a point of ${}^{\Psi}\mathfrak{S}$ there is a polynomial $p(H)$ and a neighbourhood U as before so that $p(H)(N(s, H)F, F')$ is analytic on U for all F and F'. Moreover there is an integer n so that $(N(s, H)F, F')$ vanishes identically if the order of F or of F' is greater than n. Finally, if

$${}^{\Psi}P' = {}^*N \backslash P' \cap {}^*S$$

then

$$(7.\text{b}) \int_{{}^*\Theta \cap {}^{\Psi}N' \backslash {}^{\Psi}N'} E(nmk, F, H)\,dn = \sum_{s \in {}^{\Psi}\Omega^{(j)}(\mathfrak{S})} \exp(\langle {}^{\Psi}H'(m), sH \rangle + \rho({}^{\Psi}H'(m)))N(s, H)F(mk)$$

provided both sides are defined. However, if P'' is a cuspidal subgroup to which ${}^*P$ belongs and P'' does not belong to $\{P\}$ then the cuspidal component of

$$\int_{{}^*\Theta \cap {}^{\Psi}N'' \backslash {}^{\Psi}N''} E(nmk, F, H)\,dn$$

is zero.

(iii) Suppose ${}^*P_1$, with split component ${}^*\mathfrak{a}_1$, is a cuspidal subgroup belonging to some element of $\{P\}$ and ${}^*P$, with split component ${}^*\mathfrak{a}$, is a cuspidal subgroup belonging to ${}^*P_1$ and suppose $\widetilde{\mathfrak{S}}$ contains ${}^*\mathfrak{a}_1$. If P belongs to ${}^*\{P\}_1^{(i)}$ and hence to ${}^*\{P\}^{(i)}$ and F belongs to $L(S(\mathfrak{S}), \mathcal{E}(V, W))$ then $E_1(\cdot, F, \cdot)$ and $E(\cdot, F, \cdot)$ are functions on ${}^*A_1 {}^*T_1 \backslash G \times {}^{\Psi}\mathfrak{S}_1$ and ${}^*A {}^*T \backslash G \times {}^{\Psi}\mathfrak{S}$

respectively. If H belongs to ${}^{\Psi}\mathcal{E}$ let $H = H^* + {}^{\Psi}H$ where ${}^*H$ belongs to the complexification of the orthogonal complement of ${}^*\mathfrak{a}$ in ${}^*\mathfrak{a}_1$ and ${}^{\Psi}H$ belongs to ${}^{\Psi}\mathcal{E}_1$; if H belongs to

$$\bigcup_{{}^*\mathfrak{a}_1 \subseteq \mathfrak{a}^{(j)}} \bigcup_{s \in {}^{\Psi}\Omega_1(\mathfrak{a}^{(i)}, \mathfrak{a}^{(j)})} s^{-1}({}^{\Psi}U^{(j)})$$

then

$$\text{(7.c)} \quad E(g, F, H) = \Sigma_{{}^*\Delta_1 \backslash {}^*\Delta} \exp(\langle {}^*H_1(\gamma g), {}^*H \rangle + \rho_0({}^*H_1(\gamma g))) E_1(\gamma g, F, {}^{\Psi}H)$$

if $E_1(\cdot, F, \cdot)$ is analytic at ${}^{\Psi}H$. Here $\rho_0({}^*H_1(g))$ is the value of ρ at the projection of ${}^*H_1(g)$ on the orthogonal complement of ${}^*\mathfrak{a}$. The convergence of (7.c) is implied by the remarks following the proof of Lemma 4.1. Moreover if P' belongs to ${}^*\{P\}_1^{(j)}$ and s belongs to ${}^{\Psi}\Omega_1^{(j)}(\mathcal{E})$ then

$$N(s, H) = N_1(s, {}^{\Psi}H)$$

(iv) Suppose ${}^*P_1$ and ${}^*P_2$ are cuspidal subgroups with the split component ${}^*\mathfrak{a}$ which both belong to elements of $\{P\}$ and suppose $\tilde{\mathcal{E}}$ contains ${}^*\mathfrak{a}$. Suppose P_1 belongs to ${}^*\{P\}_1^{(i)}$ and P_2 belongs to ${}^*\{P\}_2^{(i)}$ and suppose there is a γ in Γ so that $\gamma {}^*P_1 = {}^*P_2\gamma$ and $\gamma P_1 = P_2\gamma$. If H belongs to ${}^{\Psi}\mathcal{E}$ let $D(H)$ be the map from $\mathcal{E}(V_1, W)$ to $\mathcal{E}(V_2, W)$ and D be the map from functions on ${}^*A_1 {}^*T_1\backslash G$ to functions on ${}^*A_2 {}^*T_2\backslash G$ which were defined in Section 4; then if F belongs to $L(S(\mathcal{E}), \mathcal{E}(V, W))$

$$\text{(7.d)} \qquad DE_1(g, F, H) = E_2(g, dD(H)F, H)$$

Moreover if P_1' and P_2' belong to ${}^*\{P\}_1^{(j)}$ and ${}^*\{P\}_2^{(j)}$ respectively and there is a δ in Γ with

$$\delta P_1' = P_2'\delta$$

and

$$\delta^* P_1 = {}^* P_2 \delta$$

so that the map $D(H)$ from $\mathcal{E}(V_1', W)$ to $\mathcal{E}(V_2', W)$ is defined for all H in ${}^\Psi\mathfrak{S}$, and if s belongs to ${}^\Psi\Omega^{(j)}(\mathfrak{S})$ then

(7.e) $$(N_1(s, H)F, F') = (N_2(s, H)(dD(H)F), dD(-s\overline{H})F')$$

for all F and F'.

(v) If k is in K then

$$\lambda(k)E(g, F, H) = E(g, \lambda(k)F, H)$$

and if f belongs to $\mathcal{C}(W, W)$ then

$$\lambda(f)E(g, F, H) = E(g, d(\pi(f, H))F, H)$$

Suppose that $\mathfrak{S} = \mathfrak{a}^{(i)}$. Then $S(\mathfrak{S})$ is just the space of constants so that, for all P in $\{P\}^{(i)}$, the map $F \longrightarrow F(1)$ defines an isomorphism of $L(S(\mathfrak{S}), \mathcal{E}(V, W))$ with $\mathcal{E}(V, W)$. If ${}^*P$ is a cuspidal subgroup with split component ${}^*\mathfrak{a}$ which belongs to some element of $\{P\}$, if ${}^*\mathfrak{a}$ is contained in $\mathfrak{a}^{(i)}$, if P belongs to ${}^*\{P\}^{(i)}$, and if F belongs to $L(S(\mathfrak{S}), \mathcal{E}(V, W))$ we let

$$E(g, F, H) = \sum_{\Delta \backslash {}^*\Delta} \exp(\langle H(\gamma g), H\rangle + \rho_0(H(\gamma g))\Phi(\gamma g)$$

if H belongs to ${}^\Psi\mathfrak{U}^{(i)}$. Here $\Phi = F(1)$ and $\rho_0(H(g))$ is the value of ρ at the projection of $H(g)$ on the orthogonal complement of ${}^*\mathfrak{a}$. This collection of functions certainly defines an Eisenstein system and, as remarked before, all the other Eisenstein systems of interest to us will be obtained from systems of this type by taking residues. Let us see explicitly how this is done.

Suppose that $\mathfrak{S}$ is a subspace of $\mathfrak{a} = \mathfrak{a}^{(i)}$ defined by equations of the same form as before and suppose that $\phi(\cdot)$ is a function meromorphic on all of $\mathfrak{S}$ whose singularities lie along hyperplanes of the form $\alpha(H) = \mu$ where α is a real linear function $\mathfrak{a}$ and μ is a complex number. Suppose we have a hyperplane $\mathfrak{s}$, not necessarily a singular hyperplane of $\phi(\cdot)$, of this form and choose a real unit normal H_0 to $\mathfrak{s}$. Then we can define a meromorphic function $\operatorname{Res}_{\mathfrak{s}} \phi(\cdot)$ on $\mathfrak{s}$ by

$$\operatorname{Res}_{\mathfrak{s}} \phi(H) = \frac{\delta}{2\pi i} \int_0^1 \phi(H + \delta e^{2\pi i\Theta} H_0) d(e^{2\pi i\Theta})$$

if δ is so small that $\phi(H + zH_0)$ has no singularities for $0 < |z| < 2\delta$. It is easily verified that the singularities of $\operatorname{Res}_{\mathfrak{s}} \phi(\cdot)$ lie on the intersections with $\mathfrak{s}$ of the singular hyperplanes of $\phi(\cdot)$ different from $\mathfrak{s}$. Now suppose we have an Eisenstein system $\{E(\cdot, \cdot, \cdot)\}$ belonging to $\mathfrak{S}$ and suppose $\mathfrak{s}$ is a hyperplane of $\mathfrak{S}$ defined by an equation of the form $\alpha(H) = \mu$ where α is a positive root of $\mathfrak{a}$. We now define an Eisenstein system belonging to $\mathfrak{s}$. Suppose that ${}^*P$ is a cuspidal subgroup belonging to some element of $\{P\}$ and suppose that the split component ${}^*\mathfrak{a}$ of ${}^*P$ is contained in $\tilde{\mathfrak{s}}$. Then ${}^*\mathfrak{a}$ is also contained in $\tilde{\mathfrak{S}}$ so that if P belongs to ${}^*\{P\}^{(i)}$ there is a function $E(\cdot, \cdot, \cdot)$ defined on

$${}^*A\,{}^*T\backslash G \times L(S(\mathfrak{S}), \mathcal{L}(V, W)) \times {}^{\Psi}\mathfrak{S}$$

If g is in G and $\Phi(\cdot)$ is a function on $\mathfrak{a}_c$ with values in $\mathcal{L}(V, W)$ which is defined and analytic in a neighbourhood of H in ${}^{\Psi}\mathfrak{a}^{(i)}$ then $\operatorname{Res}_{\Psi_{\mathfrak{s}}} E(g, d\Phi(\cdot), \cdot)$ is defined in a neighbourhood of H in $\mathfrak{s}$. Let

$$d\Phi(H + zH_0) = \Sigma_{x=0}^{\infty} \frac{z^x}{x!} d(D(H_0^x)\Phi(H))$$

and let

$$E(g, F, H + zH_0) = \Sigma_{y=-\infty}^{\infty} z^y E_y(g, F, H)$$

if F belongs to $L(S(\mathfrak{S}), \mathcal{L}(V, W))$. Of course only a finite number of terms with negative y actually occur. Then

$$\operatorname{Res}_{\Psi_{\mathfrak{s}}} E(g, d\Phi(H), H) = \Sigma_{x+y=-1} \frac{1}{x!} E_y(g, d(D(H_0^x)\Phi(H)), H)$$

If F belongs to $L(S(\mathfrak{s}), E(V, W))$ we set

$$\operatorname{Res}_{\mathfrak{s}} E(g, F, H) = \Sigma_{x+y=-1} \frac{1}{x!} E_y(g, H_0^x \vee F, H)$$

We must verify that the collection of functions $\operatorname{Res}_{\mathfrak{s}} E(\cdot, \cdot, \cdot)$ is an Eisenstein system belonging to $\mathfrak{s}$. Condition (i) is easily verified. If $^*P$ and P are as above and if P' belongs to $^*\{P\}^{(j)}$ then

$$\int_{{}^*\Theta \cap {}^{\Psi}N' \backslash {}^{\Psi}N'} \operatorname{Res}_{\mathfrak{s}} E(nmk, F, H) = \Sigma_{x+y=-1} \frac{1}{x!} \int_{{}^*\Theta \cap {}^{\Psi}N' \backslash {}^{\Psi}N'} E_y(nmk, H_0^k \vee F, H)dn$$

Suppose that, for s in ${}^{\Psi}\Omega^{(j)}(\mathfrak{a})$,

$$N(s, H + zH_0) = \Sigma_{v=-\infty}^{\infty} z^v N_v(s, H)$$

then

$$\int_{{}^*\Theta \cap {}^{\Psi}N' \backslash {}^{\Psi}N'} E_y(nmk, H_0^x \vee F, H)dn$$

is equal to the sum over s in ${}^{\Psi}\Omega^{(j)}(\mathfrak{a})$ of

$$\exp(\langle {}^{\Psi}H'(m), sH\rangle + \rho({}^{\Psi}H'(m))\Sigma_{u+v=y} \frac{1}{u!} \langle {}^{\Psi}H'(m), sH_0\rangle^u (N_v(s, H)(H_0^x \vee F))(mk)$$

If for t in ${}^{\Psi}\Omega^{(j)}(\mathfrak{s})$ we let $\operatorname{Res}_{\mathfrak{s}} N(t, H)$ be that linear transformation from $L(S(\mathfrak{s}), \mathcal{E}(V, W))$ to $S(\mathfrak{s}_t) \otimes \mathcal{E}(V', W))$, where

$$\mathfrak{s}_t = \{-(\overline{tH}) \mid H \in \mathfrak{s}\}$$

which sends F in $L(S(\mathfrak{s}), \mathcal{E}(V, W))$ to

$$\Sigma_s \Sigma_{x+u+v=-1} \frac{1}{x!\,u!} (sH_0)^u \otimes N_v(s, H)(H_0^x \vee F)$$

where the outer sum is over those s in ${}^{\Psi}\Omega^{(j)}(\mathfrak{a})$ whose restriction to $\mathfrak{s}$ equals t, then

$$\int_{{}^*\Theta \cap {}^{\Psi}N' \backslash {}^{\Psi}N'} \operatorname{Res}_{\mathfrak{s}} E(nmk, F, H)dn$$

is equal to

$$\sum_{t \in {}^{\Psi}\Omega^{(j)}(\mathfrak{T})} \exp(\langle {}^{\Psi}H'(m), tH \rangle + \rho({}^{\Psi}H'(m)))(\mathrm{Res}_{\mathfrak{T}} N(t, H)F)(mk)$$

It is now an easy matter to complete the verification of condition (ii). It should be remarked that if $\Phi(\cdot)$ is a function with values in $\mathcal{E}(V, W)$ which is defined and analytic in a neighbourhood of H in ${}^{\Psi}\mathfrak{a}^{(i)}$ and if $\Psi(\cdot)$ is a function with values in $\mathcal{E}(V', W)$ which is defined and analytic in an open set of ${}^{\Psi}\mathfrak{a}^{(j)}$ containing

$$\{-s\overline{H} \mid s \in {}^{\Psi}\Omega^{(j)}(\mathfrak{S})\}$$

then

$$\mathrm{Res}_{{}^{\Psi}\mathfrak{T}} \left\{ \sum_{s \in {}^{\Psi}\Omega^{(j)}(\mathfrak{S})} (N(s, H)d\Phi(H), d\Psi(-s\overline{H})) \right\}$$

is equal to

$$\sum_{t \in {}^{\Psi}\Omega^{(j)}(\mathfrak{T})} \mathrm{Res}_{\mathfrak{T}} (N(t, H)d\Phi(H), d\Psi(-s\overline{H}))$$

The conditions of (iii), (iv), and (v) are also verified easily.

There is a lemma which should be proved before we leave the subject of residues. It appears rather complicated because it is stated in such a form that it is directly applicable in the proof of Theorem 7.1, which is the only place it is used; however, it is essentially a simple consequence of the usual residue theorem. If $\mathfrak{S}$ is a subspace of $\mathfrak{a} = \mathfrak{a}^{(i)}$, for some i with $1 \leq i \leq r$, defined by equations of the usual form and if $\{E(\cdot, \cdot, \cdot)\}$ is an Eisenstein system belonging to $\mathfrak{S}$ then a hyperplane $\mathfrak{T}$ of $\mathfrak{S}$ will be called a singular hyperplane of the Eisenstein system if there is a cuspidal subgroup ${}^{*}P$ whose split component ${}^{*}\mathfrak{a}$ is contained in $\mathfrak{T}$ and a cuspidal subgroup P contained in ${}^{*}\{P\}^{(i)}$ so that the projection of $\mathfrak{T}$ on ${}^{\Psi}\mathfrak{S}$ is either a singular hyperplane of $E(\cdot, F, H)$ for some F in $L(S(\mathfrak{S}), \mathcal{E}(V, W))$ or a singular hyperplane of $(N(s, H)F, F')$ for some F in $L(S(\mathfrak{S}), \mathcal{E}(V, W))$, some P' in ${}^{*}\{P\}^{(j)}$, some s in ${}^{\Psi}\Omega^{(j)}(s)$, and some F'

in $L(S(\mathfrak{S}_s), \mathcal{E}(V', W))$. Only a finite number of singular hyperplanes of the Eisenstein system meet each compact subset of $\mathfrak{S}$. Let

$$\hat{\mathfrak{S}} = X(\mathfrak{S}) + (\tilde{\mathfrak{S}} \cap \mathfrak{n})$$

If Z is a point in $\hat{\mathfrak{S}}$ and a is a positive number or infinity let

$$U(\mathfrak{S}, Z, a) = \{Z + iH \mid H \in \tilde{\mathfrak{S}} \cap \mathfrak{n}, \; \|H\| < a\}$$

and if ${}^*\mathfrak{n}$ is a distinguished subspace of $\mathfrak{n}$ which is contained in $\tilde{\mathfrak{S}}$ let $U({}^\Psi\mathfrak{S}, Z, a)$ be the projection of $U(\mathfrak{S}, Z, a)$ on ${}^\Psi s$. Let a be a positive number and let Z_1 and Z_2 be two distinct points in $\hat{\mathfrak{S}}$. If $0 \le x \le 1$ let

$$Z(x) = xZ_1 + (1-x)Z_2$$

and suppose that there is a number x_0 with $0 < x_0 < 1$ so that no singular hyperplane of the Eisenstein system meets the closure of $U(\mathfrak{S}, Z(x), a)$ if $x \neq x_0$ and so that any singular hyperplane which meets the closure of $U(\mathfrak{S}, Z(x_0), a)$ is defined by an equation of the form $\langle H, Z_2 - Z_1\rangle = \mu$ where μ is a complex number. If ${}^*P$, P, P' and $\mathfrak{S}$ are given we want to consider

$$\text{(7.f)} \quad \frac{1}{(2\pi i)^m}\int_{U({}^\Psi\mathfrak{S}, Z_2, a)} E(g, d\Phi(H), H)dH - \frac{1}{(2\pi i)^m}\int_{U({}^\Psi\mathfrak{S}, Z_1, a)} E(g, d\Phi(H), H)dH$$

as well as the sum over s in $\Omega^{(j)}(\mathfrak{S})$ of

$$\text{(7.g)} \left\{\frac{1}{(2\pi i)^m}\int_{U({}^\Psi\mathfrak{S}, Z_2, a)} (N(s,H)d\Phi(H), d\Psi(-s\overline{H}))dH - \frac{1}{(2\pi i)^m}\int_{U({}^\Psi\mathfrak{S}, Z_1, a)} (N(s,H)d\Phi(H), d\Psi(-s\overline{H}))dH\right\}$$

$\Phi(H)$ is a function with values in $\mathcal{E}(V, W)$ which is defined and analytic in a neighbourhood of the closure of $\bigcup_{0\le x\le 1} U({}^\Psi\mathfrak{S}, Z, a)$ in ${}^\Psi\mathfrak{n}^{(i)}$ and $\Psi(H)$ is a function with values in $\mathcal{E}(V', W)$ which is defined and analytic in a neighbourhood of

$$\bigcup_{s\in{}^\Psi\Omega^{(j)}(\mathfrak{S})} \bigcup_{0\le x\le 1} U({}^\Psi\mathfrak{S}_s, -s\overline{Z}, a)$$

in ${}^{\Psi}\mathfrak{w}^{(j)}$. The dimension of ${}^{\Psi}\mathfrak{S}$ is m.

Choose coordinates $z = (z_1, \ldots, z_m)$ on ${}^{\Psi}\mathfrak{S}$ so that $H(z)$ belongs to $\hat{\mathfrak{S}}$ if and only if z is real, so that

$$\langle H(z), H(w)\rangle = \Sigma_{k=1}^{m} z_k w_k$$

so that ${}^{\Psi}Z_1$, the projection of Z_1 on ${}^{\Psi}\mathfrak{S}$, is equal to $H(0, \ldots, 0)$, and so that ${}^{\Psi}Z_2 = H(0, \ldots, 0, c)$ with some positive number c. Set $w = (0, \ldots, c)$. If $a' = (a^2 - \|\operatorname{Im} X(\mathfrak{S})\|^2)^{\frac{1}{2}}$ the above differences are equal to

$$(7.h)\quad \left(\frac{1}{2\pi}\right)^m \int_{|y|<a'} \phi(w + iy)dy_1 \ldots dy_m - \left(\frac{1}{2\pi}\right)^m \int_{|y|<a'} \phi(iy)dy_1 \ldots dy_m$$

with $\phi(z)$ equal to $E(g, d\Phi(H(z)), H(z))$ or to

$$\Sigma_{s\in {}^{\Psi}\Omega^{(j)}(\mathfrak{S})} (N(s, H(z))d\Phi(H(z)), d\Psi(-sH(\bar{z})))$$

Choose $b > a'$ so that no singular hyperplane of $\phi(\cdot)$ intersects

$$\{(iy_1, \ldots, x + iy_m) \mid \Sigma_{k=1}^{m} |y_i|^2 < b^2, \ 0 \le x \le c, \ x \neq x_0 c\}$$

and so that any singular hyperplane which intersects

$$\{(iy_1, \ldots, iy_{m-1}, x_0 c + iy_m) \mid \Sigma_{k=1}^{m} |y_k|^2 < b^2\}$$

is defined by an equation of the form $z_m = \mu$. Choose, in the $m-1$ dimensional coordinate space, a finite set of half-open rectangles $J^{(\ell)}$, $1 \le \ell \le n$, defined by

$$\alpha_k^{(\ell)} < y_k < \beta_k^{(\ell)}, \quad 1 \le k \le m-1$$

and for each ℓ a positive number γ^{ℓ} so that

$$\bigcup_{\ell=1}^{n} \{(y_1, \ldots, y_{m-1}, y_m) \mid (y_1, \ldots, y_{m-1}) \in J^{\ell}, \ |y_m| < \gamma^{\ell}\}$$

contains the closed ball of radius a' and is contained in the open ball of radius b.

The expression (7.h) differs from

$$(7.i)\ \Sigma_{\ell=1}^{n} \frac{1}{(2\pi)^{m-1}} \int_{a_1^\ell}^{\beta_1^\ell} dy_1 \ldots \int_{a_{m-1}^\ell}^{\beta_{m-1}^\ell} dy_{m-1} \frac{1}{2\pi i} \left\{ \int_{c-i\gamma^k}^{c+i\gamma^k} - \int_{-i\gamma^k}^{i\gamma^k} \phi(iy_1, \ldots, iy_{m-1}, z_m) dz_m \right\}$$

by the sum of two integrals. Each of these integrals is of the form

$$(7.k) \qquad \frac{1}{(2\pi i)^m} \int_U \phi(z) dz_1 \wedge \ldots \wedge dz_m$$

where U is an open subset of a real oriented subspace of the coordinate space which is of dimension m and is contained in

$$\{z = (z_1, \ldots, z_m) \mid \|\operatorname{Im} z\| > a'\}$$

If $z_m = \mu_j$, $1 \leq j \leq p$ are the singular hyperplanes of $\phi(z)$ which meet

$$\{(iy_1, \ldots, iy_{m-1}, x_0 c + iy_m) \mid \Sigma_{i=1}^m |y_i|^2 < b\}$$

and $\phi_j(z_1, \ldots, z_{m-1})$ is the residue of $\phi(z_1, \ldots, z_{m-1}, z_m)$ at μ_j the sum (7.i) differs from

$$\Sigma_{j=1}^p \Sigma_{\ell=1}^n \frac{1}{(2\pi)^{m-1}} \int_{J^\ell} \phi_j(iy_1, \ldots, iy_{m-1}) dy_1 \ldots dy_{m-1}$$

by an integral of the form (7.k). The latter sum differs from

$$\Sigma_{j=1}^p \frac{1}{(2\pi)^{m-1}} \int_{|y|<a_j'} \phi_j(iy_1, \ldots, iy_{m-1}) dy_1 \ldots dy_{m-1}$$

with $a_j' = ((a')^2 - (\operatorname{Im} \mu_j)^2)^{\frac{1}{2}}$ by a sum of the form

$$\Sigma_{j=1}^p \frac{1}{(2\pi i)^{m-1}} \int_{U_j} \phi_j(z) dz_1 \wedge \ldots \wedge dz_{m-1}$$

where U_j is an open subset of a real oriented subspace of dimension $m-1$ of the hyperplane $z_m = \mu_j$ which is contained in

$$\{z = (z_1, \ldots, z_{m-1}, z_m) \mid \|\operatorname{Im} z\| > a'\}$$

Let $\mathfrak{s}_1, \ldots, \mathfrak{s}_n$ be the singular hyperplanes of the Eisenstein system which meet the closure of $U(\mathfrak{S}, Z(x_0), a)$. If none of the $\tilde{\mathfrak{s}}_\ell$, $1 \leq \ell \leq n$ contain ${}^*\mathfrak{a}$ then the expression (7.f) is equal to a sum of integrals of the form

$$(7.\ell) \qquad \frac{1}{(2\pi i)^{m'}} \int_{U'} E'(g, d\Phi(H), H)\,dH$$

where U' is an open subset of some real subspace of dimension m' of the space ${}^\Psi\mathfrak{s}$, the projection on ${}^\Psi\mathfrak{a}^{(i)}$ of $\mathfrak{s}$, which is $\mathfrak{S}$ itself or a singular hyperplane of the Eisenstein system such that $\tilde{\mathfrak{s}}$ contains ${}^*\mathfrak{a}$, and is contained in $\{H \mid \|\operatorname{Im} H\| > a\}$ and $E'(\cdot, \cdot, \cdot)$ is $E(\cdot, \cdot, \cdot)$ or $\operatorname{Res}_{\mathfrak{s}} E(\cdot, \cdot, \cdot)$. If ${}^*\mathfrak{a}$ is contained in one, and hence all, of the $\tilde{\mathfrak{s}}_\ell$ then the expression (7.f) differs from

$$\Sigma_{\ell=1}^{n} \frac{1}{(2\pi i)^{m-1}} \int_{U({}^\Psi\mathfrak{s}_\ell, W_\ell, a)} \operatorname{Res}_{\mathfrak{s}_\ell} E(g, d\Phi(H), H)\,dH$$

where W_ℓ is a point in $X(\mathfrak{s}_\ell) + (\mathfrak{s}_\ell \cap \mathfrak{a}^{(i)})$ such that $\operatorname{Re} W_\ell$ is in the convex hull of $\operatorname{Re} Z_1$ and $\operatorname{Re} Z_2$, by a sum of integrals of the form (7.ℓ). A similar assertion is valid for the expression (7.g). The last sum is replaced by

$$\Sigma_{\ell=1}^{n} \Sigma_{t \in {}^\Psi\Omega^{(j)}(\mathfrak{s}_\ell)} \frac{1}{(2\pi i)^{m-1}} \int_{U({}^\Psi\mathfrak{s}_\ell, W_\ell, a)} (\operatorname{Res}_{\mathfrak{s}} N(t, H)\,d\Phi(H), d\Psi(-t\overline{H}))\,dH$$

and the integrals (7.ℓ) are replaced by

$$(7.m) \qquad \frac{1}{(2\pi i)^{m'}} \int_{U'} (N'(t, H)\,d\Phi(H), d\Psi(-t\overline{H}))\,dh$$

with t in ${}^\Psi\Omega^{(j)}(\mathfrak{s}')$ and $N'(t, H)$ equal to $N(t, H)$ if $t' = s$ and to $\operatorname{Res}_{\mathfrak{s}'} N(t, H)$ if $\mathfrak{s}'$ is a singular hyperplane. The lemma we need is a refinement of these observations. In stating it we keep to our previous notation.

LEMMA 7.1. <u>Suppose that for every positive number</u> a <u>there is given a non-empty</u>

open convex subset V(a) of

$$X(\mathfrak{S}) + (\tilde{\mathfrak{S}} \cap \mathfrak{a}^{(1)})$$

so that no singular hyperplane intersects the closure of $U(\mathfrak{S}, W, a)$ if W belongs to V(a) and so that $V(a_1)$ contains $V(a_2)$ if a_1 is less than a_2. Let Z be a given point in $X(\mathfrak{S}) + (\tilde{\mathfrak{S}} \cap \mathfrak{a}^{(i)})$ and if W belongs to $X(\mathfrak{S}) + (\tilde{\mathfrak{S}} \cap \mathfrak{a}^{(i)})$ let

$$W(x) = (1-x)Z + xW$$

Then there is a subset T of the set S of singular hyperplanes, and for each $\mathfrak{s}$ in T a distinguished unit normal, and for each $a > 0$ a non-empty open convex subset W(a) of V(a), and, for each $\mathfrak{s}$, a non-empty open convex subset $V(\mathfrak{s}, a)$ of $X(\mathfrak{s}) + (\tilde{\mathfrak{s}} \cap \mathfrak{a}^{(i)})$ so that, for any ${}^*P$, P, and P' such that $\tilde{\mathfrak{S}}$ contains ${}^*\mathfrak{a}$, any W in V(a), any choice of $W(\mathfrak{s})$ in $V(\mathfrak{s}, a)$, and any $\varepsilon > 0$ such that no element of T meets the closure of $U(\mathfrak{S}, W(x), a)$ if $0 < x \leq \varepsilon$, the difference between

$$\frac{1}{(2\pi i)^m} \int_{U({}^{\Psi}\mathfrak{S}, W, a)} E(g, d\Phi(H), H)dH$$

and, if $0 < x \leq \varepsilon$,

$$\frac{1}{(2\pi i)^m} \int_{U({}^{\Psi}\mathfrak{S}, W(x), a)} E(g, d\Phi(H), H)dH + \Sigma \frac{1}{(2\pi i)^{m-1}} \int_{U({}^{\Psi}\mathfrak{s}, W(\mathfrak{s}), a)} \mathrm{Res}_{\mathfrak{s}} E(g, d\Phi(H), H)dH$$

is a sum of integrals of the form (7.ℓ). In the above expression the second sum is over those $\mathfrak{s}$ in T such that $\tilde{\mathfrak{s}}$ contains ${}^*\mathfrak{a}$. Moreover the difference between

$$\Sigma_{s \in {}^{\Psi}\Omega^{(j)}(\mathfrak{S})} \frac{1}{(2\pi i)^m} \int_{U({}^{\Psi}\mathfrak{S}, W, a)} (N(s, H)d\Phi(H), d\Psi(-s\overline{H}))dH$$

and the sum of

$$\sum_{s \in \Psi_{\Omega}^{(j)}(\mathfrak{S})} \frac{1}{(2\pi i)^{m}} \int_{U(\Psi_{\mathfrak{S}},\, W(x),\, a)} (N(s, H)d\Phi(H),\ d\Psi(-s\overline{H}))dH$$

and

$$\sum\sum_{t \in \Psi_{\Omega}^{(j)}(t)} \frac{1}{(2\pi i)^{m-1}} \int_{U(\Psi_{\mathfrak{T}},\, W(\mathfrak{T}),\, a)} (\mathrm{Res}_{\mathfrak{T}}\, N(t, H)d\Phi(H),\ d\Psi(-t\overline{H}))dH$$

is a sum of integrals of the form (7.m). *The sets* U' *appearing in the integrals of the form* (7.ℓ) *and* (7.m) *can be taken to be such that* $\{\mathrm{Re}\, H \mid H \in U'\}$ *lies in the convex hull of* $\mathrm{Re}\, Z$ *and* $\{\mathrm{Re}\, H \mid H \in V(a)\}$. *The sets* $V(\mathfrak{T}, a)$ *can be so chosen that* $\{\mathrm{Re}\, H \mid H \in V(\mathfrak{T}, a)\}$ *lies in the interior of the convex hull of* $\mathrm{Re}\, Z$ *and* $\{\mathrm{Re}\, H \mid H \in V(a)\}$, *and* $V(\mathfrak{T}, a_1)$ *contains* $V(\mathfrak{T}, a_2)$ *if* a_1 *is less than* a_2, *and no singular hyperplane of the Eisenstein system belonging to* $\mathfrak{T}$ *meets the closure of* $U(\mathfrak{T}, W, a)$ *if* W *lies in* $V(\mathfrak{T}, a)$. *If no singular hyperplane meets the closure of* $U(\mathfrak{S}, Z, a)$ *the conclusions are valid when* $x = 0$.

We have not troubled to be explicit about the conditions on the functions $\Phi(\cdot)$ and $\Psi(\cdot)$. They will become clear. Replacing $V(a)$ by $V(N(a))$ where $N(a)$ is the integer such that

$$N(a) - 1 < a \leq N(a)$$

we can suppose that

$$V(a) = V(N(a))$$

Let $P(a)$ be the set of hyperplanes of $\hat{\mathfrak{S}}$ which are the projections on $\hat{\mathfrak{S}}$ of those elements of S which meet $\{H \mid \|\mathrm{Im}\, H\| \leq a\}$. If N is a positive integer the set of points W in $V(N)$ such that the interior of the segment joining Z and W does not contain a point belonging to two distinct hyperplanes in $P(N)$ is a non-empty open subset of $V(N)$. Let $W(N)$ be a non-empty convex open subset of this set and let

$$W(a) = W(N(a))$$

if $a > 0$. If the sets are chosen inductively it can be arranged that $W(N_1)$ contains $W(N_2)$ if N_1 is less than N_2. Let $T(a)$ be the set of singular hyperplanes whose projection on $\hat{\mathfrak{b}}$ separates Z and $W(a)$ and let

$$T = \bigcup_{a>0} T(a)$$

If $\mathfrak{s}$ belongs to T and $\mathfrak{s}$ intersects

$$\{H \mid \|\operatorname{Im} H\| \leq a\}$$

so that $\mathfrak{s}$ belongs to $T(a)$ let $V(\mathfrak{s}, a)$ be the inverse image in $\hat{\mathfrak{s}}$ of the intersection of the projection of $\mathfrak{s}$ on $\hat{\mathfrak{b}}$ with the convex hull of Z and $W(a)$; let the distinguished normal to $\mathfrak{s}$ be the one which points in the direction of $W(a)$. If $\mathfrak{s}$ does not intersect $\{H \mid \|\operatorname{Im} H\| \leq a\}$ let b be the smallest number such that $\mathfrak{s}$ intersects $\{H \mid \|\operatorname{Im} H\| \leq b\}$ and set

$$V(\mathfrak{s}, a) = V(\mathfrak{s}, b)$$

In proving the lemma it may be assumed that $W(\mathfrak{s})$ is the inverse image in $\hat{\mathfrak{s}}$ of the intersection of the projection of $\mathfrak{s}$ on $\hat{\mathfrak{b}}$ with the line joining Z and W. Choosing a polygonal path $Z_0, Z_1, \ldots, Z_n$ from $W(x)$ to W which lies in the convex hull of Z and $W(a)$, which meets no element of $P(a)$ except the projections of the elements of $T(a)$ and these only once and in the same point as the line joining Z and W, and which is such that no point Z_j, $1 \leq j \leq n$ lies on any element of $P(a)$ and such that any line segment of the path crosses at most one element of $P(a)$ and crosses that in a normal direction, and observing that the difference between the integrals over $U({}^{\Psi}\mathfrak{b}, W, a)$ and $U({}^{\Psi}\mathfrak{b}, W(x), a)$ is equal to the sum of the differences between the integrals over $U({}^{\Psi}\mathfrak{b}, Z_j, a)$ and $U({}^{\Psi}\mathfrak{b}, Z_{j-1}, a)$, $1 \leq j \leq n$, we see that the lemma is a consequence of the discussion preceding it. To conform to the definition of an Eisenstein system we have to remove those $\mathfrak{t}$ for which all the functions $\operatorname{Res}_{\mathfrak{t}} E(\cdot, \cdot, \cdot)$ vanish.

Unfortunately this lemma on residues is not sufficient for our needs; it must be supplemented by another which we state informally but, for an obvious reason, do not prove. If $\mathfrak{S}$ is as above and ε and a are positive numbers let

$$C(\mathfrak{S}, \varepsilon, a) = \{X(\mathfrak{S}) + H \mid H \in \tilde{\mathfrak{S}}, \ \|\operatorname{Re} H\| < \varepsilon, \ \|\operatorname{Im}(X(\mathfrak{S}) + H)\| < a\}$$

If U is an open set on the sphere of radius ε in $\tilde{\mathfrak{S}} \cap \mathfrak{n}^{(i)}$ then

$$\{xX(\mathfrak{S}) + (1-x)Z \mid 0 < x < 1, \ Z \in U\}$$

will be called a cone of radius ε and centre $X(\mathfrak{S})$. Suppose that, just as in the lemma, we are given an Eisenstein system belonging to $\mathfrak{S}$. Suppose that for every $a > 0$ there are two non-empty convex cones $V_i(\mathfrak{S}, \varepsilon(a), a)$, $i = 1, 2$, of radius $\varepsilon(a)$ and centre $X(\mathfrak{S})$ such that no singular hyperplane meets the closure of $U(\mathfrak{S}, W, a)$ if W belongs to $V_i(\mathfrak{S}, \varepsilon(a), a)$ and such that

$$V_i(\mathfrak{S}, \varepsilon(a_1), a_1) \supseteq V_i(\mathfrak{S}, \varepsilon(a_2), a_2)$$

if $a_1 \leq a_2$. Suppose that, for all a, every singular hyperplane which meets the closure of $C(\mathfrak{S}, \varepsilon(a), a)$ meets the closure of $U(\mathfrak{S}, X(\mathfrak{S}), a)$. Then there is a subset T of the set of singular hyperplanes such that $\operatorname{Re} X(\mathfrak{S}) = X(\mathfrak{t})$ for all $\mathfrak{t}$ in T, and for each $\mathfrak{t}$ in T a distinguished unit normal to $\mathfrak{t}$, and, for each $a > 0$, two non-empty convex cones $W_i(\mathfrak{S}, \varepsilon(a), a)$ of radius $\varepsilon(a)$ and centre $X(\mathfrak{t})$ such that

$$W_i(\mathfrak{S}, \varepsilon(a), a) \subseteq V_i(\mathfrak{S}, \varepsilon(a), a)$$

and, for each $\mathfrak{t}$ in T, an open convex cone $V(\mathfrak{t}, \varepsilon(a), a)$ of radius $\varepsilon(a)$ and centre $X(\mathfrak{t})$ so that if, for some $a > 0$, W_i belongs to $W_i(\mathfrak{S}, \varepsilon(a), a)$ and $W(\mathfrak{t})$, $\mathfrak{t} \in T$, belongs to $V(\mathfrak{t}, \varepsilon(a), a)$ then the difference between the sum over s in $\Psi_\Omega^{(j)}(\mathfrak{s})$ of

$$\frac{1}{(2\pi i)^m}\left\{\int_{U(\Psi\mathfrak{s},\, W_1,\, a)} (N(s,H)d\Phi(H),\, d\Psi(-s\overline{H}))dH - \int_{U(\Psi\mathfrak{s},\, W_2,\, a)} (N(s,H)d\Phi(H),\, d\Psi(-s\overline{H}))dH\right.$$

and

$$\Sigma\Sigma_{t\in\Psi_\Omega^{(j)}(\mathfrak{t})} \frac{1}{(2\pi i)^{m-1}} \int_{U(\Psi\mathfrak{t},\, W(\mathfrak{t}),\, a)} (\mathrm{Res}_{\mathfrak{t}}\, N(t,\, H)d\Phi(H),\ d\Psi(-t\overline{H}))dH$$

is a sum of integrals of the form (7.m). It is clear that one again has some control over the location of the sets U' which occur. Moreover if $\mathfrak{t}$ is in T any singular hyperplane of the associated Eisenstein system which meets the closure of $C(\mathfrak{t},\ \varepsilon(a),\ a)$ meets the closure of $U(\mathfrak{t},\, X(\mathfrak{t}),\ a)$ and we can assume that if W lies in $V(\mathfrak{t},\ \varepsilon(a),\ a)$ then no such hyperplane meets the closure of $U(\mathfrak{t}\ ,\ W,\ a)$.

Suppose that for each i, $1 \leq i \leq r$, we are given a collection $S^{(i)}$ of distinct affine subspaces of dimension m which are defined by equations of the usual form. Let

$$S = \bigcup_{i=1}^{r} S^{(i)}$$

and suppose that for each $\mathfrak{s}$ in S we are given an Eisenstein system belonging to $\mathfrak{s}$. In order to appreciate Theorem 7.7 we have to have some understanding of the relations which the functions in this collection of Eisenstein systems may satisfy and of the conditions under which the relations must be satisfied. The next four lemmas provide us with the necessary understanding. In other words Theorem 7.7 can be regarded, if one is thinking only of the Eisenstein series, as asserting that all Eisenstein series satisfy certain conditions and we are about to show that all Eisenstein series satisfying these conditions satisfy functional equations.

If $\mathfrak{s}$ is a subspace of $\mathfrak{a}^{(i)}$ and $\mathfrak{t}$ is a subspace of $\mathfrak{a}^{(j)}$ defined by equations of the usual form and if ${}^*\mathfrak{a}$ is a distinguished subspace of both $\mathfrak{a}^{(i)}$ and $\mathfrak{a}^{(j)}$ which is contained in $\tilde{\mathfrak{s}}$ and $\tilde{\mathfrak{t}}$ we let $\Psi_\Omega(\mathfrak{s}\ ,\ \mathfrak{t}\ )$ be the set of distinct linear transformations in $\Psi\Omega^{(j)}(\mathcal{E})$ such that $\mathcal{E}_s = \mathcal{F}$. Two linear transformations of $\Omega(\mathfrak{a}^{(i)}, \mathfrak{a}^{(j)})$ which have the same effect on every element of $\mathcal{E}$ have the same effect on every element of the space $\mathcal{E}'$ spanned by $\mathcal{E}$ and zero and on

$$\bar{\mathcal{E}}' = \{H \mid \bar{H} \in \mathcal{E}'\}$$

Thus $\Psi\Omega(\mathcal{E}, \mathcal{F})$ can also be regarded as a set of linear transformations from $\mathcal{E}'$ to $\bar{\mathcal{F}}'$ or from $\bar{\mathcal{E}}'$ to $\mathcal{F}'$. Such a convention is necessary in order to make some of the expressions below meaningful. Suppose that for every element $\mathcal{E}$ of the collection S there is an element s^o of $\Omega(\mathcal{E}, \mathcal{E})$ which fixes each element of $\tilde{\mathcal{E}}$. Certainly s^o is unique. If $^*\mathfrak{a}$ is a distinguished subspace of $\mathfrak{h}$ let

$$^*S^{(i)} = \{\mathcal{E} \in S^{(i)} \mid {}^*\mathfrak{a} \subseteq \tilde{\mathcal{E}}\}$$

and let

$$^*S = \bigcup_{i=1}^{r} {}^*S^{(i)}$$

Two elements $\mathcal{E}$ and $\mathcal{F}$ of $^*S$ are said to be equivalent if $\Psi\Omega(\mathcal{E}, \mathcal{F})$ is not empty.

LEMMA 7.2. Suppose that for each i, $1 \leq i \leq r$, $S^{(i)}$ is a collection of distinct affine subspaces, of dimension m, of $\mathfrak{a}^{(i)}$, defined by equations of the form $\alpha(H) = \mu$ where α is a positive root of $\mathfrak{a}^{(i)}$ and μ is a complex number, such that only a finite number of the elements of $S^{(i)}$ meet each compact subset of $\mathfrak{a}^{(i)}$. Suppose that if $\mathcal{E}$ belongs to $S^{(i)}$ and $\mathfrak{a}$ is the orthogonal complement of the distinguished subspace of largest dimension which is contained in $\hat{\mathcal{E}}$ then $\operatorname{Re} X(\mathcal{E})$ belongs to $^+\mathfrak{a}$ and lies in a fixed compact subset of $\mathfrak{a}^{(i)}$ and suppose that for each $\mathcal{E}$ in S the set $\Omega(\mathcal{E}, \mathcal{E})$ contains an element which leaves each point of $\tilde{\mathcal{E}}$ fixed. Finally suppose that if $\mathcal{E}$ is in S there is given an Eisenstein system belonging to $\mathcal{E}$ and that if $^*P$ is a cuspidal subgroup, with split component $^*\mathfrak{a}$, if P belongs to $^*\{P\}^{(i)}$, P' belongs to $^*\{P\}^{(j)}$, $\mathcal{E}$ belongs to $^*S^{(i)}$, and s belongs to $\Psi_\Omega^{(j)}(\mathcal{E})$ then $N(s, H)$ vanishes identically unless s belongs to $\Psi\Omega(\mathcal{E}, \mathcal{F})$ for some $\mathcal{F}$ in $^*S^{(j)}$. Then S is finite and for each $\mathcal{E}$ in S the point $X(\mathcal{E})$ is real. Moreover, for any choice of $^*\mathfrak{a}$, every equivalence class in $^*S$ contains an element $\mathcal{E}$ such that $\tilde{\mathcal{E}}$ is the complexification of a distinguished subspace of $\mathfrak{h}$.

There is another lemma which must be proved first.

LEMMA 7.3. Suppose that ϕ is a function in $\mathcal{L}(\{P\}, \{V\}, W)$ and suppose that there is an integer N so that if P belongs to $\{P\}$ and $\{p_y \mid 1 \leq y \leq x\}$ is a basis for the polynomials on $\mathfrak{a}$, the split component of P, of degree at most N then there are distinct points $H_1, \ldots, H_u$ in $\mathfrak{a}_c$ and functions $\Phi_{x,y}$, $1 \leq x \leq u$, $1 \leq y \leq v$ in $\mathcal{E}(V, W)$ so that

$$(7.n)\quad \int_{\Gamma\cap N\backslash N} \phi(ng)dn = \Sigma_{x=1}^{u} \exp(\langle H(g), H_x\rangle + \rho(H(g)))\left\{\Sigma_{y=1}^{v} p_y(H(g))\Phi_{x,y}(g)\right\}$$

If

$$\Sigma_{y=1}^{v} p_y(H(g))\Phi_{x,y}$$

does not vanish identically then H_x is real.

If we agree that an empty sum is zero then we can suppose that

$$\Sigma_{y=1}^{v} p_y(H(g))\Phi_{x,y}(g)$$

is never identically zero. The lemma will be proved by induction on the rank of the elements in $\{P\}$. If that rank is zero there is nothing to prove so suppose it is a positive number q and the lemma is true for families of cuspidal subgroups of rank $q-1$. If P belongs to $\{P\}$ and $P' = \gamma P \gamma^{-1}$, γ in Γ, then

$$\int_{\Gamma\cap N'\backslash N'} \phi(n'\gamma g)dn' = \int_{\Gamma\cap N\backslash N} \phi(ng)dn$$

so that the right side of (7.n) is equal to

$$\Sigma_{x=1}^{u'} \exp(\langle H'(\gamma g), H'_x \rangle + \rho(H'(\gamma g)))\Sigma_{y=1}^{v} p_y(H(g))\Phi'_{x,y}(g)$$

Since $H'(\gamma g) = H(g) + H'(\gamma)$ the sets $\{H_1, \ldots, H_u\}$ and $\{H'_1, \ldots, H'_{u'}\}$ are the same. Thus for $1 \leq i \leq r$ the set

$$\bigcup_{P \in \{P\}^{(i)}} \{H_1, \ldots, H_u\} = F_i$$

is finite.

If P belongs to $\{P\}^{(i)}$ let X_y be that element of $S(\{0\})$ such that $p_{X_y} = p_y$ and let

$$T_x = \Sigma_{y=1}^{v} X_y \otimes \Phi_{x,y}$$

If ψ belongs to $\mathfrak{D}(V, W)$ it follows from the relation (7.a) that

$$(\phi, \psi^\wedge) = \Sigma_{x=1}^{u}(T_x, d\Psi(-\overline{H}_x))$$

If

$$f(\cdot) = (f_1(\cdot), \ldots, f_r(\cdot))$$

is such that $\lambda(f)$ can be defined as in Section 6 then

$$(\lambda(f)\phi, \psi^\wedge) = (\phi, \lambda(f^*)\psi^\wedge)$$

is equal to

$$\Sigma_{x=1}^{u}(T_x, d(f_i^*\Psi)(-\overline{H}_x))$$

In particular, if for each i, f_i vanishes to a sufficiently high order at each point of F_i then $\lambda(f)\phi = 0$. If H belongs to F_i and H' belongs to F_j and there is no s in $\Omega(\mathfrak{a}^{(i)}, \mathfrak{a}^{(j)})$ such that $sH = H'$ then we can choose an $f(\cdot)$ so that $f_i(H) \neq f_j(H')$. Consequently we can find $f^{(1)}, \ldots, f^{(w)}$ so that

$$\phi = \Sigma_{x=1}^{w} \lambda(f^{(x)})\phi$$

and $\lambda(f^{(x)})\phi$ satisfies the same conditions as ϕ except that if H belongs to $F_i^{(x)}$, the analogue of F_i, and H' belongs to $F_j^{(x)}$ then $H' = sH$ for some s in $\Omega(\mathfrak{a}^{(i)}, \mathfrak{a}^{(j)})$. Since it is enough to prove the lemma for each $\lambda(f^{(x)})\phi$ we assume that ϕ already satisfies this extra condition. Let $c(f)$ be the value of f_i at one and hence every point in F_i. Since $\lambda(f)$ is normal $\lambda(f)\phi = c(f)\phi$ and $\lambda(f^*)\phi = \overline{c}(f)\phi = c(f^*)\phi$. Thus, if H belongs to F_i, $f_i(H) = f_i(-\overline{H})$ and there is an s in $\Omega(\mathfrak{a}^{(i)}, \mathfrak{a}^{(i)})$ so that $sH = \overline{H}$ and $\langle H, H\rangle$ is real.

To prove the lemma we need only show that for some P in $\{P\}$ one of $H_1, \ldots, H_u$ is real. It is not difficult to see that for each P in $\{P\}$ the points $-\mathrm{Re}\,H_x$, $1 \leq x \leq u$, belong to ${}^{+}\mathfrak{a}$. We forego describing the proof in detail because in all applications we shall make of the lemma it will be apparent that this is so. Let

$$\mu_x = \max_{1 \leq k \leq q} \{\langle \alpha_{,k}, \alpha_{,k}\rangle^{-\frac{1}{2}} \alpha_{,k}(\mathrm{Re}\,H_x)\}$$

and if $\{H_1, \ldots, H_u\}$ is not empty let

$$\mu(P) = \max_{1 \leq x \leq u} \mu_x$$

If ϕ does not vanish identically choose P_0 so that

$$\mu_0 = \mu(P_0) \geq \mu(P)$$

for all P in $\{P\}$; the number μ_0 is negative. Let

$$\|\alpha_{,\ell_0}\| \mu(P_0) = \alpha_{,\ell_0}(\mathrm{Re}\,H_{x_0})$$

and let ${}^{*}P$ be the cuspidal subgroup belonging to P_0 with split component

$${}^{*}\mathfrak{a} = \{H \in \mathfrak{a}_0 \mid \alpha_{\ell,}(H) = 0,\ \ell \neq \ell_0\}$$

It follows without difficulty from Lemma 4.2 that if $\{q_y \mid 1 \le y \le v'\}$ is a basis for the polynomials on ${}^*\mathfrak{n}$ of degree at most N then there are distinct points ${}^*H_x$, $1 \le x \le u'$, in ${}^*\mathfrak{n}_c$ and functions ϕ_{xy} on ${}^*A\,{}^*T\backslash G$ so that

$$\int_{\Gamma \cap {}^*N \backslash {}^*N} \phi(ng)dn$$

is equal to

$$\Sigma_{x=1}^{u'} \exp(\langle {}^*H(g), {}^*H_x \rangle + \rho({}^*H(g)))\left\{\Sigma_{y=1}^{v'} q_y({}^*H(g))\phi_{x,y}(g)\right\}$$

It follows from formula (3.d) that if P is an element of ${}^*\{P\}$ and $g = amk$ with a in ${}^*A$, m in ${}^*M$, and k in K then

$$\Sigma_{x=1}^{u} \exp(\langle H(g), H_x \rangle + \rho(H(g)))\left\{\Sigma_{y=1}^{v} p_y(H(g))\Phi_{x,y}(g)\right\}$$

is equal to

$$\Sigma_{x=1}^{u'} \exp(\langle {}^*H(g), {}^*H_x \rangle + \rho({}^*H(g)))\left\{\Sigma_{y=1}^{v'} q_y({}^*H(g)) \int_{{}^*\Theta \cap {}^\Psi N \backslash {}^\Psi N} \phi_{xy}(nmk)dn\right\}$$

Applying this relation to P_0 we see that if the indices are chosen appropriately we can suppose that the projection of H_{x_0} on ${}^*\mathfrak{n}_c$ is ${}^*H_1$. Let

$$\int_{{}^*\Theta \cap {}^\Psi N \backslash {}^\Psi N} \phi_{1y}(nmk)dn$$

equal

$$\Sigma_{x=1}^{u''} \exp(\langle {}^\Psi H(m), {}^\Psi H_x \rangle + \rho({}^\Psi H(m)))\left\{\Sigma_{z=1}^{v''} r_z({}^\Psi H(m))\Phi_{x,y,z}(g)\right\}$$

where $\{r_z \mid 1 \le z \le v''\}$ is a basis for the polynomials of degree at most N on ${}^\Psi\mathfrak{n}$, the orthogonal complement of ${}^*\mathfrak{n}$ in $\mathfrak{n}$, and $\Phi_{x,y,z}$ belongs to $\mathcal{E}(V, W)$. We suppose that for each x there is a y so that

$$\Sigma_{z=1}^{v''} r_z({}^\Psi H(m))\Phi_{x,y,z}$$

does not vanish identically. If we show that $-\mathrm{Re}({}^{\Psi}H_x)$ belongs to ${}^{+}({}^{\Psi}\mathfrak{n})$ for $1 \le x \le u''$ it will follow from the corollary to Lemma 5.1 that $\phi_{1,y}$ is square integrable. It is then obvious that it belongs to ${}^{*}\mathcal{L}(\{P\}, \{V\}, W)$. The induction assumption implies that ${}^{\Psi}H_x$ is real for $1 \le x \le u''$. In particular we can choose x_1 so that

$$H_{x_0} = {}^{*}H_1 + {}^{\Psi}H_{x_1}$$

Since

$$\langle H_{x_0}, H_{x_0}\rangle = \langle {}^{*}H_1, {}^{*}H_1\rangle + \langle {}^{\Psi}H_{x_1}, {}^{\Psi}H_{x_1}\rangle$$

is real the number $\langle {}^{*}H_1, {}^{*}H_1\rangle$ is real and ${}^{*}H_1$ is either real or purely imaginary. It is not purely imaginary since $\alpha_{,\ell_0}(\mathrm{Re} H_{x_0}) = \alpha_{,\ell_0}(\mathrm{Re}\, {}^{*}H_1)$. Consequently H_{x_0} is real. To show that $-\mathrm{Re}({}^{\Psi}H_x)$ belongs to ${}^{+}({}^{\Psi}\mathfrak{n})$ we have to show that $\alpha_{,\ell}(\mathrm{Re}\, {}^{\Psi}H_x) < 0$ if $\ell \neq \ell_0$. Certainly

$$\alpha_{,\ell}(\mathrm{Re}({}^{*}H_1 + {}^{\Psi}H_x)) \le \langle \alpha_{,\ell}, \alpha_{,\ell}\rangle^{\frac{1}{2}} \mu_0$$

and

$$\alpha_{,\ell_0}(\mathrm{Re}({}^{*}H_1 + {}^{\Psi}H_x)) = \alpha_{,\ell_0}(\mathrm{Re}\, {}^{*}H_1) = \langle \alpha_{,\ell_0}, \alpha_{,\ell_0}\rangle^{\frac{1}{2}} \mu_0$$

Thus

$$\alpha_{,\ell}({}^{\Psi}H_x) \le \langle \alpha_{,\ell}, \alpha_{,\ell}\rangle^{\frac{1}{2}}\{\mu_0 - \langle \alpha_{,\ell}, \alpha_{,\ell}\rangle^{-\frac{1}{2}} \langle \alpha_{,\ell_0}, \alpha_{,\ell_0}\rangle^{-\frac{1}{2}} \langle \alpha_{,\ell}, \alpha_{,\ell_0}\rangle \mu_0\} < 0$$

if $\ell \neq \ell_0$.

Suppose that, for each P in $\{P\}^{(i)}$ and $1 \le x \le u$, T_x has the same meaning as above. It has been observed that

$$\Sigma_{x=1}^{u}(T_x, df_i\Psi(-\overline{H}_x)) = \Sigma_{x=1}^{u} \overline{f}_i(-\overline{H}_x)(T_x, d\Psi(-\overline{H}_x))$$

if $f(\cdot) = (f_1(\cdot), \ldots, f_r(\cdot))$, if, for each i, $f_i(\cdot)$ is a bounded analytic function on D_i, and if $f_j(sH) = f_i(H)$ if s belongs to $\Omega(\mathfrak{a}^{(i)}, \mathfrak{a}^{(j)})$. It is clear that the equality must also be valid for any function $f(\cdot)$ such that $f_i(\cdot)$ is analytic in a neighbourhood of

$$\bigcup_{j=1}^{r} \bigcup_{s \in \Omega(\mathfrak{a}^{(j)}, \mathfrak{a}^{(i)})} \{-s\overline{H} \mid H \in F_j\}$$

Indeed for any such function

$$(T_x, df_i \Psi(-\overline{H}_x)) = \overline{f}_i(-\overline{H}_x)(T_x, d\Psi(-\overline{H}_x))$$

We turn to the proof of Lemma 7.2. Let C be an equivalence class in ${}^*S$ and choose $\mathfrak{S}$ in C so that $\widetilde{\mathfrak{S}}$ contains a distinguished subspace of the largest possible dimension. Replace, if necessary, ${}^*\mathfrak{a}$ by this larger space and suppose that this distinguished subspace is ${}^*\mathfrak{a}$ itself. Of course the equivalence class to which $\mathfrak{S}$ belongs may become smaller but this is irrelevant. Suppose $\mathfrak{S}$ lies in ${}^*S^{(i)}$ and let ${}^\Psi\mathfrak{S}$ be the projection of $\mathfrak{S}$ on the orthogonal complement of ${}^*\mathfrak{a}$. A point H in ${}^\Psi\mathfrak{S}$ which does not lie on a singular hyperplane of any of the functions $E(\cdot, \cdot, \cdot)$ which are defined on ${}^\Psi\mathfrak{S}$ and which is such that if s_1 and s_2 are in ${}^\Psi\Omega^{(j)}(\mathfrak{S})$ for some j then $s_1H = s_2H$ only if $s_1 = s_2$ will be called a general point of ${}^\Psi\mathfrak{S}$. There is at least one cuspidal subgroup ${}^*P$ with ${}^*\mathfrak{a}$ as split component and one element of ${}^*\{P\}^{(i)}$ so that for some F in $L(S(\mathfrak{S}), \mathcal{E}(V, W))$ the function $E(g, F, H)$ on ${}^*A{}^*T\backslash G \times {}^\Psi\mathfrak{S}$ does not vanish identically. Suppose that the general point H lies in $U({}^\Psi\mathfrak{S}, X(\mathfrak{S}), \infty)$. If P' belongs to ${}^*\{P\}^{(j)}$ then

$$\int_{{}^*\Theta \cap {}^\Psi N' \backslash {}^\Psi N'} E(nmk, F, H)$$

is equal to

$$\sum_{s \in {}^\Psi\Omega^{(j)}(\mathfrak{S})} \exp(\langle H'(g), sH\rangle + \rho(H'(g)))N(s, H)F(g)$$

$N(s, H)$ is zero unless s belongs to ${}^{\Psi}\Omega(\mathfrak{s}, \mathfrak{t})$ for some $\mathfrak{t}$ in ${}^{*}S^{(j)}$. Moreover if $\mathfrak{t}$ belongs to C and ${}^{*}S^{(j)}$ the largest distinguished subspace which $\mathfrak{t}$ contains is ${}^{*}\mathfrak{a}$. Thus if $N(s, H)F$ is not zero then

$$-\mathrm{Re}(sH) = \mathrm{Re}(X_{\mathfrak{s}_s})$$

belongs to ${}^{+}({}^{\Psi}\mathfrak{a}^{(j)})$ if ${}^{\Psi}\mathfrak{a}^{(j)}$ is the orthogonal complement of ${}^{*}\mathfrak{a}$ in $\mathfrak{a}^{(j)}$. Lemma 7.3 implies that sH is real for all such s. If $\tilde{\mathfrak{s}}$ were not the complexification of ${}^{*}\mathfrak{a}$ we could choose an H which was not real so that $E(g, F, H)$ did not vanish identically and obtain a contradiction. Consequently ${}^{\Psi}\mathfrak{s} = \{X(\mathfrak{s})\}$ and $X(\mathfrak{s})$ is real. If $\mathfrak{s}$ and $\mathfrak{t}$ are equivalent then $X(\mathfrak{t})$ is real if and only if $X(\mathfrak{s})$ is; so it has only to be shown that S is finite. This of course follows immediately from the assumptions of the lemma and the fact that

$$X(s) = \mathrm{Re}\, X(s)$$

for all s in S.

Suppose ${}^{*}P$ with the split component ${}^{*}\mathfrak{a}$ is a cuspidal subgroup belonging to one of the elements of $\{P\}$. Let $P^{(i,k)}$, $1 \leq k \leq m_i$, be a complete set of representatives for the elements of ${}^{*}\{P\}^{(i)}$ and let

$$\mathcal{E}^{(i)} = \oplus_{k=1}^{m_i} \mathcal{E}(V^{(i,k)}, W)$$

If S is as above and $\mathfrak{s}$ belongs to ${}^{*}S^{(i)}$, $\mathfrak{t}$ belongs to ${}^{*}S^{(j)}$, and s belongs to $\Omega(\mathfrak{s}, \mathfrak{t})$ let $M(s, H)$ be that linear transformation from $L(S(\mathfrak{s}), \mathcal{E}^{(i)})$ to $S(\mathfrak{t}) \otimes \mathcal{E}^{(j)}$ such that if F belongs to $L(S(\mathfrak{s}), E(\mathcal{V}^{(i,k)}, W))$ then the component of $M(s, H)F$ in $S(\mathfrak{t}) \otimes \mathcal{E}(V^{(j,\ell)}, W)$ is $N(s, H)F$. Of course $N(s, H)$ depends on $P^{(i,k)}$ and $P^{(j,\ell)}$. If C is an equivalence class in ${}^{*}S$ choose $\mathfrak{s}$ in C so that $\tilde{\mathfrak{s}} = \mathfrak{a}_c$ where, if $\mathfrak{s}$ belongs to ${}^{*}S^{(i)}$, $\mathfrak{a}$ is a distinguished subspace of $\mathfrak{a}^{(i)}$. Let

$$\Omega(\mathfrak{S}, C) = \bigcup_{t \in C} {}^{\Psi}\Omega(\mathfrak{S}, \mathfrak{t})$$

and let $\Omega_0(\mathfrak{S}, C)$ be the set of elements in $\Omega(\mathfrak{S}, C)$ which leave each point of $\tilde{\mathfrak{S}}$ fixed. Let s^o be the linear transformation in $\Omega(\mathfrak{S}, \mathfrak{S})$ which induces the identity on $\tilde{\mathfrak{S}}$. If $\mathfrak{t}_1$ and $\mathfrak{t}_2$ belong to C then every element of $\Omega(\mathfrak{t}_1, \mathfrak{t}_2)$ can be written as a product ts^os^{-1} with s in $\Omega(\mathfrak{S}, \mathfrak{t}_1)$ and t in $\Omega(\mathfrak{S}, \mathfrak{t}_2)$. If H is in ${}^{\Psi}\mathfrak{S}$ we form the two matrices

$$M(H) = (M(ts^os^{-1}, ss^oH));\ s,\ t \in \Omega(\mathfrak{S}, C)$$

$$M = (M(ts^os^{-1}, ss^oH));\ s,\ t \in \Omega_0(\mathfrak{S}, C)$$

The first matrix is a meromorphic function of H; the second is a constant. If s belongs to $\Omega(\mathfrak{S}, C)$ there is a unique j_s so that $\mathfrak{S}_s$ belongs to ${}^*S^{(j_s)}$. The matrix M(H) can be regarded as a linear transformation from

$$\Sigma_{s \in \Omega(\mathfrak{S}, C)} L(S(\mathfrak{S}_s), \mathcal{E}^{(j_s)})$$

to

$$\Sigma_{s \in \Omega(\mathfrak{S}, C)} S(\mathfrak{S}_s) \otimes \mathcal{E}^{(j_s)}$$

It has a finite dimensional range and the dimension of its range is its rank. A similar remark applies to M. We shall see that the functional equations for all the Eisenstein series are a consequence of the following lemma.

LEMMA 7.4. <u>Suppose that, for</u> $1 \leq i \leq r$, $S^{(i)}$ <u>is the collection of Lemma</u> 7.2 <u>and suppose that for any</u> ${}^*P$, <u>any</u> $\mathfrak{S}$ <u>in</u> ${}^*S^{(i)}$, <u>any</u> $\mathfrak{t}$ <u>in</u> ${}^*S^{(j)}$, <u>any</u> P <u>in</u> ${}^*\{P\}^{(i)}$, <u>any</u> P' <u>in</u> ${}^*\{P\}^{(j)}$, <u>and any</u> s <u>in</u> ${}^{\Psi}\Omega(\mathfrak{S}, \mathfrak{t})$ <u>the functions</u> N(s, H) <u>and</u> $N^*(s^{-1}, -s\bar{H})$ <u>are equal.</u> <u>If</u> ${}^*P$ <u>with the split component</u> ${}^*\mathfrak{a}$ <u>is given, if</u> C <u>is an equivalence class in</u> ${}^*S$, <u>if</u> $\mathfrak{S}$ <u>belongs to</u> C <u>and</u> $\tilde{\mathfrak{S}}$ <u>is the complexification of a distinguished subspace of</u> $\mathfrak{h}$, <u>then, if</u> $M(\cdot)$ <u>is defined at</u> H, <u>the rank of</u> M(H) <u>is the same as the rank of</u> M.

If $M(\cdot)$ is defined at H and if $sH = tH$ for some s and t in $\Omega(\mathcal{E}, C)$ implies $sH' = tH'$ for all H' in $\Psi_{\mathcal{E}}$ then H is said to be a general point of $\Psi_{\mathcal{E}}$. Since the rank of $M(H)$ is never less than the rank of M it is enough to show that at a general point the rank of $M(H)$ is no greater than the rank of M. If $\mathfrak{a}$ belongs to ${}^*S^{(j)}$ and

$$F = \oplus_{\ell=1}^{m} F_\ell$$

belongs to

$$L(S(\mathfrak{a}), \mathcal{E}^{(j)}) = \oplus_{\ell=1}^{m_i} L(S(\mathfrak{a}), \mathcal{E}(V^{(j,\ell)}, W))$$

let

$$E(g, F, H) = \Sigma_{\ell=1}^{m_i} E(g, F_\ell, H)$$

If $F = \oplus F_s$ belongs to

$$\oplus_{s\in\Omega(\mathcal{E}, C)} L(S(\mathcal{E}_s), \mathcal{E}^{(j_s)})$$

and H belongs to $\Psi_{\mathcal{E}}$ let

$$E(g, F, H) = \Sigma_s E(g, F_s, ss^o H)$$

Suppose that H is a general point and for some such F the function $E(\cdot, F, H)$, which is defined, is zero. If m belongs to ${}^*M$ and k belongs to K then

$$\int_{{}^*_{\Theta} \cap \Psi_{N^{(j,\ell)}} \backslash \Psi_{N^{(j,\ell)}}} E(nmk, F, H)\,dn$$

is equal to

$$\Sigma_t \exp(\langle H^{(j,\ell)}(m), tH\rangle + \rho(H^{(j,\ell)}(m))) \left\{\Sigma_{s\in\Omega(\mathcal{E}, C)} \Phi_{t,s}^{(j,\ell)}(mk)\right\} \tag{7.o}$$

where the outer sum is over those t in $\Omega(\mathcal{E}, C)$ such that $j_t = j$ and $\Phi_{t,s}^{(j,\ell)}$ is the function on ${}^*AT^{(j,\ell)}\backslash G$ associated to the projection of $M(ts^o s^{-1}, ss^o H)F_s$ on

$$S(\mathfrak{S}_t) \otimes \mathcal{E}(V^{(j,\ell)}, W)$$

Since H is a general point it follows that

$$\Sigma_{s \in \Omega(\mathfrak{S}, C)} \Phi^{(j,\ell)}_{t,s}$$

is zero; consequently

$$\Sigma_{s \in \Omega(\mathfrak{S}, C)} M(ts^o s^{-1}, ss^o H) F_s = 0$$

for all t in $\Omega(\mathfrak{S}, C)$.

If the dimension of ${}^*\mathfrak{a}$ is m, the dimension of the elements of S, there is nothing to prove. We treat the case that the dimension of ${}^*\mathfrak{a}$ is $m-1$ first. Let H be a general point of ${}^{\Psi}\mathfrak{S}$ and suppose that $\operatorname{Im} H \neq 0$ and

$$\operatorname{Re} H = X(\mathfrak{S}) + H'$$

with $\alpha(H')$ small and positive if α is the unique simple root of ${}^{\Psi}\mathfrak{a}$. As usual $\tilde{\mathfrak{S}} = \mathfrak{a}_c$ and ${}^{\Psi}\mathfrak{a}$ is the orthogonal complement of ${}^*\mathfrak{a}$ in $\mathfrak{a}$. Let us show that if

$$F = \oplus_{s \in \Omega(\mathfrak{S}, C)} F_s$$

is such that

$$\Sigma_{s \in \Omega(\mathfrak{S}, C)} M(ts^o s^{-1}, ss^o H) F_s$$

is zero for t in $\Omega_0(\mathfrak{S}, C)$ then it is zero for all t. Lemma 7.3 implies that $E(\cdot, F, H)$ can not belong to ${}^*\mathcal{L}(\{P\}, \{V\}, W)$ and be different from zero so we show that $E(\cdot, F, H)$ belongs to ${}^*\mathcal{L}(\{P\}, \{V\}, W)$. In the expression (7.o) the sum can be replaced by a sum over the elements t of the complement of $\Omega_0(\mathfrak{S}, C)$ in $\Omega(\mathfrak{S}, C)$ such that $j_t = j$. The corollary to Lemma 5.1 can be applied if it is shown that, for all such t, $-\operatorname{Re}(tH)$ belongs to ${}^+({}^{\Psi}\mathfrak{a}^{(j)})$ provided $\alpha(H')$ is sufficiently small. Since $-\operatorname{Re}(tH)$ is close to $X(\mathfrak{S}_t)$ this is perfectly obvious if

${}^*\mathfrak{n}$ is the largest distinguished subspace contained in $\tilde{\mathfrak{G}}_t$. If it is not then $\mathfrak{G}_t$ is the complexification of a distinguished subspace $\mathfrak{n}'$ of $\mathfrak{n}^{(j)}$. If α' is the unique simple root of the orthogonal complement of ${}^*\mathfrak{n}$ in $\mathfrak{n}'$ then it follows from Lemma 2.13 that $\alpha'(tH')$ is negative. Lemma 2.5 implies that $-tH$ belongs to ${}^+({}^\Psi\mathfrak{n}^{(j)})$.

Since the set of points satisfying the condition of the previous paragraph is open it is enough to prove that the rank of $M(H)$ is no greater than the rank of M when H is in this set. Every element

$$G = \oplus_{t\in\Omega(\mathfrak{G},\, C)} G_t$$

in the range of $M(H)$ is of the form

$$G_t = \Sigma_{s\in\Omega(\mathfrak{G},\, C)} M(ts^o s^{-1},\ ss^o H) F_s$$

with F_s in $L(S(\mathfrak{G}_s),\ \mathfrak{E}^{(j_s)})$. The map

$$G \longrightarrow \oplus_{t\in\Omega_0(\mathfrak{G},\, C)} \oplus G_t$$

is an injection of the range of $M(H)$ into

$$\oplus_{t\in\Omega_0(\mathfrak{G},\, C)} \oplus S(\mathfrak{G}_t) \otimes \mathfrak{E}^{(j_t)}$$

It is sufficient to show that the image is contained in the range of M. If not there would be a set

$$\{F'_t \mid t \in \Omega_0(\mathfrak{G},\ C)\}$$

so that

$$\Sigma_{t\in\Omega_0(\mathfrak{G},\, C)} \Sigma_{s\in\Omega_0(\mathfrak{G},\, C)} (M(ts^o s^{-1},\ ss^o H) F_s,\ F'_t) = 0$$

for all sets $\{F_s \mid s \in \Omega_0(\mathfrak{E}, C)\}$ and

$$\Sigma_{t\in\Omega_0(\mathfrak{E},C)}\Sigma_{s\in\Omega(\mathfrak{E},C)}(M(ts^o s^{-1}, ss^o H)F_s, F'_t) \neq 0$$

for some set $\{F_s \mid s \in \Omega(\mathfrak{E}, C)\}$. However, the first relation is independent of H so, replacing H by $-s^o\overline{H}$ and using the relation

$$M(ts^o s^{-1}, -s\overline{H}) = M^*(ss_0 t^{-1}, ts_0 H)$$

we deduce that

$$\Sigma_{t\in\Omega_0(\mathfrak{E},C)} M(ss^o t^{-1}, ts^o H)F'_t = 0$$

for all s in $\Omega_0(\mathfrak{E}, C)$ and hence for all s and all H. But the complex conjugate of the expression on the left of the second relation is

$$\Sigma_{s\in\Omega(\mathfrak{E},C)}\left\{\Sigma_{t\in\Omega_0(\mathfrak{E},C)}(M(ss^o t^{-1}, ts^o(-s^o\overline{H}))F'_t, F_s)\right\}$$

and must be zero.

The general case will be treated by induction. Suppose that the dimension of ${}^*\mathfrak{n}$ is n with n less than m-1 and that the assertion of the lemma is valid if the dimension of ${}^*\mathfrak{n}$ is greater than n. Let $\Omega'(\mathfrak{E}, C)$ be the set of all s in $\Omega(\mathfrak{E}, C)$ such that $\tilde{\mathfrak{E}}$ contains a distinguished subspace which is larger than ${}^*\mathfrak{n}$ and let

$$M'(H) = (M(ts^o s^{-1}, ss^o H));\ s, t \in \Omega'(\mathfrak{E}, C)$$

We first show that the rank of M(H) is no larger than the rank of M'(H). It is enough to show this when H is a general point in $U({}^\Psi\mathfrak{E}, X(\mathfrak{E}), \infty)$ which is not real. The argument is then very much like the one just presented. Indeed if

$$F = \oplus_{s\in\Omega(\mathfrak{E},C)} F_s$$

and

$$\Sigma_{s\in\Omega(\mathcal{E}, C)} M(ts^{o}s^{-1}, ss^{o}H)F_{s} = 0$$

for all t in $\Omega'(\mathcal{E}, C)$ then $E(\cdot, F, H)$ is zero because $-\mathrm{Re}(sH) = \mathrm{Re}\,X(\mathcal{E}_s)$ lies in $^{+}({}^{\Psi}\mathfrak{a}^{(j)})$ if $\mathfrak{a}_s$ belongs to $^{*}S^{(j)}$ and s does not belong to $\Omega'(\mathcal{E}, C)$. Consequently this equality is valid for all t. As before the restriction of the map

$$\oplus_{s\in\Omega(\mathcal{E}, C)} G_s \longrightarrow \oplus_{s\in\Omega'(\mathcal{E}, C)} G_s$$

to the range of $M(H)$ can be shown to be an injection into the range of $M'(H)$. It remains to show that the rank of $M'(H)$ is no larger than the rank of M.

Suppose $^{*}P_1$ is a cuspidal subgroup with split component $^{*}\mathfrak{a}_1$ belonging to an element of $\{P\}$. Suppose also that $^{*}P$ belongs to $^{*}P_1$ and that $^{*}\mathfrak{a}$ is properly contained in $^{*}\mathfrak{a}_1$. For each i, $1 \leq i \leq r$, $^{*}\{P\}_1^{(i)}$ is a subset of $^{*}\{P\}^{(i)}$. Let $P_1^{(i,k)}$, $1 \leq k \leq m_i$, be a complete set of representatives for the conjugacy classes in $^{*}\{P\}_1^{(i)}$. It may as well be supposed that $P_1^{(i,k)}$ is conjugate to $P^{(i,k)}$, $1 \leq k \leq m_i'$. The elements of C which belong to $^{*}S_1$ break up into a number of equivalence classes $C_1, \ldots, C_u$. In each C_x, $1 \leq x \leq u$, choose an $\mathcal{E}_x$ so that $\tilde{\mathcal{E}}_x$ is a distinguished subspace of $\mathfrak{f}$. For each x fix s_x in $\Omega(\mathcal{E}, \mathcal{E}_x)$ and let $\Omega(\mathcal{E}, C_x)$ be the set of all s in $\Omega(\mathcal{E}, C)$ such that $ss^{o}s_x^{-1}$ belongs to $\Omega(s_x, C_x)$ and let $\Omega_0(\mathcal{E}, C_x)$ be the set of all s such that $ss^{o}s_x^{-1}$ belongs to $\Omega_0(\mathcal{E}_x, C_x)$. The induction assumption will be used to show that if F_s, $s \in \Omega(\mathcal{E}, C_x)$, belongs to

$$\oplus_{\ell=1}^{m'_{j_s}} L(S(\mathcal{E}_s), \mathcal{E}(V^{(j_s,\ell)}, W))$$

if H is a general point of $^{\Psi}\mathcal{E}$, and if

$$\Sigma_{s\in\Omega(\mathcal{E}, C_x)} M(ts^{o}s^{-1}, ss^{o}H)F_{s} = 0$$

for all t in $\Omega_0(\mathfrak{S}, C_x)$ then this relation is valid for all t in $\Omega(\mathfrak{S}, C)$. It is sufficient to establish this when $s_x s^o H$ belongs to the intersection of $^{\Psi}\mathfrak{S}$ and

$$\bigcup_{{}^*\mathfrak{a}_1 \subseteq \mathfrak{a}^{(j)}} \bigcup_{s \in {}^{\Psi}\Omega_1(\mathfrak{a}^{(i)}, \mathfrak{a}^{(j)})} s^{-1}({}^{\Psi}\mathfrak{U}^{(j)})$$

where i is such that $\mathfrak{S}_x \in {}^*S_1^{(i)}$. If

$$F_s = \oplus_{\ell=1}^{m_{j_s}^!} F_s^{\ell}, \quad t = ss^o s_x^{-1}$$

and $D(H)$ is the linear transformation from $\mathcal{E}(V^{(j_s, \ell)}, W)$ to $\mathcal{E}(V_1^{(j_s, \ell)}, W)$ defined in Section 4 let $G_t^{\ell} = dD(H)F_s^{\ell}$ and let

$$G_t = \oplus_{\ell=1}^{m_{j_s}^!} G_t^{\ell}$$

The relation (7.e) and the last part of condition (iii) for an Eisenstein system imply that

$$\Sigma_{s \in \Omega(\mathfrak{S}_x, C_x)} M_1(ts_x^o s^{-1}, ss_x^o H_x) G_s = 0$$

for all t in $\Omega_0(\mathfrak{S}_x, C_x)$ if H_x is the projection of

$$s_x^o s_x H = s_x s^o H$$

on the orthogonal complement of ${}^*\mathfrak{a}_1$. According to the induction assumption the relation must then be valid for all t in $\Omega(\mathfrak{S}_x, C_x)$. Consequently

$$\Sigma_{s \in \Omega(\mathfrak{S}_x, C_x)} E_1(g, G_s, ss_x^o H_x) = 0$$

The relations (7.c) and (7.d) imply that

$$\Sigma_{s \in \Omega(\mathfrak{S}, C_x)} E(g, F_s, ss^o H) = \Sigma_{s \in \Omega(\mathfrak{S}_x, C_x)} E(g, G_s, ss_x H) = 0$$

We obtain the assertion by appealing to the remarks made when we started the proof.

Suppose that for each s in $\Omega(\mathfrak{a}, C_x)$ we are given F_s in

$$\oplus_{k=1}^{m'_{j_s}} L(S(\mathfrak{a}_s), \mathcal{E}(V^{(j_s, \ell)}, W))$$

It will also be necessary to know that we can find for each s in $\Omega_0(\mathfrak{a}, C_x)$ an element F'_s of

$$\Sigma_{k=1}^{m'_{j_s}} L(S(\mathfrak{a}_s), \mathcal{E}(V^{(j_s, \ell)}, W))$$

so that

$$\Sigma_{s \in \Omega(\mathfrak{a}, C_x)} M(ts^{o}s^{-1}, ss^{o}H)F_s = \Sigma_{s \in \Omega_0(\mathfrak{a}, C_x)} M(ts^{o}s^{-1}, ss^{o}H)F'_s$$

for all t in $\Omega(\mathfrak{a}, C)$. If G_s is defined for s in $\Omega(\mathfrak{a}_x, C_x)$ as before the induction assumption guarantees the existence of a set

$$\{G'_s \mid s \in \Omega_0(\mathfrak{a}_x, C_x)\}$$

so that

$$\Sigma_{s \in \Omega(\mathfrak{a}_x, C_x)} M_1(ts_x^{o}s^{-1}, ss_x^{o}H_x)G_s = \Sigma_{s \in \Omega_0(\mathfrak{a}_x, C_x)} M_1(ts_x^{o}s^{-1}, ss_x^{o}H_x)G'_s$$

for all t in $\Omega(\mathfrak{a}_x, C_x)$. We need only choose a set

$$\{F'_s \mid s \in \Omega_0(\mathfrak{a}, C_x)\}$$

which is related to $\{G'_s\}$ the way $\{F_s\}$ is related to $\{G_s\}$.

Let

$$M'_0(H) = (M(ts^{o}s^{-1}, ss^{o}H)), \quad s \in \Omega_0(\mathfrak{a}, C), \; t \in \Omega'(\mathfrak{a}, C)$$

Choosing ${}^*\mathfrak{a}_1$ so that its complexification is $\tilde{\mathfrak{s}}$ we see that the ranks of $M_0'(H)$ and M are the same. It will now be shown that the range of $M'(H)$ is contained in the range of $M_0'(H)$ and this will complete the proof of the lemma. Suppose that $\mathfrak{t}$ in ${}^*S^{(j)}$ belongs to C, that F belongs to $L(S(\mathfrak{t}), \mathcal{E}(V^{(j,\ell)}, W))$ for some ℓ, $1 \leq \ell \leq m_j$, and that there is an r in $\Omega'(\mathfrak{s}, C)$ with $\mathfrak{s}_r = \mathfrak{t}$. Let us show that

$$\Sigma_{t\in\Omega'(\mathfrak{s},C)} M(ts^{o}r^{-1}, rs^{o}H)F$$

belongs to the range of $M_0'(H)$. Choose ${}^*\mathfrak{a}_1$ so that $\tilde{\mathfrak{t}}$ contains ${}^*\mathfrak{a}_1$ and choose ${}^*P_1$ so that ${}^*P_1$ belongs to $P^{(j,\ell)}$. If t belongs to C_x then we can choose for each s in $\Omega_0(\mathfrak{s}, C_x)$ an element F_s of

$$\oplus_{k=1}^{m'_{j_s}} L(S(\mathfrak{s}_s), \mathcal{E}(V^{(j_s,k)}, W))$$

so that

$$\Sigma_{s\in\Omega_0(\mathfrak{s},C_x)} M(ts^{o}s^{-1}, ss^{o}H)F_s = M(ts^{o}r^{-1}, rs^{o}H)F$$

for all t. We may as well assume then that $\tilde{\mathfrak{t}}$ is the complexification of a distinguished subspace of $\mathfrak{h}$.

Since the lemma is true for $n = m-1$ the set of $\mathfrak{t}$ in C such that $\tilde{\mathfrak{t}}$ is the complexification of a distinguished subspace satisfies the hypothesis of the second corollary to Lemma 2.13. The assertion will be proved by induction on the length of r. Suppose that $\mathfrak{t}'$ is another element of C such that $\tilde{\mathfrak{t}}'$ is the complexification of a distinguished subspace and suppose that $r = pt^{o}r'$ where r' belongs to $\Omega(\mathfrak{s}, \mathfrak{t}')$ and has length one less than that of r, t^{o} belongs to $\Omega(\mathfrak{t}', \mathfrak{t}')$ and leaves every element of $\tilde{\mathfrak{t}}'$ fixed, and p is a reflection in $\Omega(\mathfrak{t}', \mathfrak{t})$. Choose ${}^*\mathfrak{a}_1$ so that ${}^*\mathfrak{a}_1$ is of dimension $m-1$, is contained in $\tilde{\mathfrak{t}}$ and $\tilde{\mathfrak{t}}'$, and is such that p leaves each element of ${}^*\mathfrak{a}_1$ fixed and let ${}^*P_1$

belong to $P^{(j,\ell)}$. There is an x so that $\mathcal{D}$ and $\mathcal{D}'$ both belong to C_x. Let $\mathcal{C}_x = \mathcal{D}'$ and let $s_x = r'$. It has been shown that for each s in $\Omega_0(\mathcal{C}, C_x)$ we can choose F_s in

$$\oplus_{k=1}^{m'_{j_s}} L(S(\mathcal{C}_s), \mathcal{E}(V^{(j_s,k)}, W))$$

so that

$$\Sigma_{s\in\Omega_0(\mathcal{C}, C_x)} M(ts^{o}s^{-1}, ss^{o}H)F_s = M(ts^{o}r'^{-1}, r's^{o}H)F$$

for all $\mathcal{D}$. Since the length of each s in $\Omega_0(\mathcal{C}, C_x)$ is the same as that of r' the proof may be completed by applying the induction assumption.

COROLLARY 1. <u>Suppose the collections</u> $S^{(i)}$, $1 \leq i \leq r$, <u>and the associated Eisenstein systems satisfy the conditions of Lemmas</u> 7.2 <u>and</u> 7.3. <u>Suppose moreover that if</u> ${}^*P$ <u>is a cuspidal subgroup belonging to an element of</u> $\{P\}$, <u>if</u> $\mathfrak{a}^{(i)}$, $1 \leq i \leq r'$, <u>are the elements of</u> $\{\mathfrak{a}^{(i)} \mid 1 \leq i \leq r\}$ <u>which contains</u> ${}^*\mathfrak{a}$, <u>the split component of</u> ${}^*P$, <u>and if, for</u> $1 \leq i \leq r'$, p_i <u>is a polynomial on</u> ${}^{\Psi}\mathfrak{a}^{(i)}$, <u>the orthogonal complement of</u> ${}^*\mathfrak{a}$ <u>in</u> $\mathfrak{a}^{(i)}$, <u>and</u> $p_j(sH) = p_i(H)$ <u>for all</u> H <u>in</u> ${}^{\Psi}\mathfrak{a}^{(i)}$ <u>and all</u> s <u>in</u> ${}^{\Psi}\Omega(\mathfrak{a}^{(i)}, \mathfrak{a}^{(j)})$ <u>then for any</u> $\mathcal{C}$ <u>in</u> ${}^*S^{(i)}$, <u>any</u> P <u>in</u> ${}^*\{P\}^{(i)}$, <u>any</u> F <u>in</u> $L(S(\mathcal{C}), \mathcal{E}(V, W))$, <u>any</u> $\mathcal{D}$ <u>in</u> ${}^*S^{(j)}$, <u>any</u> P' <u>in</u> ${}^*\{P\}^{(j)}$, <u>any</u> F' <u>in</u> $L(S(\mathcal{D}), \mathcal{E}(V, W))$, <u>and any</u> s <u>in</u> $\Omega(\mathcal{C}, \mathcal{D})$

$$(N(s, H)dp_i(H)F, F') \equiv (N(s, H)F, dp_j^*(-s\bar{H})F')$$

<u>Then for any</u> ${}^*P$, $p_1(\cdot), \ldots, p_{r'}(\cdot)$, $\mathcal{C}$, P, F, $\mathcal{D}$, P', <u>and</u> s <u>as above</u>

$$E(g, dp_i(H)F, H) \equiv p_i(H)E(g, F, H)$$

<u>and</u>

$$(N(s, H)(dp_i(H)F) \equiv p_i(H)N(s, H)F$$

Of course the equalities above are not valid for literally all H in ${}^{\Psi}\mathcal{C}$;

rather the two sides are equal as meromorphic functions. It is enough to prove the equalities when H is a general point of ${}^{\Psi}\mathfrak{S}$. Since the two equalities are then equivalent it is only necessary to prove one of them. Suppose first of all that $\tilde{\mathfrak{S}}$ is the complexification of ${}^{*}\mathfrak{a}$. It was seen in the proof of Lemma 7.2 that if $H = X(\mathfrak{S})$ then $E(\cdot, F, H)$ belongs to ${}^{*}\mathfrak{L}(\{P\}, \{V\}, W)$. If ψ belongs to ${}^{*}\mathfrak{J}(V', W)$ then

$$\int_{{}^{*}\Theta\backslash {}^{*}M\times K} E(mk, F, H)\bar{\psi}^{\wedge}(mk)\,dm\,dk$$

is equal to

$$\sum_{\mathfrak{t}\in {}^{*}S^{(j)}} \sum_{s\in {}^{\Psi}\Omega(\mathfrak{S}, \mathfrak{t})} ((N(s, H)F, d\Psi(-s\bar{H}))$$

According to the remarks following the proof of Lemma 7.3

$$(N(s, H)F, d(p_j^{*}\Psi(-s\bar{H})) = p_i(H)(N(s, H)F, d\Psi(-s\bar{H}))$$

for all s. Thus for all F' in $\mathfrak{L}(S(\mathfrak{t}), E(\mathcal{V}, W))$

$$(N(s, H)dp_i(H)F, F') = (N(s, H)F, dp_j^{*}(-s\bar{H})F') = p_i(H)(N(s, H)F, F')$$

This proves the second equality in this case. Next suppose that $\tilde{\mathfrak{S}}$ is the complexification of a distinguished subspace of $\mathfrak{a}^{(i)}$. It follows from the relation (7.e) that the first equality is valid on an open set and hence on all of ${}^{\Psi}\mathfrak{S}$.

In the general case we prove the second equality. Because of the relation (7.e) it is enough to show that if C is an equivalence class in ${}^{*}S$ and if an $\mathfrak{S}$ in C such that $\tilde{\mathfrak{S}}$ is the complexification of a distinguished subspace of $\mathfrak{f}$ is chosen then for all s and t in $\Omega(\mathfrak{S}, C)$ and all F in $\mathcal{E}^{(j_s)}$

$$M(ts^{o}s^{-1}, ss^{o}H)(dp_{j_s}(ss_0H)F) = p_{j_s}(ss_0H)M(ts^{o}s^{-1}, ss^{o}H)F$$

It follows from Lemma 7.4 that if for a given s and F this relation is valid for all t in $\Omega_0(\mathfrak{S}, C)$ then it is true for all t in $\Omega(\mathfrak{S}, C)$. It has just been proved that it is valid for s in $\Omega_0(\mathfrak{S}, C)$ and t in $\Omega(\mathfrak{S}, C)$ and it remains to prove that it is valid for s in $\Omega(\mathfrak{S}, C)$ and t in $\Omega_0(\mathfrak{S}, C)$. Take such an s and t and let F belong to $\mathcal{E}^{(j_s)}$ and F' to $\mathcal{E}^{(j_t)}$; then

$$(M(ts^o s^{-1}, ss^o H)(dp_{j_s}(ss^o H)F), F') = (M(ts^o s^{-1}, ss^o H)F, dp^*_{j_t}(-t\overline{H})F')$$

which is the complex conjugate of

$$(M(ss^o t^{-1}, -t\overline{H})(dp^*_{j_t}(-t\overline{H})F'), F) = p^*_{j_t}(-t\overline{H})(M(ss^o t^{-1}, -t\overline{H})F', F)$$

Since the complex conjugate of the right hand side is

$$p_{j_s}(ss^o H)(M(ts^o s^{-1}, ss^o H)F, F')$$

we are done.

The next corollary can be obtained by an argument essentially the same as the one just given. Since it is of no great importance the proof will be omitted.

COROLLARY 2. <u>If the collections</u> $S^{(i)}$, $1 \le i \le r$, <u>and the associated Eisenstein systems satisfy the hypotheses of Lemmas</u> 7.2 <u>and</u> 7.4, <u>they are uniquely determined if for every cuspidal subgroup</u> $^*P$ <u>of rank</u> m <u>belonging to some element of</u> $\{P\}$ <u>the sets</u> $^*S^{(i)}$, $1 \le i \le r$ <u>are given and if for every</u> $\mathfrak{S}$ <u>in</u> $^*S^{(i)}$, <u>every</u> P <u>in</u> $^*\{P\}^{(i)}$, <u>and every</u> F <u>in</u> $L(S(\mathfrak{S}), \mathcal{E}(V, W))$ <u>the function</u> $E(\cdot, X(\mathfrak{S}), F)$ <u>is given</u>.

It is now necessary to find some conditions on the collections $S^{(i)}$, $1 \le i \le r$, and the associated Eisenstein systems which imply the hypotheses of Lemmas 7.2 and 7.4 and the first corollary but which are, at least in our context, easier to verify. It must be expected that they will be rather technical. For convenience, if $^*P$ and $^{\Psi}\mathfrak{V}^{(i)}$, $1 \le i \le r'$, are as in the first corollary we will denote the collection of r'-tuples $p(\cdot) = (p_1(\cdot), \ldots, p_{r'}(\cdot))$ satisfying the conditions of that corollary by ${}^{\Psi}\mathcal{J}$, the collection of r'-tuples $f(\cdot) = \{f_1(\cdot), \ldots, f_{r'}(\cdot)\}$ where $f_i(\cdot)$ is a bounded analytic function on $\{H \in {}^{\Psi}\mathfrak{a}^{(i)} \mid \|\operatorname{Re} H\| < R\}$ and $f_i(H) = f_j(sH)$ if s belongs to ${}^{\Psi}\Omega(\mathfrak{a}^{(i)}, \mathfrak{a}^{(j)})$ will be denoted by ${}^{\Psi}\mathcal{J}_0$. The number R has been introduced in Section 4.

LEMMA 7.5. *Suppose that* $S^{(i)}$, $1 \le i \le r$, *is a collection of distinct affine subspaces of* $\mathfrak{a}^{(i)}$ *which are of dimension* m *and which are defined by equations of the form* $\alpha(H) = \mu$ *where* α *is a positive root of* $\mathfrak{a}^{(i)}$ *and* μ *is a complex number and suppose that there is an Eisenstein system associated to each element of* $S = \bigcup_{i=1}^{r} S^{(i)}$. *Suppose that if* $\mathfrak{s}$ *belongs to* $S^{(i)}$ *and* $\mathfrak{a}$ *is the orthogonal complement in* $\mathfrak{a}^{(i)}$ *of the distinguished subspace of largest dimension which is contained in* $\tilde{\mathfrak{s}}$ *then* $\operatorname{Re} X(\mathfrak{s})$ *belongs to* ${}^{+}\mathfrak{a}$ *and* $X(\mathfrak{s})$ *lies in* D_i. *Suppose also that only a finite number of the elements of* $S^{(i)}$ *meet any compact subset of* $\mathfrak{a}_c^{(i)}$. *Finally suppose that if* ${}^{*}P$ *is a cuspidal subgroup with split component* ${}^{*}\mathfrak{a}$ *belonging to an element of* $\{P\}$ *and if* $\mathfrak{a}^{(i)}$, $1 \le i \le r'$, *are the elements of* $\{\mathfrak{a}^{(i)} \mid 1 \le i \le r\}$ *which contain* ${}^{*}\mathfrak{a}$ *then there is an orthogonal projection* Q, *which commutes with* $\lambda(f)$ *if* $f(\cdot)$ *belongs to* ${}^{\Psi}\mathcal{J}_0$, *of* ${}^{*}\mathcal{L}(\{P\}, \{V\}, W)$ *onto a subspace and for every positive number* a *and each* i *a polynomial* r_i *on* ${}^{\Psi}\mathfrak{a}^{(i)}$ *which does not vanish identically on* ${}^{\Psi}\mathfrak{s}$ *if* $\mathfrak{s}$ *belongs to* ${}^{*}S^{(i)}$ *so that if* P *belongs to* ${}^{*}\{P\}^{(i)}$, P' *belongs to* ${}^{*}\{P\}^{(j)}$, $\Phi'(\cdot)$ *belongs to* ${}^{*}\mathcal{H}(V, W)$,

$$\Phi(\cdot) = r_i(\cdot)\Phi'(\cdot)$$

and $\Psi(\cdot)$ *belongs to* ${}^{*}\mathcal{H}(V', W)$ *then the difference* (7.p) *between*

$$(R(\lambda, A)Q\phi^{\wedge}, \psi^{\wedge})$$

and

$$\Sigma_{\mathfrak{s} \in {}^{*}S^{(i)}} \Sigma_{s \in {}^{\Psi}\Omega^{(j)}(\mathfrak{s})} \frac{1}{(2\pi i)^m} \int_{U({}^{\Psi}\mathfrak{s}, X(\mathfrak{s}), a)} (N(s, H)d((\lambda - \langle H, H \rangle)^{-1}\Phi(H)), d\Psi(-s\overline{H}))dH$$

as well as the difference (7.q) between

$$(Q\phi^\wedge,\ R(\bar{\lambda},\ A)\psi^\wedge)$$

and

$$\Sigma_{\mathcal{S}\in {}^*S^{(i)}} \Sigma_{s\in {}^\Psi\Omega^{(j)}(\mathcal{S})} \frac{1}{(2\pi i)^m} \int_{U({}^\Psi\mathcal{S}, X(\mathcal{S}), a)} (N(s,H)d\Phi(H), d((\lambda - \langle -s\bar{H}, -s\bar{H}\rangle)^{-1}\Psi(-s\bar{H})))dH$$

are analytic for $\operatorname{Re}\lambda > R^2 - a^2$. Then for every $\mathcal{S}$ in S the set $\Omega(\mathcal{S}, \mathcal{S})$ contains an element which leaves each point of $\tilde{\mathcal{S}}$ fixed. Moreover if ${}^*P$ is a cuspidal subgroup with split component ${}^*\mathfrak{a}$, P belongs to ${}^*\{P\}^{(i)}$, P' belongs to ${}^*\{P\}^{(j)}$, $\mathcal{S}$ belongs to ${}^*S^{(i)}$, and s belongs to ${}^\Psi\Omega^{(j)}(\mathcal{S})$ then N(s, H) vanishes identically unless $\mathcal{S}_s = \mathcal{T}$ for some $\mathcal{T}$ in ${}^*S^{(j)}$ and then

$$N(s,\ H) = N^*(s^{-1},\ -s\bar{H})$$

Finally, if F belongs to $L(S(\mathcal{S}), \mathcal{E}(V, W))$, F' belongs to $L(S(\mathcal{T}), \mathcal{E}(V', W))$, and p(·) belongs to ${}^\Psi\mathfrak{I}$ then

$$(N(s,\ H)dp_i(H)F,\ F') \equiv (N(s,\ H)F,\ dp_j^*(-s\bar{H})F')$$

There is one simple assertion which is central to the proof of this lemma. We first establish it. Let a be a positive number, let ${}^*P$ be a cuspidal subgroup belonging to some element of {P}, let $\mathcal{S}$ belong to ${}^*S^{(i)}$, let P belong to ${}^*\{P\}^{(i)}$, let P' belong to ${}^*\{P\}^{(j)}$ and suppose that for each $\mathcal{S}$ in ${}^\Psi\Omega^{(j)}(\mathcal{S})$ there is a given a function M(s, H) on ${}^\Psi\mathcal{S}$ with values in the space of linear transformations from $L(S(\mathcal{S}), \mathcal{E}(V, W))$ to $S(\mathcal{S}_s) \otimes \mathcal{E}(V', W)$ such that (M(s, H)F, F') is meromorphic on ${}^\Psi\mathcal{S}$ for all F and F' and vanishes identically if the order of F or of F' is sufficiently large. Suppose that if

$$\Phi(\cdot) = r_i(\cdot)\Phi'(\cdot)$$

with $\Phi'(\cdot)$ in ${}^*\mathfrak{H}(V, W)$ and $\Psi(\cdot)$ belongs to ${}^*\mathfrak{H}(V', W)$ then

$$(M(s, H)d\Phi(H), d\Psi(-s\overline{H}))$$

is analytic on the closure of $U({}^{\Psi}\mathfrak{C}, X(\mathfrak{C}), a)$ for all s and

$$\Sigma_{s\epsilon^{\Psi}\Omega^{(j)}(\mathfrak{C})} \frac{1}{(2\pi i)^m} \int_{U({}^{\Psi}\mathfrak{C}, X(\mathfrak{C}), a)} (M(s,H)d((\lambda - \langle H, H\rangle)^{-1}\Phi(H)), d\Psi(-s\overline{H}))dH$$

is analytic for $\operatorname{Re}\lambda > R^2 - a^2$; then if

$$\operatorname{Re}\langle X(\mathfrak{C}), X(\mathfrak{C})\rangle > R^2 - a^2$$

each of the functions $M(s, \cdot)$ is identically zero. Suppose not and suppose that $M(s, H)F$ vanishes identically for all s if the order of F is greater than n but that for some s and some $F = F_0$ of order n the function $M(s, H)F_0$ does not vanish identically. There are polynomials h_k, $1 \le k \le \ell$ on ${}^{\Psi}\mathfrak{a}^{(i)}$ and functions Φ_k, $1 \le k \le \ell$, in $\mathfrak{E}(V, W)$ so that the order of

$$F_0 - d(\Sigma_k h_k(H)\Phi_k)$$

is greater than n. Let

$$\Phi'(\cdot) = f(\cdot)\left\{\Sigma_{k=1}^{\ell} h_k(\cdot)\Phi_k\right\}$$

with some scalar valued function $f(\cdot)$; then

$$(M(s,H)d((\lambda - \langle H, H\rangle)^{-1}\Phi(H), d\Psi(-s\overline{H})) = (\lambda - \langle H, H\rangle)^{-1} f(H)r_i(H)(M(s,H)F_0, d\Psi(-s\overline{H}))$$

Let

$$g(H) = \Sigma_{s\epsilon^{\Psi}\Omega^{(j)}(\mathfrak{C})} r_i(H)(M(s, H)F_0, d\Psi(-s\overline{H}))$$

Then

$$\int_{U({}^{\Psi}\mathfrak{S},\, X(\mathfrak{S}),\, a)} (\lambda - \langle H, H\rangle)^{-1} f(H)g(H)dH$$

is analytic for $\operatorname{Re}\lambda > R^2 - a^2$. Let B be the unit sphere in $\tilde{\mathfrak{S}} \cap {}^{\Psi}\mathfrak{v}^{(i)}$ and let dB be the volume element on B. Set

$$\xi(r) = \int_B f(X(\mathfrak{S}) + ir^{\frac{1}{2}}H)g(X(\mathfrak{S}) + ir^{\frac{1}{2}}H)dB$$

If $\langle \operatorname{Re} X(\mathfrak{S}), \operatorname{Re} X(\mathfrak{S})\rangle = \mu$ and $\langle \operatorname{Im} X(\mathfrak{S}), \operatorname{Im} X(\mathfrak{S})\rangle = \nu$ then

$$\zeta(\lambda) = \int_0^{a^2-\nu} (\lambda + r)^{-1}\xi(r)r^{n/2-1}dr$$

is analytic for

$$\operatorname{Re}\lambda > R^2 - a^2 - \mu + \nu$$

and the right side is negative. On the other hand if $0 < \varepsilon < a^2 - \nu$

$$\lim_{\delta \searrow 0} \frac{1}{2\pi i}\{\zeta(-\varepsilon+i\delta) - \zeta(-\varepsilon-i\delta)\} = \xi(\varepsilon)\varepsilon^{n/2-1}$$

so that

$$\xi(r) = 0$$

for $0 < r < \mu + a^2 - R^2 - \nu$ and hence for all r. Since $f(H)$ can be taken to be the product of $\exp\langle H, H\rangle$ and any polynomial we conclude that $g(H)$ vanishes identically. A simple approximation argument which has been used implicitly before allows us to conclude that

$$M(s, H)F_0 = 0$$

for all s and this is a contradiction.

Let ${}^*P$ be a cuspidal subgroup belonging to some element of $\{P\}$, let P belong to ${}^*\{P\}^{(i)}$, let P' belong to ${}^*\{P\}^{(j)}$, and let $\mathfrak{S}$ belong to ${}^*S^{(i)}$. Let

$q(\cdot)$ be a polynomial on $\mathfrak{a}^{(i)}$ which vanishes to such a high order on every element $\mathfrak{a}$ of ${}^*S^{(j)}$ different from $\mathfrak{b}$ itself that if $\mathfrak{a}$ belongs to ${}^{\psi}\Omega^{(j)}(\mathfrak{a})$ then $N(t, H) \circ dq(H)$ vanishes identically on ${}^{\psi}\mathfrak{a}$ and to such a high order on every space $\mathfrak{a}_{t}$, with $\mathfrak{a}$ in ${}^*S^{(j)}$ and t in ${}^{\psi}\Omega^{(j)}(\mathfrak{a})$, different from $\mathfrak{b}$ itself that $d^*q(-t\overline{H})N(t, H)$ vanishes identically on ${}^{\psi}\mathfrak{a}$ but which does not vanish identically on $\mathfrak{b}$ itself. Of course $d^*q(H)$ is defined by the condition that

$$(d^*q(H)T, F) = (T, dq(H)F)$$

for all T in $S(\mathfrak{a}_t) \otimes \mathcal{E}(V, W)$ and all F in $L(S(\mathfrak{a}_t), \mathcal{E}(V, W))$. In (7.p) replace $\Phi(\cdot)$ by $q(\cdot)\Phi(\cdot)$ and let

$$\Psi(\cdot) = r_j(\cdot)\Psi'(\cdot)$$

and in (7.q) replace j by i, λ by $\overline{\lambda}$, $\Phi(\cdot)$ by

$$\Psi(\cdot) = r_j(\cdot)\Psi'(\cdot)$$

and $\Psi(\cdot)$ by $q(\cdot)\Phi(\cdot)$; then subtract the complex conjugate of (7.q) from (7.p). Since the complex conjugate of $(Q\psi^\wedge, R(\lambda, A)\phi^\wedge)$ is

$$(R(\lambda, A)Q\phi^\wedge, \psi^\wedge)$$

the result is

$$\Sigma_{s \in {}^{\psi}\Omega^{(j)}(\mathfrak{b})} \frac{1}{(2\pi i)^m} \int_{U({}^{\psi}\mathfrak{b}, X(\mathfrak{b}), a)} (M(s, H)d((\lambda - \langle H, H\rangle)^{-1}\Phi(H), \Psi(-s\overline{H}))dH$$

where $M(s, H)$ equals

$$d^*r_j(-s\overline{H})N(s, H)dq(H)$$

if $\mathfrak{b}_s$ does not belong to ${}^*S^{(j)}$ and equals

$$d^*r_j(-s\overline{H})\{N(s, H) - N^*(s^{-1}, -s\overline{H})\}dq(H)$$

if $\mathfrak{S}_s$ does belong to ${}^*S^{(j)}$. Since a can be taken as large as necessary we conclude that

$$N(s, H) \equiv 0$$

if $\mathfrak{S}_s$ does not belong to ${}^*S^{(j)}$ and that

$$N(s, H) \equiv N^*(s^{-1}, -s\overline{H})$$

if $\mathfrak{S}_s$ does belong to ${}^*S^{(j)}$. If $f(\cdot)$ belongs to ${}^\Psi\mathcal{J}_0$ then

$$(R(\lambda, A)Q\lambda(f)\phi^\wedge, \psi^\wedge) = (R(\lambda, A)Q\phi^\wedge, \lambda(f^*)\psi^\wedge)$$

Thus we can also conclude that if

$$M(s, H) = \{N(s, H)df(H) - d^*(f^*(-s\overline{H}))N(s, H)\}dq(H)$$

then

$$\Sigma_{s \in {}^\Psi\Omega^{(j)}(\mathfrak{S})} \frac{1}{(2\pi i)^m} \int_{U({}^\Psi\mathfrak{S}, X(\mathfrak{S}), a)} (M(s, H)d((\lambda - \langle H, H \rangle)^{-1}\Phi(H)), \Psi(-s\overline{H}))dH$$

is analytic for $\mathrm{Re}\,\lambda > R^2 - a^2$. Consequently

$$N(s, H)df(H) = d^*(f^*(-s\overline{H}))N(s, H)$$

for $f(\cdot)$ in ${}^\Psi\mathcal{J}_0$ and hence, by a simple approximation argument, for $f(\cdot)$ in ${}^\Psi\mathcal{J}$. The first assertion of the lemma has still to be proved.

Suppose $\mathfrak{S}$ belongs to ${}^*S^{(i)}$. Let ${}^*P$ be a cuspidal subgroup belonging to some element of $\{P\}$ such that $E(\cdot, \cdot, \cdot)$ is not identically zero for some P in ${}^*\{P\}^{(i)}$. Suppose that $E(\cdot, F, \cdot) \equiv 0$ if $O(F) > n$ but $E(\cdot, F, \cdot) \not\equiv 0$ for some F with $O(F) = n$. Let $h(\cdot)$ be a polynomial on $\mathfrak{n}_c^{(i)}$ such that

$$h(H - X(s)) = p_X(H)$$

where X lies in $S(\mathfrak{S})$ and is homogeneous of degree n, and such that

$$E(\cdot,\ dh(H)F,\ \cdot) \not\equiv 0$$

for some F in $L(S(\mathfrak{S}),\ \mathcal{E}(V,\ W))$. We first show that if we take $P' = P$ then for some $\mathfrak{S}$ in ${}^{\Psi}\Omega(\mathfrak{S},\ \mathfrak{S})$ the function

$$d^*(h^*(-s\overline{H}))N(s,\ H)dh(H) \not\equiv 0$$

Suppose the contrary. Fix some positive number a with

$$\langle \operatorname{Re} X(\mathfrak{S}),\ \operatorname{Re} X(\mathfrak{S}) \rangle > R^2 - a^2$$

Choose $q(\cdot)$ as above; let $\Phi'(\cdot)$ belong to ${}^*\mathfrak{H}(V,\ W)$; and set

$$\Phi(\cdot) = r_i(\cdot)q(\cdot)h(\cdot)\Phi'(\cdot)$$

Replacing $\Psi(\cdot)$ by $\Phi(\cdot)$ in (7.q) we obtain $(R(\lambda,\ A)Q\phi^\wedge,\ Q\phi^\wedge)$, which must be analytic for $\operatorname{Re}\lambda > R^2 - a^2$. It follows from Theorem 5.10 of [21] that $Q\phi^\wedge$ belongs to the range of $E(R^2 - a^2)$. However, if this is so then for any P' in ${}^*\{P\}^{(j)}$ and any $\Psi(\cdot)$ in ${}^*\mathfrak{H}(V',\ W)$ the function $(R(\lambda,\ A)Q\phi^\wedge,\ \psi^\wedge)$ is analytic for $\operatorname{Re}\lambda > R^2 - a^2$; consequently

$$\sum_{s \in {}^{\Psi}\Omega^{(i)}(\mathfrak{S})} \frac{1}{(2\pi i)^m} \int_{U({}^{\Psi}\mathfrak{S},\, X(\mathfrak{S}),\, a)} (N(s, H)d((\lambda - \langle H, H\rangle)^{-1}\Phi(H)),\ d\Psi(-s\overline{H}))dH$$

is analytic for $\operatorname{Re}\lambda > R^2 - a^2$. Thus

$$N(s,\ H)dh(H) \equiv 0$$

for all s which is impossible. In particular there is some s in $\Omega(\mathfrak{S},\ \mathfrak{S})$. For such an s,

$$sX(\mathfrak{S}) = -\overline{X}(\mathfrak{S})$$

consequently $\langle X(\mathfrak{S}),\ X(\mathfrak{S}) \rangle$ is real. Choose Φ in $\mathcal{E}(V,\ W)$ so that

$$d^*(h^*(-s\overline{H}))N(s, H)d(h(H)\Phi) \not\equiv 0$$

for some s in ${}^{\Psi}\Omega^{(i)}(\mathfrak{S}, \mathfrak{S})$. If

$$\Phi(\cdot) = f(\cdot)r_i(\cdot)q(\cdot)h(\cdot)\Phi$$

and

$$\Psi(\cdot) = g(\cdot)r_i(\cdot)q(\cdot)h(\cdot)\Psi$$

and if $b < a$, and $\mu = \langle X(\mathfrak{S}), X(\mathfrak{S})\rangle - b^2$ then

$$((I - E(\mu)Q\phi^{\wedge}, \psi^{\wedge}) = \frac{1}{(2\pi i)^m} \sum_{s \in {}^{\Psi}\Omega^{(i)}(\mathfrak{S}, \mathfrak{S})} \int_{U({}^{\Psi}\mathfrak{S}, X(\mathfrak{S}), b)} f(H)g(-s\overline{H})\xi(s, H)dH$$

with

$$\xi(s, H) = (N(s, H)d(r_i(H)q(H)h(H)\Phi), d(r_i(H)q(H)h(H)\Psi))$$

For some s the function $\xi(s, H)$ does not vanish identically. Consequently the expression on the right is a positive semi-definite Hermitian symmetric form in $f(\cdot)$ and $g(\cdot)$ which does not vanish identically. A simple approximation argument shows that there must be an s^o in ${}^{\Psi}\Omega^{(i)}(\mathfrak{S}, \mathfrak{S})$ so that $H = -s^o H$ for all H in $U({}^{\Psi}\mathfrak{S}, X(\mathfrak{S}), b)$. Choosing b sufficiently close to a we conclude that $H = -s^o\overline{H}$ for all H in $U({}^{\Psi}\mathfrak{S}, X(\mathfrak{S}), \infty)$ and that s^o leaves every element of $\tilde{\mathfrak{S}}$ fixed.

Collections of subspaces and Eisenstein systems satisfying the conditions of Lemma 7.5 are just what we need to describe the spectral decomposition of the spaces ${}^*\mathfrak{L}(\{P\}, \{V\}, W)$. Let us see how to associate to each such collection a closed subspace of each of the spaces ${}^*\mathfrak{L}(\{P\}, \{V\}, W)$.

LEMMA 7.6. Suppose that $S^{(i)}$, $1 \le i \le r$, is a collection of distinct affine subspaces of $\mathfrak{a}^{(i)}$ and that if $\mathfrak{S}$ belongs to $S = \bigcup_{i=1}^r S^{(i)}$ there is given an Eisenstein system belonging to $\mathfrak{S}$. Suppose that S and the associated Eisenstein systems satisfy the hypotheses of Lemma 7.5. Let ${}^*P$ be a cuspidal subgroup belonging to some element of $\{P\}$ and let ${}^*\mathcal{L}'(\{P\}, \{V\}, W)$ be the closed subspace of ${}^*\mathcal{L}(\{P\}, \{V\}, W)$ generated by functions of the form $(I-E(\lambda))Q\phi^\wedge$ where λ and $\Phi(\cdot)$ are such that for some positive number a, some i, and some P in ${}^*\{P\}^{(i)}$ the inequality $R^2 - \lambda < a^2$ is satisfied and

$$\Phi(\cdot) = r_i(\cdot)\Phi'(\cdot)$$

with $\Phi'(\cdot)$ in ${}^*\mathfrak{H}(V, W)$. Let $C_1, \ldots, C_u$ be the equivalence classes in ${}^*S$ and for each x, $1 \le x \le u$, choose $\mathcal{E}_x$ in C_x so that $\tilde{\mathcal{E}}_x$ is the complexification of a distinguished subspace of $\mathfrak{f}$. If P belongs to $\{P\}^{(i)}$ and $\Phi(\cdot)$ belongs to ${}^*\mathfrak{H}(V, W)$ then

$$\sum_{s\in\Omega^{(i)}(\mathcal{E}_x, C_x)} E(g, d\Phi(ss_x^o H), ss_x^o H)$$

is analytic on $U({}^\Psi\mathcal{E}_x, X(\mathcal{E}_x), \infty)$, where

$$\Omega^{(i)}(\mathcal{E}_x, C_x) = \{s \in \Omega(\mathcal{E}_x, C_x) \,|\, j_s = i\}$$

and if ω_x is the number of elements in ${}^\Psi\Omega(\mathcal{E}_x, \mathcal{E}_x)$ then

$$\phi(g, a) = \Sigma_{x=1}^u 1/\omega_x (2\pi i)^m \int_{U({}^\Psi\mathcal{E}_x, X(\mathcal{E}_x), a)} \sum_{s\in\Omega^{(i)}(\mathcal{E}_x, C_x)} E(g, d\Phi(ss_x^o H), ss_x^o H)\,dH$$

belongs to ${}^*\mathcal{L}'(\{P\}, \{V\}, W)$ and the projection of $\phi^\wedge$ on ${}^*\mathcal{L}'(\{P\}, \{V\}, W)$ is equal to $\lim_{a\to\infty} \phi(\cdot, a)$. Moreover if P' belongs to ${}^*\{P\}^{(j)}$ and $\Psi(\cdot)$ belongs to ${}^*\mathfrak{H}(V', W)$ then

$$\sum_{t\in\Omega^{(j)}(\mathcal{E}_x, C_x)} \sum_{s\in\Omega^{(i)}(\mathcal{E}_x, C_x)} (N(ts_x^o s^{-1}, ss_x^o H)d\Phi(ss_x^o H), d\Psi(-t\overline{H}))$$

is analytic on $U({}^\Psi\mathcal{E}_x, X(\mathcal{E}_x), \infty$ and the inner product of the projection s of $\phi^\wedge$ and $\psi^\wedge$ on ${}^*\mathcal{L}'(\{P\}, \{V\}, W)$ is equal to

$$\Sigma_{x=1}^u 1/\omega_x (2\pi i)^m \int \Sigma\, \Sigma\, (N(ts_x^o s^{-1}, ss_x^o H)d\Phi(ss_x^o H), d\Psi(-t\overline{H}))dH$$

The inner sums are over $t \in \Omega^{(j)}(\mathfrak{S}_x, C_x)$ and $s \in \Omega^{(i)}(\mathfrak{S}_x, C_x)$. The integral is over $U({}^{\Psi}\mathfrak{S}_x, X(\mathfrak{S}_x), \infty)$. Suppose a is a positive number, P belongs to ${}^*\{P\}^{(i)}$,

$$\Phi(\cdot) = r_i(\cdot)\Phi'(\cdot)$$

with $\Phi'(\cdot)$ in ${}^*\mathfrak{H}(V, W)$, P' belongs to ${}^*\{P\}^{(j)}$, and $\Psi(\cdot)$ belongs to ${}^*\mathfrak{H}(V', W)$. To begin the proof of the lemma we calculate the inner product $((I-E(\lambda))Q\phi^\wedge, \psi^\wedge)$ when $\lambda > R^2 - a^2$. Choose $\beta > R^2$, $\alpha = \lambda$, and $\gamma > 0$; according to Theorem 5.10 of [21]

$$((I-E(\lambda))Q\phi^\wedge, \psi^\wedge) = \lim_{\delta \to 0} \frac{1}{2\pi i} \int_{C(\alpha, \beta\gamma, \delta)} (R(z, A)Q\phi^\wedge, \psi^\wedge)dz$$

Since (7.p) is analytic for $\operatorname{Re}\lambda > R^2 - a^2$ it follows from the first corollary to Lemma 7.4 that the right side equals

$$\lim_{\delta \to 0} \frac{1}{2\pi i} \int_{C(\alpha, \beta, \gamma, \delta)} \left\{ \Sigma\, \Sigma \frac{1}{(2\pi i)^m} \int (z - \langle H, H \rangle)^{-1} (N(s, H)d\Phi(H), d\Psi(-s\overline{H}))dH \right\} dz$$

The sums are over $\mathfrak{S} \in {}^*S^{(i)}$ and $s \in {}^{\Psi}\Omega^{(j)}(\mathfrak{S})$. The inner integral is over $U({}^{\Psi}\mathfrak{S}, X(\mathfrak{S}), a)$. Let

$$\langle X(\mathfrak{S}), X(\mathfrak{S}) \rangle = \mu(\mathfrak{S})$$

Then $a > (\mu(s) - \lambda)^{\frac{1}{2}}$; so this limit equals

$$\Sigma_{\mathfrak{S} \in {}^*S^{(i)}} \Sigma_{s \in {}^{\Psi}\Omega^{(j)}(\mathfrak{S})} \frac{1}{(2\pi i)^m} \int_{U({}^{\Psi}\mathfrak{S}, X(\mathfrak{S}), (\mu(\mathfrak{S}) - \lambda)^{\frac{1}{2}})} (N(s, H)d\Phi(H), d\Psi(-s\overline{H}))dH$$

which we prefer to write as

$$\text{(7.r)} \qquad \Sigma_{x=1}^{u} \Sigma\, \Sigma \frac{1}{\omega_x (2\pi i)^m} \int (N(ts_x^o s^{-1}, ss_x^o H)d\Phi(ss_x^o H), d\Psi(-t\overline{H}))dH$$

The inner sums are over $t \in \Omega^{(j)}(\mathfrak{S}_x, C_x)$ and $s \in \Omega^{(i)}(\mathfrak{S}_x, C_x)$. The inner integral is over $U({}^{\Psi}\mathfrak{S}_x, X(\mathfrak{S}_x), (\mu(\mathfrak{S}_x) - \lambda)^{\frac{1}{2}})$. It should perhaps be observed that if

H belongs to $U({}^{\Psi}\mathfrak{G}_x, X(\mathfrak{G}_x), \infty)$ then

$$-t\overline{H} = ts_x^o H$$

Let

$$\mathcal{H}_x(V, W) = \Sigma_{s \in \Omega^{(i)}(\mathfrak{G}_x, C_x)} L(S((\mathfrak{G}_x)_s), \mathcal{E}(V, W)), \quad 1 \leq x \leq u$$

and define $\mathcal{H}_x(V', W)$ in a similar fashion. If $F = \oplus F_s$ belongs to $\mathcal{H}_x(V, W)$ and $F' = \oplus F'_s$ belongs to $\mathcal{H}_x(V', W)$ let

$$[F, F'] = \Sigma_{t \in \Omega^{(j)}(\mathfrak{G}_x, C_x)} \Sigma_{s \in \Omega^{(i)}(\mathfrak{G}_x, C_x)} (N(ts_x^o s^{-1}, ss_x^o H)F_s, F'_t)$$

Of course $[F, F']$ depends on H. A simple approximation argument shows that, when H belongs to $U(\mathfrak{G}_x, X(\mathfrak{G}_x), \infty)$,

(7.s) $$[F, F] \geq 0; \ |[F, F']|^2 \leq [F, F][F', F']$$

At this point we need to remind ourselves of a number of simple facts from the theory of integration (cf. [5], Ch. II). Let $\mathcal{L}_x(V, W; \lambda)$ be the space of all functions

$$F(H) = \oplus F_s(H)$$

on $U({}^{\Psi}\mathfrak{G}_x, X(\mathfrak{G}_x), (\mu(\mathfrak{G}_x) - \lambda)^{\frac{1}{2}})$ with values in $\mathcal{H}_x(V, W)$ such that $[F(H), F]$ is measurable for every F in $\mathcal{H}_x(V, W)$,

$$F_{ss_x^o r}(H) = F_s(s_x^o rH)$$

for all r in ${}^{\Psi}\Omega(\mathfrak{G}_x, \mathfrak{G}_x)$, and

$$1/\omega_x(2\pi)^m \int_{U({}^{\Psi}\mathfrak{G}_x, X(\mathfrak{G}_x), (\mu(\mathfrak{G}_x)-\lambda)^{\frac{1}{2}}} [F(H), F(H)] |dH| = \|F(\cdot)\|^2$$

is finite. If two functions whose difference has norm zero are identified

$\mathcal{L}_x(V, W; \lambda)$ becomes a Hilbert space.

$$\oplus_{x=1}^{u} \mathcal{L}_x(V, W; \lambda) = \mathcal{L}(V, W; \lambda)$$

is also a Hilbert space with the dense subset

$$\mathfrak{K}(V, W; \lambda) = \left\{ \oplus_{x=1}^{u} \oplus_{s \in \Omega^{(i)}(\mathfrak{a}_x, C_x)} d\Phi(ss_x^o H) \,|\, \Phi(\cdot) = r_i(\cdot)\Phi'(\cdot),\ \Phi'(\cdot) \in {}^*\mathfrak{H}(V, W) \right\}$$

The map

$$\Phi(\cdot) \longrightarrow (I - E(\lambda))\Omega\phi^{\wedge}$$

can be extended to an isometric map of $\mathcal{L}(V, W; \lambda)$ into ${}^*\mathcal{L}'(\{P\}, \{V\}, W)$

$$F(\cdot) \longrightarrow f^{\wedge}$$

where

$$F(\cdot) = \oplus_{x=1}^{u} F_x(\cdot)$$

Let $\mathcal{L}_x(V, W)$ be the set of all functions

$$F(\cdot) = \oplus F_s(\cdot)$$

on $U({}^{\Psi}\mathfrak{a}_x, X(\mathfrak{a}_x), \infty)$ with values in $\mathcal{H}_x(V, W)$ such that $[F(H), F]$ is measurable for every F in $\mathcal{H}_x(V, W)$,

$$F_{ss_x^o r}(H) = F_s(rs_x^o H)$$

for all r in ${}^{\Psi}\Omega(\mathfrak{a}_x, \mathfrak{a}_x)$, and

$$1/\omega_x(2\pi)^m \int_{U({}^{\Psi}\mathfrak{a}_x, X(\mathfrak{a}_x), \infty)} [F(H), F(H)]\,|dH| = \|F(\cdot)\|^2$$

is finite. $\mathcal{L}_x(V, W)$ is also a Hilbert space; let

$$\oplus_{x=1}^{u} \mathcal{L}_x(V, W) = \mathcal{L}(V, W)$$

The spaces $\mathcal{L}(V, W; \lambda)$ can be regarded as subspaces of $\mathcal{L}(V, W)$ and $\cup_\lambda \mathcal{L}(V, W; \lambda)$ is dense in $\mathcal{L}(V, W)$. The map $F(\cdot) \longrightarrow f^\wedge$ can be extended to an isometric mapping of $\mathcal{L}(V, W)$ into ${}^*\mathcal{L}'(\{P\}, \{V\}, W)$. It follows readily from (7.r) that if $F(\cdot)$ belongs to $\mathcal{L}(V, W)$ and $G(\cdot)$ belongs to $\mathcal{L}(V', W)$ then

$$(f^\wedge, g^\wedge) = \Sigma_{x=1}^{u} \frac{1}{\omega_x (2\pi)^m} \int_{U({}^\Psi\mathfrak{S}_x, X(\mathfrak{S}_x), \infty)} [F(H), G(H)] |dH|$$

Let $F^x(\cdot) = \oplus F_s^x(\cdot)$, $1 \le x \le u$, be a function on $U({}^\Psi\mathfrak{S}_x, X(\mathfrak{S}_x), \infty)$ with values in $\mathcal{H}_x(V, W)$ such that

$$F^x_{ss_x^o r}(H) = F_s(rs_x^o H)$$

for all r in ${}^\Psi\Omega(\mathfrak{S}_x, \mathfrak{S}_x)$ and suppose that if $G(\cdot) = \oplus_{x=1}^u G_x(\cdot)$ belongs to $\mathfrak{K}(V, W; \lambda)$ for some λ then $[F_x(H), G_x(H)]$ is measurable for $1 \le x \le u$ and

$$\Sigma_{x=1}^u 1/\omega_x (2\pi)^m \int_{U({}^\Psi\mathfrak{S}_x, X(\mathfrak{S}_x), (\mu(\mathfrak{S}_x)-\lambda)^{\frac{1}{2}}} [F_x(H), G_x(H)] dH$$

is defined and is at most $c \| G(\cdot) \|$ where c is some constant. Then $F(\cdot)$ belongs to $\mathcal{L}(V, W)$ and its norm is at most c.

If $\Phi(\cdot)$ belongs to ${}^*\mathfrak{H}(V, W)$ and

$$F_x(H) = \oplus_{s\in\Omega^{(i)}(\mathfrak{S}_x, C_x)} d\Phi(ss_x^o H)$$

this condition is satisfied with $c = \| \Phi^\wedge \|$. If P' belongs to ${}^*\{P\}^{(j)}$ and $\Psi(\cdot)$ belongs to ${}^*\mathfrak{H}(V', W)$ it then follows from (7.s) that

$$\Sigma_{t\in\Omega^{(j)}(\mathfrak{S}_x, C_x)} \Sigma_{s\in\Omega^{(i)}(\mathfrak{S}_x, C_x)} (N(ts_x^o s^{-1}, ss_x^o H) d\Phi(ss_x^o H), d\Psi(-t\overline{H}))$$

is integrable on $U({}^\Psi\mathfrak{S}_x, X(\mathfrak{S}_x), \infty)$. However, it is a meromorphic function with singularities which lie along hyperplanes of the form $\alpha(H) = \mu$ so it is integrable over $U({}^\Psi\mathfrak{S}_x, X(\mathfrak{S}_x), \infty)$ only if it is analytic on this set. Applying the map

$F(\cdot) \longrightarrow f^\wedge$ to the above element of $\mathfrak{L}(V, W)$ we obtain a function ϕ' in $^*\mathfrak{L}'(\{P\}, \{V\}, W)$. To prove the final assertion of the lemma it is sufficient to show that ϕ' is the projection of $\phi^\wedge$ on $^*\mathfrak{L}'(\{P\}, \{V\}, W)$ or that

$$((I-E(\lambda))Q\psi^\wedge, \phi^\wedge) = ((I-E(\lambda))Q\psi^\wedge, \phi')$$

whenever there is a positive number a and a P' in $^*\{P\}^{(j)}$ for some j so that $R^2 - \lambda < a^2$ and $\Psi(\cdot) = r_j(\cdot)\Psi'(\cdot)$ with $\Psi'(\cdot)$ in $^*\mathfrak{H}(V', W)$. This follows from the formula (7.r) with $\Phi(\cdot)$ and $\Psi(\cdot)$ interchanged.

Take $\Phi(\cdot)$ as in the last paragraph and suppose that for some x

$$\sum_{s\in\Omega^{(i)}(\mathfrak{a}_x, C_x)} E(\cdot, d\Phi(ss_x^o H), ss_x^o H) = E(\cdot, H)$$

is not analytic on $U({}^\Psi\mathfrak{a}_x, X(\mathfrak{a}_x), \infty)$. Let $\mathfrak{t}$ be a singular hyperplane which intersects $U({}^\Psi\mathfrak{a}_x, X(\mathfrak{a}_x), \infty)$. Select a unit normal to $\mathfrak{t}$; take an arbitrary analytic function $g(\cdot)$ on ${}^\Psi\mathfrak{a}$; and consider $\operatorname{Res}_{\mathfrak{t}}\{g(H)E(\cdot, H)\}$. If P' belongs to $^*\{P\}^{(j)}$ for some j and ψ belongs to $^*\mathfrak{D}(V, W)$ then

$$\int_{^*\Theta\backslash^* M\times K} \operatorname{Res}_{\mathfrak{t}}\{g(H)E(mk, H)\}\bar{\psi}^\wedge(mk)\,dm\,dk$$

is equal to

$$\operatorname{Res}_{\mathfrak{t}}\left\{\int_{^*\Theta\backslash^* M\times K} g(H)E(mk, H)\bar{\psi}^\wedge(mk)\,dm\,dk\right\}$$

If ψ is the Fourier transform of $\Psi(\cdot)$ the expression in brackets is equal to

$$g(H)\sum_{t\in\Omega^{(j)}(\mathfrak{a}_x, C_x)}\ \sum_{s\in\Omega^{(i)}(\mathfrak{a}_x, C_x)} (N(ts_x^o s^{-1}, ss_x^o H)d\Phi(ss_x^o H), d\Psi(-t\bar{H}))$$

Since no singular hyperplanes of this function intersect $U({}^\Psi\mathfrak{a}_x, X(\mathfrak{a}_x), \infty)$ the residue is zero. Comparing this conclusion with Lemma 3.7 we obtain a contradiction. Suppose that $\phi'(\cdot, a)$ is the image in $^*\mathfrak{L}'(\{P\}, \{V\}, W)$ of the element

$$\oplus_{x=1}^{u} F_x(H)$$

of $\mathcal{L}(V, W)$ where

$$F_x(H) = \oplus_{s\in\Omega^{(i)}(\mathfrak{a}_x, C_x)} d\Phi(ss_x^{o}H)$$

if $\|\mathrm{Im}\,H\| < a$ and $F_x(H) = 0$ if $\|\mathrm{Im}\,H\| \geq a$. Certainly the limit of $\phi'(\cdot, a)$ as a approaches infinity is equal to the function ϕ' of the previous paragraph. To complete the proof of the lemma it has to be shown that $\phi'(\cdot, a) = \phi(\cdot, a)$. To do this we show that if P' belongs to ${}^*\{P\}^{(j)}$ for some j and ψ belongs to ${}^*\mathfrak{J}(V, W)$ then

$$(\phi'(\cdot, a)\, \psi^{\wedge}) = (\phi(\cdot, a), \psi^{\wedge})$$

Now $(\phi(\cdot, a), \psi^{\wedge})$ is equal to

$$\Sigma_{x=1}^{u} \frac{1}{\omega_x(2\pi i)^m} \int \Sigma \left\{ \int_{{}^*\Theta\backslash {}^*M\times K} E(mk, d\Phi(ss_x^{o}H), ss_x^{o}H)\overline{\psi}^{\wedge}(mk)dmdk \right\} dH$$

The outer integral is over $U({}^{\Psi}\mathfrak{a}, X(\mathfrak{a}), a)$ and the inner sum is over $s \in \Omega^{(i)}(\mathfrak{a}_x, C_x)$. Referring to (7.a) we see that if ψ is the Fourier transform of $\Psi(\cdot)$ this equals

$$\Sigma_{x=1}^{u} \frac{1}{\omega_x(2\pi i)^m} \int \Sigma_t\, \Sigma_s (N(ts_x^{o}s^{-1}, ss_x^{o}H)d\Phi(ss_x^{o}H), d\Psi(-t\overline{H}))dH$$

which is, of course, equal to $(\phi'(\cdot, a), \psi^{\wedge})$.

There is a corollary to this lemma which is of great importance to us.

COROLLARY. Let ${}^*P$ be a cuspidal subgroup belonging to some element of $\{P\}$, let P belong to ${}^*\{P\}^{(i)}$, let $\mathfrak{a}$ belong to ${}^*S^{(i)}$ and let F belong to $L(S(\mathfrak{a}), \mathcal{E}(V, W))$. If $\mathfrak{n}$ is the largest distinguished subspace which $\tilde{\mathfrak{a}}$ contains and if r is the inverse image in $\mathfrak{a}$ of a singular hyperplane of the function

$E(\cdot, F, H)$ on ${}^{\Psi}\mathfrak{S}$ which meets $U({}^{\Psi}\mathfrak{S}, X(\mathfrak{S}), \infty)$ then $\tilde{r}$ contains $\mathfrak{v}$.

Suppose that $\tilde{\mathfrak{S}}$ is the complexification of a distinguished subspace $\mathfrak{v}$ of $\mathfrak{f}$. The assertion in this case is that $E(\cdot, F, H)$ has no singular hyperplanes which meet $U({}^{\Psi}\mathfrak{S}, X(\mathfrak{S}), \infty)$ and it will be proved by induction on $\dim \mathfrak{v} - \dim {}^{*}\mathfrak{v}$. We first take this difference to be one. Let H_0 be a unit vector in ${}^{\Psi}\mathfrak{S} \cap \mathfrak{v}^{(i)}$. If X is a singular point of $E(\cdot, F, H)$ lying in $U({}^{\Psi}\mathfrak{S}, X(\mathfrak{S}), \infty)$ let

$$E(g, F, X + izH_0) = \Sigma_{k=-m}^{\infty} z^k E_k(g)$$

with $m > 0$ and $E_{-m}(g) \not\equiv 0$. If $\mathfrak{S}$ belongs to C_x choose $\mathfrak{S}_x = \mathfrak{S}$ and let $F^n(\cdot)$, for sufficiently large n, be that element of $\mathcal{L}_x(V, W)$ such that $F^n_s(H)$ vanishes identically if s is not in ${}^{\Psi}\Omega(\mathfrak{S}, \mathfrak{S})$, $F_{s^o}(H)$ equals $nz^m F$ if $H = X + izH_0$ with $1/2n < z < 1/n$ and equals zero otherwise, and $F_r(H) = F_{s^o}(rs^o H)$ if r belongs to ${}^{\Psi}\Omega(\mathfrak{S}, \mathfrak{S})$. Since, for large n,

$$\|F^n(\cdot)\|^2 = n^2/2\pi \int_{1/2n}^{1/n} z^{2m}(N(s^o, X)F, F)dz$$

$\lim_{n\to\infty} \|F^n(\cdot)\|^2 = 0$. Let f^n be the image of $F^n(\cdot)$ in ${}^{*}\mathcal{L}'(\{P\}, \{V\}, W)$. An argument like that used in the proof of the lemma shows that

$$f^n(g) = n/2\pi \int_{1/2n}^{1/n} z^m E(g, F, X + izH_0)dz$$

$$= n/2\pi \Sigma_{k=-m}^{\infty} \int_{1/2n}^{1/n} z^{m+k} E_k(g)dz$$

so that

$$\lim_{n\to\infty} f^n(g) = \frac{1}{4\pi} E_{-m}(g)$$

uniformly on compact sets. Comparing the two results we conclude that $E_{-m}(g) \equiv 0$, and this is impossible.

Suppose that dim $\mathfrak{v}$ - dim $^*\mathfrak{v}$ = n is greater than one and that the assertion is true when dim $\mathfrak{v}$ - dim $^*\mathfrak{v}$ is less than n. If $\mathfrak{f}$ in $^*S^{(j)}$ belongs to the same equivalence class as $\mathfrak{s}$, if P belongs to $^*\{P\}^{(i)}$, if F belongs to $L(S(\mathfrak{s}), \mathfrak{E}(V, W))$, if P' belongs to $^*\{P\}^{(j)}$, if F' belongs to $L(S(\mathfrak{f}), \mathfrak{E}(V, W))$, if s belongs to $^{\Psi}\Omega(\mathfrak{s}, \mathfrak{s})$, and if t belongs to $^{\Psi}\Omega(\mathfrak{s}, \mathfrak{f})$ then

$$|(N(ts^os^{-1}, ss^oH)F, F')|^2$$

is at most

$$(N(ss^os^{-1}, ss^oH)F, F)(N(ts^ot^{-1}, ts^oH)F', F')$$

which in turn equals

$$(N(s^o, ss^oH)F, F)(N(t^o, ts^oH)F', F')$$

if H belongs to $U(^{\Psi}\mathfrak{s}, X(\mathfrak{s}), \infty)$. Consequently if a singular hyperplane of the function $(N(ts^os^{-1}, ss^oH)F, F')$ meets $U(^{\Psi}\mathfrak{s}, X(\mathfrak{s}), \infty)$ it must be a singular hyperplane of $(N(t^o, ts^oH)F', F')$. This fact will be used to show that, if for some F in $L(S(\mathfrak{s}), \mathfrak{E}(V, W))$ the hyperplane $^{\Psi}\mathfrak{H}$ meets $U(^{\Psi}\mathfrak{s}, X(\mathfrak{s}), \infty)$ and is a singular hyperplane of $E(\cdot, F, H)$, then for some j, some $\mathfrak{f}$ in $^*S^{(j)}$ such that the largest distinguished subspace contained in $\tilde{\mathfrak{f}}$ is larger than $^*\mathfrak{v}$, some P' in $^*\{P\}^{(j)}$, some F' in $L(S(\mathfrak{f}), \mathfrak{E}(V', W))$, and some t in $\Omega(\mathfrak{s}, \mathfrak{f})$ the function $(N(t^o, ts^oH)F', F')$ has $^{\Psi}\mathfrak{H}$ as a singular hyperplane. Suppose not, and let $\mathfrak{H}$ be the inverse image of $^{\Psi}\mathfrak{H}$ in s. Select a unit normal to $\mathfrak{H}$ and consider the function $\operatorname{Res}_{\mathfrak{H}} E(\cdot, \cdot, \cdot)$ defined on

$$^*A\,^*T\backslash G \times L(S(\mathfrak{H}), E(V, W)) \times {}^{\Psi}\mathfrak{H}$$

If r belongs to $^{\Psi}\Omega^{(j)}(\mathfrak{H})$ for some j, then $\operatorname{Res}_{\mathfrak{H}} N(r, \cdot)$ is zero unless there is a $\mathfrak{f}$ in $^*S^{(j)}$ such that the largest distinguished subspace contained in $\tilde{\mathfrak{f}}$ is $^*\mathfrak{v}$, and a t in $^{\Psi}\Omega(\mathfrak{s}, \mathfrak{f})$ so that r is the restriction to $\mathfrak{H}$ of t. Then

$$\mathrm{Re}\{-r(X(\mathfrak{X}))\} = -r(X(\mathfrak{S})) = X(\mathfrak{t})$$

belongs to ${}^{+}(\Psi_{\mathfrak{V}}^{(j)})$. It follows from the corollary to Lemma 5.1 that if F is in $L(S(\mathfrak{X}), \mathcal{E}(V, W))$ and $\mathrm{Res}_{\mathfrak{X}}\, E(\cdot, F, H)$ is defined at H in $U(\Psi_{\mathfrak{X}}, X(\mathfrak{X}), \infty)$, then it belongs to ${}^{*}\mathcal{L}(\{P\}, \{V\}, W)$. Choosing such an H which is not real we contradict Lemma 7.3.

Let us, for brevity, call those $\mathfrak{S}$ in S such that $\tilde{\mathfrak{S}}$ is the complexification of a distinguished subspace principal. To complete the induction and to prove the lemma for those elements of S which are not principal we will use the functional equations proved in Lemma 7.4. Let C be an equivalence class in ${}^{*}S$ and choose a principal element $\mathfrak{S}$ in C. If $\mathfrak{S}_1$ is in ${}^{*}S^{(i)}$ and belongs to C and if P in ${}^{*}\{P\}^{(i)}$ is given we can choose the set of representatives $P^{(i,k)}$, $1 \le k \le m_i'$, so that it contains P. Choose

$$F^y = \oplus_{s \in \Omega_0(\mathfrak{S}, C)} F_s^y$$

with F_s^y in $L(S(\mathfrak{S}_s), \mathcal{E}^{(j_s)})$ so that the set of vectors $\oplus_{t \in \Omega(\mathfrak{S}, C)} F_t^y(H)$ with

$$F_t^y(H) = \Sigma_{s \in \Omega_0(\mathfrak{S}, C)} M(ts^o s^{-1}, ss^o H) F_s^y$$

is a basis for the range of $M(H)$ when $M(H)$ is defined. There are elements

$$G^y = \oplus_{s \in \Omega_0(\mathfrak{S}, C)} G_s^y$$

so that if $\{G_t \mid t \in \Omega(s, C)\}$ belongs to the range of $M(H)$ and

$$G_t = \Sigma_{y=1}^{v} c_y F_t^y$$

for all t then

$$c_y = \Sigma_{s \in \Omega_0(\mathfrak{S}, C)} (G_s, G_s^y)$$

If F belongs to $L(S(\mathfrak{S}_1), \mathcal{E}(V^{(i,k)}, W))$ for some k, s_1 belongs to ${}^{\Psi}\Omega(\mathfrak{S}, \mathfrak{S}_1)$, and

$$\Sigma_{y=1}^{v} c_y(H)F_t^y(H) = M(ts^o s_1^{-1}, s_1 s^o H)F$$

for all t then

$$E(g, F, s_1 s^o H) = \Sigma_{y=1}^{v} c_y(H)\left\{\Sigma_{s\in\Omega_0(\mathfrak{S}, C)} \mathcal{E}(g, F_s^y, ss^o H)\right\} \tag{7.t}$$

Suppose that $\mathfrak{S}_1$ is principal and that $\mathfrak{r}_1$ is the inverse image in $\mathfrak{S}_1$ of a singular hyperplane of $E(\cdot, F, \cdot)$. Let $\mathfrak{t}$ in ${}^*S^{(j)}$ be an element of C which contains a distinguished subspace ${}^*\mathfrak{a}_1$ larger than ${}^*\mathfrak{a}$ such that for some P' in ${}^*\{P\}^{(j)}$, some F' in $L(S(\mathfrak{t}), \mathcal{E}(V', W))$, and some t_1 in ${}^{\Psi}\Omega(\mathfrak{S}_1, \mathfrak{t})$ the function $(N(t^o, t_1 s_1^o H)F', F')$ has ${}^{\Psi}\mathfrak{r}_1$ as a singular hyperplane. Since $N(t^o, t_1 s_1^o H)$ depends only on the projection of $t_1 s_1^o H$ on the orthogonal complement of ${}^*\mathfrak{a}_1$ the hyperplane $t_1 s_1^o(\tilde{\mathfrak{r}}_1)$ contains ${}^*\mathfrak{a}_1$. There is a principal element in ${}^*S_1$ which is equivalent to $\mathfrak{t}$; it may be supposed that we have chosen $\mathfrak{S}$ to be this element. Choose a t in $\Omega(\mathfrak{S}, \mathfrak{t})$ which leaves every element of ${}^*\mathfrak{a}_1$ fixed. Let us take s_1 to be $t_1^{-1} t^o t$. Then $E(g, F, s_1 s^o H)$ has a singular hyperplane ${}^{\Psi}\mathfrak{r}$ which meets $U({}^{\Psi}\mathfrak{S}, X(\mathfrak{S}), \infty)$ such that $\tilde{\mathfrak{r}}$ contains ${}^*\mathfrak{a}_1$. As usual the inverse image $\mathfrak{r}$ of ${}^{\Psi}\mathfrak{r}$ in $\mathfrak{S}$ is written as $X(\mathfrak{r}) + \tilde{\mathfrak{r}}$. Now

$$c_y(H) = \Sigma_{t\in\Omega_0(\mathfrak{S}, C)}(M(ts^o s_1^{-1}, s_1 s^o H)F, G_t^y)$$

and is thus analytic on $U({}^{\Psi}\mathfrak{S}, X(\mathfrak{S}), \infty)$. Consequently for some s in $\Omega_0(\mathfrak{S}, C)$ the function $E(\cdot, F_s^y, ss^o H)$ has $\mathfrak{r}$ for a singular hyperplane. In other words we can suppose that if s belongs to ${}^*S^{(k)}$, there is a P in ${}^*\{P\}^{(k)}$, and an F in $L(S(\mathfrak{S}), \mathcal{E}(V, W))$ so that $E(\cdot, F, \cdot)$ has a singular hyperplane ${}^{\Psi}\mathfrak{r}$ which meets $U({}^{\Psi}\mathfrak{S}, X(\mathfrak{S}), \infty)$ and is such that $\tilde{\mathfrak{r}}$ contains ${}^*\mathfrak{a}_1$. Let ${}^*P_1$ be the cuspidal subgroup with split component ${}^*\mathfrak{a}_1$ which belongs to P and let

$E_1(\cdot, \cdot, \cdot)$ be the associated function on

$$^*A_1 {}^*T_1 \backslash G \times L(S(\mathfrak{E}), \mathfrak{E}(V, W)) \times {}^\Psi\mathfrak{E}'$$

if ${}^\Psi\mathfrak{E}'$ is the projection of $\mathfrak{E}$ on the orthogonal complement of ${}^*\mathfrak{a}_1$. It follows from the relation (7.c) that $E_1(\cdot, F, \cdot)$ must have a singular hyperplane in ${}^\Psi\mathfrak{E}'$ which meets $U({}^\Psi\mathfrak{E}', X(\mathfrak{E}), \infty)$ and this contradicts the induction assumption.

The general case now follows readily from the relation (7.t). Indeed a singular hyperplane of the function $E(\cdot, F, H)$ defined on ${}^\Psi\mathfrak{E}_1$, the projection of $\mathfrak{E}_1$ on the orthogonal complement of ${}^*\mathfrak{a}$, can meet $U({}^\Psi\mathfrak{E}, X(\mathfrak{E}), \infty)$ only if it is a singular hyperplane of $(M(ts^o s_1^{-1}, H)F, G_t)$ for some s_1 in ${}^\Psi\Omega(\mathfrak{E}, \mathfrak{E}_1)$ and some t in $\Omega_0(\mathfrak{E}, C)$ and hence, by (7.s), a singular hyperplane of $(M(s_1^o, H)F, F)$. Since $M(s_1^o, H)$ depends only on the projection of H on the orthogonal complement of the largest distinguished subspace contained in $\tilde{\mathfrak{E}}_1$ the corollary follows.

The principal assertion of the paper can now be formulated as follows.

THEOREM 7.1. <u>There are</u> $q+1$ <u>unique collections</u> $S_m = \bigcup_{i=1}^r S_m^{(i)}$ <u>of affine spaces of dimension</u> m <u>and unique Eisenstein systems, one belonging to each element of</u> S_m, $0 \leq m \leq q$, <u>which satisfy the hypotheses of Lemma</u> 7.5 <u>so that if</u> ${}^*P$ <u>is a cuspidal subgroup belonging to some element of</u> $\{P\}$ <u>and</u> ${}^*\mathfrak{L}_m(\{P\}, \{V\}, W)$ <u>is the closed subspace of</u> ${}^*\mathfrak{L}(\{P\}, \{V\}, W)$ <u>associated to</u> S_m <u>by Lemma</u> 7.6 <u>then, if</u> ${}^*q$ <u>is the dimension of</u> ${}^*\mathfrak{a}$,

$${}^*\mathfrak{L}(\{P\}, \{V\}, W) = \Sigma_{m={}^*q}^{q} \, {}^*\mathfrak{L}_m(\{P\}, \{V\}, W)$$

<u>and</u> ${}^*\mathfrak{L}_{m_1}(\{P\}, \{V\}, W)$ <u>is orthogonal to</u> ${}^*\mathfrak{L}_{m_2}(\{P\}, \{V\}, W)$ <u>if</u> $m_1 \neq m_2$.

We will use induction on m to establish the existence of these collections and the associated Eisenstein system. Let us first describe the form the induction step takes, then show how to start the induction, and then carry it out in general.

Let m be an integer with $0 \leq m \leq q$ and suppose that we have defined the collections $S_n^{(i)}$, $1 \leq i \leq r$ for all $n > m$ and that if $n_1 \neq n_2$ the spaces $^*\mathcal{L}_{n_1}(\{P\}, \{V\}, W)$ and $^*\mathcal{L}_{n_2}(\{P\}, \{V\}, W)$ are orthogonal. Suppose that for $1 \leq i \leq r$ we have also defined a collection $S^{(i)}$ of distinct affine subspaces of $\mathfrak{a}_c^{(i)}$ of dimension m and a collection $T^{(i)}$ of not necessarily distinct affine subspaces of $\mathfrak{a}_c^{(i)}$ of dimension $m-1$ and that we have associated an Eisenstein system to each element of $S^{(i)}$ and $T^{(i)}$. Suppose that every space in $S^{(i)}$ or $T^{(i)}$ meets D_i and that only a finite number of the elements of $S^{(i)}$ or $T^{(i)}$ meet each compact subset of $\mathfrak{a}_c^{(i)}$. In particular then if $\mathfrak{s}$ belongs to $S^{(i)}$ or $T^{(i)}$ the point $X(\mathfrak{s})$ lies in D_i; we assume also that $\operatorname{Re} X(\mathfrak{s})$ belongs to $^+\mathfrak{a}$ if $\mathfrak{a}$ is the orthogonal complement of the largest distinguished subspace contained in $\tilde{\mathfrak{s}}$ and to the closure of $^+\mathfrak{a}(\mathfrak{s})$ if $\mathfrak{a}(\mathfrak{s})$ is the orthogonal complement of $\tilde{\mathfrak{s}}$ in $\mathfrak{a}^{(i)}$. Recall that $^+\mathfrak{a}(\mathfrak{s})$ has been defined in the discussion preceding Lemma 2.6. If $\mathfrak{s}$ belongs to $S^{(i)}$ it is said to be of type A if for every positive number a we have defined a non-empty convex cone $V(\mathfrak{s}, a)$ with centre $X(\mathfrak{s})$ and radius $\varepsilon(\mathfrak{a})$ so that if a_1 is less than a_2 then $V(\mathfrak{s}, a_1)$ contains $V(\mathfrak{s}, a_2)$, so that every singular hyperplane of the associated Eisenstein system which meets the closure of the cylinder $C(\mathfrak{s}, \varepsilon(a), a)$ meets the closure of $U(\mathfrak{s}, X(\mathfrak{s}), a)$ but no singular hyperplane meets the closure of $U(\mathfrak{s}, Z, a)$ if Z belongs to $V(\mathfrak{s}, a)$, and so that the closure of $V(\mathfrak{s}, a)$ is contained in D_i. An element $\mathfrak{t}$ of $T^{(i)}$ is said to be of type B if it satisfies, in addition to these conditions, the condition we now describe. Let P belong to $\{P\}^{(i)}$, let $^*P$ belong to P, and let F belong to $L(S(\mathfrak{t}), \mathcal{E}(V, W))$. If $\mathfrak{a}$ is the largest distinguished subspace which $\tilde{\mathfrak{t}}$ contains and if $\mathfrak{r}$ is the inverse image in $\mathfrak{t}$ of a singular hyperplane of the function $E(\cdot, F, \cdot)$, which is defined on $^{\Psi}\mathfrak{t}$, which meets $U(^{\Psi}\mathfrak{t}, X(\mathfrak{t}), \infty)$ then $\tilde{\mathfrak{r}}$ contains $\mathfrak{a}$. If $\mathfrak{t}$ lies in $T^{(i)}$ it is said to be of type C if for every positive number a we have defined a non-empty open convex subset $V(\mathfrak{t}, a)$ of $\tilde{\mathfrak{t}}$ so that if a_1 is less than a_2 then $V(\mathfrak{t}, a_1)$ contains $V(\mathfrak{t}, a_2)$, so that no singular hyperplane meets the closure of $U(\mathfrak{t}, Z, a)$ if Z belongs to $V(\mathfrak{t}, a)$, and so that $\{\operatorname{Re} Z \mid Z \in V(\mathfrak{t}, a)\}$ is contained in the interior of the convex hull of $(\mathfrak{n}^{(i)})^+$ and the closure of $^+\mathfrak{n}(\mathfrak{t})$. We assume that every element of $S^{(i)}$ is of type A and every element of $T^{(i)}$ is of type B or C.

Suppose that $^*P$ is a cuspidal subgroup belonging to some element of $\{P\}$ and let Q be the projection of $^*\mathfrak{L}(\{P\}, \{V\}, W)$ onto the orthogonal complement of

$$\Sigma_{n=m+1}^{q}\, {}^*\mathfrak{L}_n(\{P\}, \{V\}, W)$$

We suppose that Q is zero if m is less than $^*q$ but that if $m \geq {}^*q$ then for any P in $^*\{P\}^{(i)}$, any $\Phi(\cdot)$ in $^*\mathfrak{H}(V, W)$, any P' in $^*\{P\}^{(j)}$, and $\Psi(\cdot)$ in $^*\mathfrak{H}(V', W)$ and any positive number a the difference between $(R(\lambda, A)Q\phi^\wedge, \psi^\wedge)$ and the sum of

$$\Sigma_{\mathfrak{s} \in {}^*S^{(i)}} \Sigma_{s \in {}^\Psi\Omega^{(j)}(\mathfrak{s})} \frac{1}{(2\pi i)^{m'}} \int (N(s, H)d((\lambda - \langle H, H\rangle)^{-1}\Phi(H)), d\Psi(-s\overline{H}))dH$$

and

$$\Sigma_{\mathfrak{t} \in {}^*T^{(i)}} \Sigma_{t \in {}^\Psi\Omega^{(j)}(\mathfrak{t})} \frac{1}{(2\pi i)^{m'-1}} \int (N(t, H)d((\lambda - \langle H, H\rangle)^{-1}\Phi(H)), d\Psi(-s\overline{H}))dH$$

is analytic for $\operatorname{Re}\lambda > R^2 - a^2$ if $Z(\mathfrak{s})$ belongs to $V(\mathfrak{s}, a)$ and $Z(\mathfrak{t})$ belongs to $V(\mathfrak{t}, a)$. The integrals are over $U({}^\Psi\mathfrak{s}, Z(\mathfrak{s}), a)$ and $U({}^\Psi\mathfrak{t}, Z(\mathfrak{t}), a)$ respectively. The integer m' equals $m - {}^*q$. We also suppose that the difference between $(Q\phi^\wedge, R(\overline{\lambda}, A)\psi^\wedge)$ and the sum of

$$\Sigma_{\mathfrak{s} \in {}^*S^{(i)}} \Sigma_{s \in {}^\Psi\Omega^{(j)}(\mathfrak{s})} \frac{1}{(2\pi i)^{m'}} \int (N(s, H)d\Phi(H), d((\overline{\lambda} - \langle -s\overline{H}, -s\overline{H}\rangle)^{-1}\Psi(-s\overline{H})))dH$$

and

$$\Sigma_{\mathfrak{t} \in {}^*T^{(i)}} \Sigma_{t \in {}^\Psi\Omega^{(j)}(\mathfrak{t})} \frac{1}{(2\pi i)^{m'-1}} \int (N(t, H)d\Phi(H), d((\overline{\lambda} - \langle -s\overline{H}, -s\overline{H}\rangle)^{-1}\Psi(-s\overline{H})))dH$$

is analytic for $\operatorname{Re}\lambda > R^2 - a^2$. The integrals are again over $U({}^\Psi\mathfrak{s}, Z(\mathfrak{s}), a)$

and $U({}^\psi\mathfrak{t}, Z(\mathfrak{t}), a)$.

It is an easy matter to verify that the collections $S^{(i)}$ satisfy the conditions of Lemma 7.5. First of all Lemma 7.6, with m replaced by $n > m$, make it obvious that $\lambda(f)$ commutes with Q if $f(\cdot)$ belongs to ${}^\psi\mathfrak{I}_0$; so it is only necessary to verify that for each ${}^*P$ and each positive number a there are polynomials $r_i(\cdot)$, $1 \leq i \leq r'$, on ${}^\psi\mathfrak{V}^{(i)}$ for which (7.p) and (7.q) have the required property. Since there is only a finite number of $\mathfrak{t}$ in ${}^*T^{(i)}$ for which $U({}^\psi\mathfrak{t}, Z(\mathfrak{t}), a)$ is not empty a polynomial $r_i(\cdot)$ can be chosen so that, for all P and P', all such $\mathfrak{t}$, and all t in ${}^\psi\Omega^{(j)}(\mathfrak{t})$, the function $N(t, H)dr_i(H)$ vanishes identically on ${}^\psi\mathfrak{t}$ but so that $r_i(\cdot)$ does not vanish identically on ${}^\psi\mathfrak{S}$ if $\mathfrak{S}$ belongs to ${}^*S^{(i)}$. It may also be supposed that if $\mathfrak{S}$ belongs to $S^{(i)}$ and $\mathfrak{S}$ intersects

$$\{H \in D_i \mid \|\operatorname{Im} H\| \leq a\}$$

then, for all P and P' and all s in ${}^\psi\Omega^{(j)}(\mathfrak{S})$, the function $N(s, H)dr_i(H)$ on ${}^\psi\mathfrak{S}$ has no singular hyperplanes which meet

$$\{H \in D_i \mid \|\operatorname{Im} H\| \leq a\}$$

The conditions of the last paragraph imply that with such polynomials the conditions of Lemma 7.5 are satisfied. To see this one has to use the argument preceding Lemma 7.1 in the way that Lemma 7.1 is used below. We will take $S_m^{(i)}$ to be $S^{(i)}$.

We must now examine the expression

$$(7.u)\quad \sum_{\mathfrak{S} \in {}^*S^{(i)}} \sum_{s \in {}^\psi\Omega^{(j)}(\mathfrak{S})} \frac{1}{(2\pi i)^{m'}} \int_{U({}^\psi\mathfrak{S}, Z(\mathfrak{S}), a)} (N(s, H)d\Phi(H), d\Psi(-s\overline{H}))dH$$

Since the set $S = \bigcup_{i=1}^{r} S^{(i)}$ is finite we may suppose that, for each positive a, the cones $V(\mathfrak{S}, a)$, $\mathfrak{S} \in S$, all have the same radius $\varepsilon(a)$. We may also suppose that $\varepsilon(a)$ is such that for each a and each $\mathfrak{S}$ there is a cone $W(\mathfrak{S}, a)$ with centre $X(\mathfrak{S})$ and radius $\varepsilon(a)$ so that if s belongs to $\Omega(\mathfrak{S}, \mathfrak{t})$ for some $\mathfrak{t}$ in S and

Z belongs to $ss^{o}(W(\mathcal{S}, a))$ there is no singular hyperplane of the Eisenstein system associated to $\mathcal{T}$ which meets the closure of $U(\mathcal{T}, Z, a)$. It may also be supposed that if $\mathcal{V}$ and $\mathcal{S}$ belong to S the collections

$$\{ss^{o}(W(\mathcal{S}, a)) \mid s \in \Omega(\mathcal{S}, \mathcal{T})\}$$

and

$$\{rr^{o}(W(\mathcal{V}, a)) \mid r \in \Omega(\mathcal{V}, \mathcal{T})\}$$

are the same if $\Omega(\mathcal{V}, \mathcal{S})$ is not empty and that if $s \in \Omega(\mathcal{S}, \mathcal{T})$ and $t \in \Omega(\mathcal{S}, \mathcal{T})$ then

$$ss^{o}(W(\mathcal{S}, a)) \cap ts^{o}(W(\mathcal{S}, a)) \neq \phi$$

implies $s = t$. Choose for each $\mathcal{S}$ in S and each s in $\Omega(\mathcal{S}, \mathcal{S})$ a point $Z(\mathcal{S}, s)$ in $ss^{o}(W(\mathcal{S}, a))$. According to the remarks following the proof of Lemma 7.1 there is a collection $T_1^{(i)}$ of $m-1$ dimensional affine spaces and for each $\mathcal{T}$ in $T_1^{(i)}$ an Eisenstein system belonging to $\mathcal{T}$ and a cone $V(\mathcal{T}, a)$, with centre $X(t)$ and some radius $\delta(a)$, so that, for all ${}^{*}P$, (7.u) is equal to the sum of

$$(7.v) \qquad \Sigma\,\Sigma\,\Sigma \frac{1}{\omega(\mathcal{S})(2\pi i)^{m}} \int_{U({}^{\Psi}\mathcal{S},\, Z(\mathcal{S}, r),\, a)} (N(s, H)d\Phi(H), d\Psi(-s\overline{H}))dH$$

and

$$(7.w) \quad \sum_{\mathcal{T} \in {}^{*}T_i^{(i)}} \sum_{t \in {}^{\Psi}\Omega^{(j)}(\mathcal{T})} \frac{1}{(2\pi i)^{m'-1}} \int_{U({}^{\Psi}\mathcal{T},\, Z(\mathcal{T}),\, a)} (N(t, H)d\Phi(H), d\Psi(-t\overline{H}))dH$$

with $Z(\mathcal{T})$ in $V(\mathcal{T}, a)$, and a sum of terms of the same type as (7.m). The sums in (7.v) are over $\mathcal{S} \in {}^{*}S^{(i)}$, $r \in \Omega(\mathcal{S}, \mathcal{S})$, and $s \in {}^{\Psi}\Omega^{(j)}(\mathcal{S})$ and the number of elements in $\Omega(\mathcal{S}, \mathcal{S})$ is $\omega(\mathcal{S})$. We can certainly suppose that, with the cones $V(\mathcal{T}, a)$, the elements of $T_1^{(i)}$ are all of type B. The supplementary condition on elements of type B must of course be verified but that is not difficult. We can also suppose that the sets U' occurring in the terms of the form (7.m) all lie in

$$\{H \mid \|\mathrm{Re}\, H\| < R, \ \|\mathrm{Im}\, H\| \geq a\}$$

This implies that if $\Phi(H)$ is replaced by $(\lambda - \langle H, H \rangle)^{-1}\Phi(H)$ or $\Psi(H)$ is replaced by $(\bar{\lambda} - \langle H, H \rangle)^{-1}\Psi(H)$ the difference between (7.u) and the sum of (7.v) and (7.w) is analytic for $\mathrm{Re}\,\lambda > R^2 - a^2$. If $\mathfrak{f}$ belongs to $T_1^{(i)}$ we will also have to know that if $\mathfrak{n}$ is the orthogonal complement of the largest distinguished subspace contained in $\tilde{\mathfrak{f}}$ then $\mathrm{Re}\, X(\mathfrak{f})$ lies in ${}^+\mathfrak{n}$ and that if $\mathfrak{n}(\mathfrak{f})$ is the orthogonal complement in $\mathfrak{n}^{(i)}$ of $\tilde{\mathfrak{f}}$ then $\mathrm{Re}\, X(\mathfrak{f})$ lies in the closure of ${}^+\mathfrak{n}(\mathfrak{f})$. The space $\mathfrak{f}$ is a singular hyperplane of some $\mathfrak{s}$ in $S^{(i)}$ such that $\mathfrak{f}$ meets $U(\mathfrak{s}, X(\mathfrak{s}), \infty)$; consequently $\mathrm{Re}\, X(\mathfrak{f}) = X(\mathfrak{s})$. The first point follows from the corollary to Lemma 7.6 because according to it we can assume that the largest distinguished subspaces contained in $\tilde{\mathfrak{s}}$ and $\tilde{\mathfrak{f}}$ are the same. If α is a positive root of $\mathfrak{n}^{(i)}$ let H_α be such that

$$\langle H, H_\alpha \rangle = \alpha(H)$$

for all H in $\mathfrak{n}^{(i)}$. The second point follows from the observation that the closures of ${}^+\mathfrak{n}(\mathfrak{s})$ and ${}^+\mathfrak{n}(\mathfrak{f})$ are the non-negative linear combinations of the elements H_α where α varies over the positive roots which vanish on $\tilde{\mathfrak{s}}$ and $\tilde{\mathfrak{f}}$ respectively.

Let $C_1, \ldots, C_u$ be the equivalence classes in ${}^*S$ and for each x choose a principal element $\mathfrak{s}_x$ in C_x. Let r_x^y, $1 \leq y \leq v_x$, be a subset of $\Omega(\mathfrak{s}_x, \mathfrak{s}_x)$ such that every element of $\Omega(\mathfrak{s}_x, \mathfrak{s}_x)$ can be written in the form $s s_x^o r_x^y$ with a unique y and a unique s in ${}^\Psi\Omega(\mathfrak{s}_x, \mathfrak{s}_x)$. Choose for each x a point Z_x in $W(\mathfrak{s}_x, a)$ and if $\mathfrak{s}$ belongs to C_x and s belongs to $\Omega(\mathfrak{s}, \mathfrak{s})$ let

$$Z(\mathfrak{s}, s) = t s_x^o(Z_x)$$

if t is the unique element of $\Omega(\mathfrak{s}_x, \mathfrak{s})$ such that

$$ss^o(W(\mathfrak{G}, a)) = ts^o_x(W(\mathfrak{G}_x, a))$$

The expression (7.v) is equal to

$$\Sigma_{x=1}^{u}\Sigma_{y=1}^{v_x} \frac{1}{\omega(\mathfrak{G}_x)(2\pi i)^{m'}} \int \Sigma\,\Sigma\,(N(ts^o_x s^{-1}, ss^o_x H)d\Phi(ss^o_x H), d\Psi(-t\overline{H}))dH$$

The integral is taken over $U({}^{\Psi}\mathfrak{G}_x, r^y_x(Z_x), a)$ and the sums are over $t \in \Omega^{(j)}(\mathfrak{G}_x, C_x)$ and $s \in \Omega^{(i)}(\mathfrak{G}_x, C_x)$. It follows from Lemma 7.6 that each of these integrands is analytic in the closure of $C(\mathfrak{G}_x, \varepsilon(a), a)$; consequently the argument used in the proof of Lemma 7.1 shows that the sum is equal to

$$\Sigma_{x=1}^{u} \frac{1}{\omega_x(2\pi i)^{m'}} \int \Sigma\,\Sigma\,(N(ts^o_x s^{-1}, ss^o_x H)d\Phi(ss^o_x H), d\Psi(-t\overline{H}))dH \tag{7.x}$$

The integral is here taken over $U({}^{\Psi}\mathfrak{G}_x, X(\mathfrak{G}_x), a)$ plus a sum of terms of the form

$$\frac{1}{\omega_x(2\pi i)^{m'}} \int_{U'} \Sigma\,\Sigma\,(N(ts^o_x s^{-1}, ss^o_x H)d\Phi(ss^o_x H), d\Psi(-t\overline{H}))dH$$

where U' is an open subset of an m'-dimensional oriented real subspace of ${}^{\Psi}\mathfrak{G}_x$ which lies in

$$\{H \in {}^{\Psi}\mathfrak{G}_x \mid \|\operatorname{Re} H\| < R, \ \|\operatorname{Im} H\| \geq a\}$$

In any case if $\Phi(H)$ is replaced by $(\lambda - \langle H, H\rangle)^{-1}\Phi(H)$ or $\Psi(H)$ is replaced by $(\overline{\lambda} - \langle H, H\rangle)^{-1}\Psi(H)$ the difference between (7.x) and (7.v) is analytic for $\operatorname{Re}\lambda > R^2 - a^2$. If $\phi^\wedge_m$ is the projection of $\phi^\wedge$ on ${}^*\mathfrak{L}_m(\{P\}, \{V\}, W)$ it follows readily from Lemma 7.6 that the difference between $(R(\lambda, A)\phi^\wedge_m, \psi^\wedge)$ and (7.x) with $\Phi(H)$ replaced by $(\lambda - \langle H, H\rangle)^{-1}\Phi(H)$ is analytic for $\operatorname{Re}\lambda > R^2 - a^2$ and that the difference between $(\phi^\wedge_m, R(\overline{\lambda}, A)\psi^\wedge)$ and (7.x) with $\Psi(H)$ replaced by $(\overline{\lambda} - \langle H, H\rangle)^{-1}\Psi(H)$ is analytic for $\operatorname{Re}\lambda > R^2 - a^2$. In conclusion, if Q' is the projection of ${}^*\mathfrak{L}(\{P\}, \{V\}, W)$ on the orthogonal complement of

$$\Sigma_{n=m}^{q}\ {}^{*}\mathcal{L}_{m}(\{P\}, \{V\}, W)$$

and $R^{(i)}$ is the union of $T_1^{(i)}$ and $T^{(i)}$ then the difference between $(R(\lambda, A)Q'\phi^{\wedge}, \psi^{\wedge})$ and

$$\Sigma\Sigma \frac{1}{(2\pi i)^{m'-1}} \int (N(r, H)d((\lambda - \langle H, H\rangle)^{-1}\Phi(H)), d\Psi(-s\overline{H}))dH$$

and the difference between $(Q\phi^{\wedge}, R(\overline{\lambda}, A)\psi^{\wedge})$ and

$$\Sigma\Sigma \frac{1}{(2\pi i)^{m'-1}} \int (N(r, H)d\Phi(H), d((\overline{\lambda} - \langle -r\overline{H}, -r\overline{H}\rangle)^{-1}\Psi(-r\overline{H}))dH$$

are analytic for $\operatorname{Re}\lambda > R^2 - a^2$. The sums in the two displayed expressions are over $\mathcal{X} \in {}^{*}R^{(i)}$, $r \in {}^{\Psi}\Omega^{(j)}(\mathcal{X})$, and the integral is taken over $U({}^{\Psi}\mathcal{X}, Z(\mathcal{X}), a)$. In particular if $m = {}^{*}q$ these sums are empty so that $(R(\lambda, A)Q'\phi^{\wedge}, \psi^{\wedge})$ is entire and, hence, $Q'\phi^{\wedge} = 0$. Consequently

$${}^{*}\mathcal{L}(\{P\}, \{V\}, W) = \oplus_{m={}^{*}q}^{q}\ {}^{*}\mathcal{L}_{m}(\{P\}, \{V\}, W)$$

We observed after defining an Eisenstein system that, for $1 \le i \le r$, we could define in a simple manner an Eisenstein system belonging to $\mathfrak{v}^{(i)}$. If $R^{(i)} = \{\mathfrak{v}^{(i)}\}$ and if for all positive numbers a we take $V(\mathfrak{v}^{(i)}, a)$ to be $\{H \in \mathfrak{U}^{(i)} \cap \mathfrak{v}^{(i)} \mid \|H\| < R\}$ then it follows readily from the relation (4.p) that the difference between $(R(\lambda, A)\phi^{\wedge}, \psi^{\wedge})$ and

$$\Sigma\Sigma \frac{1}{(2\pi i)^{q'}} \int (N(s, H)d((\lambda - \langle H, H\rangle)^{-1}\Phi(H)), d\Psi(-s\overline{H}))dH$$

and the difference between $(\phi^{\wedge}, R(\overline{\lambda}, A)\psi^{\wedge})$ and

$$\Sigma\Sigma \frac{1}{(2\pi i)^{q'}} \int (N(s, H)d\Phi(H), d((\overline{\lambda} - \langle -s\overline{H}, -s\overline{H}\rangle)^{-1}\Psi(-s\overline{H}))dH$$

are analytic for $\operatorname{Re}\lambda > R^2 - a^2$ if $Z(\mathcal{X})$ belongs to $V(\mathcal{X}, a)$. The ranges of summation and integration are the same as above, and the integer q' equals $q - {}^{*}q$.

We now change notations so that $m-1$ or q is m and show that from the collections $R^{(i)}$ we can construct collections $S^{(i)}$ and $T^{(i)}$ which satisfy the induction assumption. Apart from the uniqueness this will complete the proof of the lemma. The construction is such that the analytic conditions on the associated Eisenstein systems are manifest so only the less obvious geometrical conditions will be verified. Suppose that $\mathfrak{s}$ belongs to $R^{(i)}$ and is of type C; since $\{\operatorname{Re} H \mid H \in V(\mathfrak{s}, a)\}$ lies in the interior of the convex hull of $(\mathfrak{a}^{(i)})^+$ and the closure of ${}^+\mathfrak{a}(\mathfrak{s})$ there is an open cone with centre $X(\mathfrak{s})$ whose projection on $\mathfrak{a}^{(i)}$ lies in the interior of this convex hull. We tentatively let $S^{(i)}$ be the set of distinct affine subspaces $\mathfrak{t}$ of $\mathfrak{a}^{(i)}$ such that $\mathfrak{t} = \mathfrak{s}$ for some $\mathfrak{s}$ in $R^{(i)}$. For each $\mathfrak{t}$ in $S^{(i)}$ and each positive number a we choose a non-empty convex open cone $V(\mathfrak{t}, a)$ with centre $X(\mathfrak{t})$ and radius $\delta(a)$ so that $V(\mathfrak{t}, a_1)$ contains $V(\mathfrak{t}, a_2)$ if a_1 is less than a_2, so that if $\mathfrak{t} = \mathfrak{s}$ and $\mathfrak{s}$ belongs to $R^{(i)}$ then every singular hyperplane of the Eisenstein system associated to $\mathfrak{s}$ which meets the closure of the cylinder $C(\mathfrak{t}, \delta(a), a)$ meets the closure of $U(\mathfrak{t}, X(\mathfrak{t}), a)$ but so that no such hyperplane meets the closure of $U(\mathfrak{t}, Z, a)$ if Z belongs to $V(\mathfrak{t}, a)$, and so that the closure of $V(\mathfrak{t}, a)$ lies in D_i. If $\mathfrak{t} = \mathfrak{s}$ with $\mathfrak{s}$ in $R^{(i)}$ and of type C we further demand that $\{\operatorname{Re} H \mid H \in V(\mathfrak{t}, a)\}$ lie in the interior of the convex hull of $(\mathfrak{t}^{(i)})^+$ and the closure of ${}^+\mathfrak{a}(\mathfrak{t})$.

Suppose that $\mathfrak{s}$ belongs to $R^{(i)}$ and is of type C. Choose the unique $\mathfrak{t}$ in $S^{(i)}$ such that $\mathfrak{s} = \mathfrak{t}$. Suppose that Y belongs to $V(\mathfrak{t}, a)$, that Z belongs to $V(\mathfrak{s}, a)$, and that the segment joining $\operatorname{Re} Y$ and $\operatorname{Re} Z$ meets the projection on $\mathfrak{a}^{(i)}$ of a singular hyperplane $\mathfrak{h}$ of the Eisenstein system belonging to $\mathfrak{s}$. We have observed that the closure of ${}^+\mathfrak{a}(\mathfrak{s})$ is contained in the closure of ${}^+\mathfrak{a}(\mathfrak{h})$. Thus $\{\operatorname{Re} H \mid H \in V(\mathfrak{s}, a)\}$ and $\{\operatorname{Re} H \mid H \in V(\mathfrak{t}, a)\}$ lie in the interior of the convex hull of $(\mathfrak{a}^{(i)})^+$ and the closure of ${}^+\mathfrak{a}(\mathfrak{h})$. The intersection of the convex hull of these two sets with $\{\operatorname{Re} H \mid H \in \mathfrak{h}\}$ also lies in this set. Take a point in this set, which is not empty, and project it on $\mathfrak{a}(\mathfrak{h})$; the result is

$\mathrm{Re}\, X(\mathfrak{s})$. Thus $\mathrm{Re}\, X(\mathfrak{s})$ lies in the interior of the convex hull of the closure of $^{+}\mathfrak{a}(\mathfrak{s})$ and the projection of $(\mathfrak{a}^{(i)})^{+}$ on $\mathfrak{a}(\mathfrak{s})$. If α is a positive root of $\mathfrak{a}(\mathfrak{s})$, if H lies in $(\mathfrak{a}^{(i)})^{+}$, and if H' is the projection of H on $\mathfrak{a}(\mathfrak{s})$ then $\alpha(H') = \alpha(H)$ which is positive. Thus $\mathrm{Re}\, X(\mathfrak{s})$ lies in the interior of the convex hull of $\mathfrak{a}^{+}(\mathfrak{s})$ and the closure of $^{+}\mathfrak{a}(\mathfrak{s})$. This is $^{+}\mathfrak{a}(\mathfrak{s})$ itself. If $\beta_{1}, \ldots, \beta_{p}$, are the simple roots of $\mathfrak{a}(\mathfrak{s})$ then

$$\mathrm{Re}\, X(\mathfrak{s}) = \Sigma_{j=1}^{p}\, b_j H_{\beta_j},$$

with $b_j > 0$. Let $\alpha_{1}, \ldots, \alpha_{q}$, be the simple roots of $\mathfrak{a}^{(i)}$ and let

$$\beta_{j}, = \Sigma_{k=1}^{q}\, b_{jk}\alpha_{k},$$

with $b_{jk} \geq 0$. If

$$\Sigma_{j=1}^{p}\, b_j b_{j\ell} = 0$$

for some ℓ then $b_{j\ell} = 0$ for all j and $\tilde{\mathfrak{s}}$ contains the distinguished subspace of $\mathfrak{a}^{(i)}$ defined by $\alpha_{k}(H) = 0$, $k \neq \ell$. It follows readily that if $\mathfrak{a}$ is the orthogonal complement of the largest distinguished subspace which $\tilde{\mathfrak{s}}$ contains then $\mathrm{Re}\, X(\mathfrak{s})$ lies in $^{+}\mathfrak{a}$.

The elements of $T^{(i)}$ will arise in two ways. Suppose that $\mathfrak{X}$ belongs to $R^{(i)}$ and is of type C. Choose the unique $\mathfrak{S}$ in $S^{(i)}$ such that $\mathfrak{X} = \mathfrak{S}$. As a consequence of Lemma 7.1 we can choose a collection $T^{(i)}_{(\mathfrak{X})}$ of affine subspaces of $\mathfrak{a}^{(i)}$ of dimension $m-1$ and a collection of Eisenstein systems, one belonging to each element of $T^{(i)}_{(\mathfrak{X})}$ so that the difference between

$$\Sigma_{\mathfrak{X} \in \Psi_{\Omega}^{(j)}(\mathfrak{X})} \frac{1}{(2\pi i)^{m'}} \int_{U(\Psi_{\mathfrak{X}},\, Z(\mathfrak{X}),\, a)} (N(r, H)\, d\Phi(H),\, d\Psi(-s\overline{H}))\, dH$$

and the sum of

$$\Sigma_{s \in \Psi_{\Omega}^{(j)}(\mathfrak{S})} \frac{1}{(2\pi i)^{m'}} \int_{U(\Psi_{\mathfrak{S}},\, Z(\mathfrak{S}),\, a)} (N_{\mathfrak{X}}(s, H)\, d\Phi(H),\, d\Psi(-s\overline{H}))\, dH$$

and

$$\sum_{\mathfrak{f} \in {}^*T^{(i)}(\mathfrak{r})} \sum_{t \in {}^\Psi\Omega^{(j)}(\mathfrak{f})} \frac{1}{(2\pi i)^{m'-1}} \int_{U({}^\Psi\mathfrak{f},\, Z(\mathfrak{f}),\, a)} (N(t, H) d\Phi(H), d\Psi(-s\overline{H}))dH$$

is a sum of integrals of the form (7.m). Of course $Z(\mathfrak{r})$ belongs to $V(\mathfrak{r}, a)$, $Z(\mathfrak{s})$ belongs to $V(\mathfrak{s}, a)$, and $Z(\mathfrak{f})$ belongs to a suitably chosen $V(\mathfrak{f}, a)$. Referring to the previous paragraph we see that $\mathfrak{f}$, with the given $V(\mathfrak{f}, a)$, may be supposed to be of type C. The meaning of the function $N_{\mathfrak{r}}(s, H)$ is clear.

Suppose that $\mathfrak{r}$ belongs to $R^{(i)}$ and is of type B. Choose the unique $\mathfrak{s}$ in $S^{(i)}$ such that $\mathfrak{r} = \mathfrak{s}$. Appealing now to the remarks following the proof of Lemma 7.1 we obtain the same conclusions as above except that the elements of $T^{(i)}(\mathfrak{r})$ are of type B. We let

$$T^{(i)} = \bigcup_{\mathfrak{r} \in R^{(i)}} T^{(i)}(\mathfrak{r})$$

If $\mathfrak{s}$ belongs to $S^{(i)}$ we associate to $\mathfrak{s}$ the Eisenstein system obtained by adding together the Eisenstein systems belonging to those $\mathfrak{r}$ in $R^{(i)}$ such that $\mathfrak{r} = \mathfrak{s}$. If the sum is not an Eisenstein system, that is, if it vanishes identically, we remove $\mathfrak{s}$ from $S^{(i)}$. The collections $S^{(i)}$ and $T^{(i)}$ satisfy the induction assumptions.

The proof of the uniqueness will merely be sketched. We apply the second corollary to Lemma 7.4. Suppose that the collections S_m, $0 \leq m \leq q$, of affine spaces together with an associated collection of Eisenstein systems satisfy the conditions of the theorem. Let ${}^*P$ be a cuspidal subgroup of rank m belonging to some element of $\{P\}$. If P belongs to ${}^*\{P\}^{(i)}$ and $\Phi(\cdot)$ belongs to ${}^*\mathcal{H}(V, W)$ the projection of $\phi^\wedge$ on the subspace of ${}^*\mathcal{L}(\{P\}, \{V\}, W)$ spanned by eigenfunctions of the operator A is uniquely determined and is equal to

$$\sum_{\mathfrak{s} \in {}^*S^{(i)}} E(\cdot, d\Phi(X(\mathfrak{s})), X(\mathfrak{s}))$$

It follows readily that the points $X(\mathcal{E})$, $\mathcal{E} \in {}^{*}S^{(i)}$, and the functions $E(\cdot, F, X(\mathcal{E}))$, $F \in L(S(\mathcal{E}), \mathcal{E}(V, W))$, are uniquely determined.

References

1. Borel, A., Density properties for certain subgroups of semi-simple groups without compact components, Ann. of Math. (2) 72 (1960).

2. ——————, Ensembles fondamentaux pour les groupes arithmétiques, Colloque sur la théorie des groupes algébrique, Brussels, 1962.

3. ——————, Some finiteness properties of adèle groups over number fields, Inst. Hautes Études Sci. Publ. Math. 16 (1963).

4. Chevalley, C., Sur certains groupes simples, Tôhoku Math. J. (2) 1 (1955).

5. Dixmier, J., Les algèbres d'opérateurs dans l'espace hilbertien, Paris (1957).

6. Gelfand, I. M., Automorphic functions and the theory of representations, Proc. Int. Congress of Math., Stockholm, 1962.

7. —————— and I. I. Pjateckii-Shapiro, Unitary representations in homogeneous spaces with discrete stationary groups, Soviet Math. Dokl. 3 (1962).

8. —————— and ——————, Unitary representations in a space G/Γ where G is a group of $n \times n$ real matrices and Γ is a subgroup of integer matrices, Soviet Math. Dokl. 3 (1962).

9. Harish-Chandra, On some applications of the universal enveloping algebra of a semi-simple Lie algebra, Trans. Amer. Math. Soc. 70 (1956).

10. ——————, Representations of semi-simple Lie groups, III, Trans. Amer. Math. Soc. 76 (1954).

11. ——————, Representations of semi-simple Lie groups, IV, Amer. J. Math. 77 (1955).

12. ——————, On a lemma of F. Bruhat, J. Math. Pures Appl. (9) 35 (1956).

13. ——————, Fourier transforms on a semi-simple Lie algebra, I, Amer. J. Math. 89 (1957).

14. ——————, Automorphic Forms on a semi-simple Lie group, Proc. Nat. Acad. Sci. U.S.A. 45 (1959).

15. Jacobson, N., Lie algebras, New York, 1962.

16. Mostow, G. D., Fully reducible subgroups of algebraic groups, Amer. J. Math. 78 (1956).

17. Selberg, A., Harmonic analysis and discontinuous groups, J. Indian Math. Soc. 20 (1956).

18. ——————, On discontinuous groups in higher-dimensional symmetric spaces, Contributions to Function Theory, Bombay, 1960.

19. ———————, Discontinuous groups and harmonic analysis, Proc. Int. Congress of Math., Stockholm, 1962.

20. Shimizu, H., On discontinuous groups operating on the product of the upper half planes, Ann. of Math. (2) 77 (1963).

21. Stone, M. H., Linear transformations in Hilbert space and their applications to analysis, New York, 1932.

22. Weil, A., On discrete subgroups of Lie groups (II), Ann. of Math. (2) 75 (1962).

23. Whitney, H., Elementary structure of real algebraic varieties, Ann. of Math. (2) 66 (1957).

Appendix I

DIRICHLET SERIES ASSOCIATED WITH QUADRATIC FORMS

1. The object of this paper is to describe and prove the functional equations for some Dirichlet series suggested by Selberg in [6]. In that paper he introduces invariant differential operators on the space of positive definite $m \times m$ matrices; it is unnecessary to describe the operators explicitly now. The series considered here arise when one attempts to construct eigenfunctions of these differential operators which are invariant under the unimodular substitutions $T \longrightarrow UTU'$. U is integral and has determinant ± 1. As Selberg observes, if $s = (s_1, \ldots, s_m)$ is a complex m-tuple and $s_{m+1} = 0$ then

$$\omega(T, S) = |T|^{\frac{m+1}{4}} \prod_{k=1}^{m} |T|_k^{s_k - s_{k+1} - \frac{1}{2}}$$

is an eigenfunction of the invariant differential operators. $|T|_k$ is the sub-determinant formed from the first k rows and columns of T. Since the differential operators are invariant, if A is a non-singular $m \times m$ matrix $\omega(A'TA, s)$ is also an eigenfunction with the same eigenvalues. In particular, if A is a sub-diagonal matrix with diagonal elements ± 1 then $\omega(ATA', s) = \omega(T, S)$. Consequently the function

$$(1) \qquad \Omega(T, s) = \Sigma_{\{U\}} \omega(UTU', s)$$

is, at least formally, an eigenfunction which is invariant under unimodular substitutions. The sum is over a set of representatives of right-cosets of the group, V, of sub-diagonal matrices in the group of unimodular matrices. The series converges when $\operatorname{Re}(s_{k+1} - s_k) > \frac{1}{2}$, $k = 1, \ldots, m-1$. One hopes to obtain eigenfunctions for other values of s by continuing $\Omega(T, S)$ analytically. If this is possible it is natural to expect that $\Omega(T, s)$ satisfies some functional equations. The form of these equations is suggested by the eigenvalues of the differential operators corresponding to the eigenfunction $\omega(T, s)$ for they are symmetric functions of s. To be precise, if

$$a(t) = t(t-1)\pi^{-t}\Gamma(t)\zeta(2t)$$

and

(2) $$\Psi(T, s) = \prod_{i>j} a(\tfrac{1}{2} + s_i - s_j)\Omega(T, s)$$

then $\Psi(T, s)$ is an entire symmetric function of s.

Similar series may be obtained from the modular group and the generalized upper half-plane. If $Z = X + iY$ with $Y > 0$, the functions

$$\chi(Z, s) = \omega(Y, s_1 + \frac{m+1}{4}, \ldots, s_m + \frac{m+1}{4})$$

are eigenfunctions of the invariant differential operators. Moreover $\chi(Z, s)$ is invariant under the group, N, of modular transformations of the form

$$\begin{bmatrix} A & B \\ 0 & A'^{-1} \end{bmatrix}$$

with A in V. Form the function

(3) $$X(Z, s) = \Sigma_{\{M\}}\chi(M(Z), s)$$

The sum is over a set of representatives of right cosets of N in the modular group. The series converges when $\mathrm{Re}(s_{k+1} - s_k) > \frac{1}{2}$, $k = 1, \ldots, m-1$, and $\mathrm{Re}(s_1) > \frac{1}{2}$. Let

(4) $$\Phi(Z, s) = \prod_{i>j} a(\tfrac{1}{2} + s_i - s_j)a(\tfrac{1}{2} + s_i + s_j)\prod_i a(\tfrac{1}{2} + s_i)X(Z, s)$$

$\Phi(Z, s)$ may be analytically continued to an entire symmetric function of s. Moreover

$$\Phi(Z, \pm s_1, \ldots, \pm s_m) = \Phi(Z, s_1, \ldots, s_m)$$

So Φ is invariant under the Weyl group of the symplectic group just as Ψ is invariant under the Weyl group of the special linear group.

Professor Bochner suggested the possibility of defining analogous functions for any algebraic number field. In order to do this I describe alternative definitions of the series (1) and (3). For this some elementary algebraic facts are required and it is convenient to state these for an arbitrary algebraic number field, k, of finite degree over the rationals.

Let z_m be the m-dimensional coordinate space over k. The elements of z_m are taken to be row vectors. All modules over $\mathfrak{o}$, the ring of integers of k are to be finitely generated and to be contained in z_m. Such a module, $\mathfrak{n}$, is said to be of rank k if the subspace z of z_m generated by $\mathfrak{n}$ is of dimension k. The rank of a module will often be indicated by a subscript. In the following $\mathfrak{m}$ will denote some fixed module in z_m of rank m. A submodule $\mathfrak{n}$ of $\mathfrak{m}$ is said to be primitive (with respect to $\mathfrak{m}$) if $\mathfrak{n} = z \cap \mathfrak{m}$. If $\mathfrak{n}_k$ is a submodule of $\mathfrak{m}$ the quotient space z_m/z may be identified with z_{m-k} and the image of $\mathfrak{m}$ is a module $\mathfrak{m}'$ in z_{m-k}. If $\mathfrak{n}_k$ is primitive the kernel of the mapping $\mathfrak{m} \longrightarrow \mathfrak{m}'$ is $\mathfrak{n}_k$. It is known that there is a submodule $\mathfrak{p}$ of $\mathfrak{m}$ which maps onto $\mathfrak{m}'$ such that $\mathfrak{m} = \mathfrak{n}_k \oplus \mathfrak{p}$.

Now suppose that k is the rational field and that $\mathfrak{m}$ consists of the elements of z_m with integral coordinates. If $U = (u'_1, \ldots, u'_m)'$ is a unimodular matrix with rows $u_1, \ldots, u_m$ let $\mathfrak{n}_k$ be the submodule of $\mathfrak{m}$ consisting of integral linear combinations of $u_1, \ldots, u_k$. $\mathfrak{n}_k$ is clearly of rank k and it is primitive. For, let $U^{-1} = (w_1, \ldots, w_n)$ then if $u = \Sigma_{i=1}^{k} a_i u_i$ is integral $uw_j = a_j$, $1 \leq j \leq k$, is integral. So to each unimodular U there is associated an ascending chain $\mathfrak{n}_1 \subset \ldots \subset \mathfrak{n}_m$ of primitive submodules. If U and V give rise to the same chain then

$$u_1 = a_{11}v_1$$
$$u_2 = a_{21}v_1 + a_{22}v_2$$
$$\vdots$$
$$u_m = a_{m1} + \dots + a_{mm}v_m$$

with integral a_{ij}; or $U = AV$ with

$$A = \begin{bmatrix} a_{11} & & \\ a_{21} & a_{22} & \\ \vdots & & \\ a_{m1} & \dots & a_{mm} \end{bmatrix}$$

Comparing determinants one sees that A is unimodular. Consequently U and V belong to the same right-coset of V. Conversely let $\mathfrak{n}_1 \subset \mathfrak{n}_2 \dots \subset \mathfrak{n}_m$ be an ascending chain of primitive submodules. There is a vector u_1 such that $\mathfrak{n}_1$ consists of integral multiples of u_1. Let $\mathfrak{n}_1 \oplus \mathfrak{z}_1$ be the decomposition of $\mathfrak{m}$ described above. Then $\mathfrak{n}_2 \cap \mathfrak{z}_1$ is of rank 1 and consists of integral multiples of a vector u_2. The elements of $\mathfrak{n}_2$ are integral linear combinations of u_1 and u_2. Continuing in this manner one obtains vectors $u_1, \dots, u_m$ such that $\mathfrak{n}_k$ consists of integral linear combinations of $u_1, \dots, u_k$. Moreover the matrix $(u_1', \dots, u_m')'$ is unimodular since $u_1, \dots, u_m$ span $\mathfrak{n}_m = \mathfrak{m}$. Thus there is a one-to-one correspondence between right-cosets of V and ascending chains of primitive submodules.

It remains to describe $\omega(UTU', s)$ in terms of the chain. Suppose once again that k is an arbitrary algebraic number field. For convenience in calculating, the kth exterior product of z_m is taken to be $z_{\binom{m}{k}}$ and the coordinates of the kth exterior product of the vectors $a_1, \dots, a_k$ are the $k \times k$

subdeterminants of the matrix $(\alpha_1', \ldots, \alpha_k')'$. If $\mathfrak{n}$ is a module in z_m then $\mathfrak{n}^k$ is the module in $z_{\binom{m}{k}}$ generated by the kth exterior products of vectors in $\mathfrak{n}$. If $\mathfrak{n}_k$ is of rank k it is often convenient to write $\mathfrak{n}^k$ instead of $\mathfrak{n}_k^k$; in this case $\mathfrak{n}^k$ is of rank 1.

Now if $U = (u_1', \ldots, u_m')'$ is a unimodular matrix and $\mathfrak{n}_1 \subset \ldots \subset \mathfrak{n}_m$ the associated chain of submodules $\mathfrak{n}^k$ consists of integral multiples of $u^k = u_1 \wedge \ldots \wedge u_k$, the exterior product of $u_1, \ldots, u_k$. Moreover, if T^k is the $\binom{m}{k} \times \binom{m}{k}$ matrix formed from the $k \times k$ subdeterminates of T then, by the general Lagrange identity,

$$|UTU'|_k = u^k T^k u^{k'}$$

Since $u^k T^k u^{k'}$ depends only on T and $\mathfrak{n}_k$ it may be written $T\{\mathfrak{n}_k\}$. Then

$$\omega(U'TU, s) = T\{\mathfrak{n}_m\}^{\frac{m+1}{4}} \prod_{k=1}^{m} T\{\mathfrak{n}_k\}^{s_k - s_{k+1} - \frac{1}{2}}$$
$$= T\{\mathfrak{m}\}^{\frac{m+1}{4}} \prod_{k=1}^{m} \{\mathfrak{n}_k\}^{s_k - s_{k+1} - \frac{1}{2}}$$

and

$$\Omega(T, s) = T\{\mathfrak{m}\}^{\frac{m+1}{4}} \Sigma \prod_{k=1}^{m} T\{\mathfrak{n}_k\}^{s_k - s_{k+1} - \frac{1}{2}}$$

The sum is over all ascending chains of primitive submodules of the module of integral vectors.

Now let k be an algebraic number field of degree n over the rationals. Let $k_1, \ldots, k_n$ be the conjugates of k; as usual k_i is real if $1 \le i \le r_1$ and complex if $r_1 < i \le n$; moreover $k_{i+r_2} = \overline{k}_i$, $r_1 < i \leqq r_1 + r_2$. Let T be the n-tuple $(T_1, \ldots, T_n)$. T_i, $1 \le i \le r_1$, is a positive definite $m \times m$ symmetric matrix; T_i, $r_1 < i \le n$, is a positive definite $m \times m$ Hermitian matrix; and $T_{i+r_2} = \overline{T}_i$, $r_1 < i \le r_1 + r_2$. If $\mathfrak{n}$ is a module of rank 1 in z_m let α be a non-zero vector in $\mathfrak{n}$ and let $\mathfrak{N} = \{a \in k \,|\, a\alpha \in \mathfrak{n}\}$. $\mathfrak{N}$ is an ideal in k and $\mathfrak{n} = \mathfrak{N}\alpha$. Let

$$T\{\mathfrak{n}\} = N^2 a \prod_{k=1}^{n} a_k T_k \bar{a}'_k$$

a_k is of course the kth conjugate of a. $T\{\mathfrak{n}\}$ is independent of the vector a chosen. If $\mathfrak{n}_k$ is of rank k set $T\{\mathfrak{n}_k\} = T^k\{\mathfrak{n}^k\}$; $T^k = (T^k_1, \ldots, T^k_n)$. Finally, if $\mathfrak{m}$ is a finitely generated module in z_m of rank m set

$$\Omega(T, \mathfrak{m}, s) = T\{\mathfrak{m}\}^{\frac{m+1}{4}} \Sigma \prod_{k=1}^{m} T\{\mathfrak{n}_k\}^{s_k - s_{k+1} - \frac{1}{2}} \tag{1'}$$

This sum is over all ascending chains, $\mathfrak{n}_1 \subset \mathfrak{n}_2 \subset \ldots \subset \mathfrak{n}_m$, of primitive submodules of m. Let

$$a(t) = t(t-1)\pi^{-nt_2 - 2r_2 t}{}_2\Delta^t \Gamma(t)^{r_1} \Gamma(2t)^{r_2} \zeta(2t)$$

Δ is the absolute value of the discriminant of k and $\zeta(\cdot)$ is the zeta-function of k. Then set

$$\Psi(T, \mathfrak{m}, s) = \prod_{i>j} a(\tfrac{1}{2} + s_i - s_j)\Omega(T, \mathfrak{m}, s) \tag{2'}$$

THEOREM 1. (i) <u>The series</u> (1') <u>converges if</u> $\mathrm{Re}(s_{k+1} - s_k) > \frac{1}{2}$, $k = 1, \ldots, m-1$.

(ii) $\Psi(T, \mathfrak{m}, s)$ <u>may be analytically continued to an entire symmetric function of</u> s.

In order to carry out an induction on m it is necessary to add

(iii) <u>If</u> $s = \sigma + i\tau$, <u>then</u> $|\Psi(T, \mathfrak{m}, s)| \leq f(\sigma)\prod_{i \neq j}(|s_i - s_j| + 1)$.

Of course f depends on T and $\mathfrak{m}$ but no attempt is made here to determine precise estimates for Ψ.

Now consider the series (3). If $M(Z) = X_1 + iY_1$ and

$$M = \begin{pmatrix} A & B \\ C & D \end{pmatrix}$$

then $Y_1 = (C\overline{Z} + D)'^{-1} Y (CZ + D)^{-1}$ so $Y_1^{-1} = (CZ + D) Y^{-1} (CZ + D)^*$. Moreover

$$\omega(Y_1, s_1, \ldots, s_m) = \omega'(Y_1^{-1}, -s_m, \ldots, -s_1)$$

and if E is the matrix $(\delta_{i, n+1-i})$,

$$\omega'(Y, s) = \omega(EYE, s)$$

Consequently the series (3) may be written

$$\Sigma\, \omega'\left((CZ + D)Y^{-1}(CZ + D)^*, -s_m - \frac{m+1}{4}, \ldots, -s_1 - \frac{m+1}{4}\right)$$

From an $m \times 2m$ matrix forming the lower half of a modular matrix, M, we may construct the chain $\mathfrak{n}_1 \subset \ldots \subset \mathfrak{n}_m$ of primitive lattices; $\mathfrak{n}_k$ is the lattice spanned by the last k rows of M. $\mathfrak{n}_m$ is orthogonal to itself with respect to the skew-symmetric form

$$\Sigma_{i=1}^{m} x_i y_{m+i} - y_i x_{m+1} = xJy'$$

Two modular matrices give rise to the same ascending chain if and only if they belong to the same right coset of N.

Conversely, given such an ascending chain of lattices, let $\{u_1, \ldots, u_k\}$ span $\mathfrak{n}_k$. Then it is possible to choose $v_1, \ldots, v_m$ so that $v_i J u_j' = \delta_{ij}$. Suppose $v_1, \ldots, v_p$ have been chosen. Select v_{p+1} so that $v_{p+1} J u_j' = \delta_{p+1, j}$ and then subtract a suitable linear combination of $u_1, \ldots, u_p$ so that $v_{p+1} J v_j' = 0$, $j = 1, \ldots, p + 1$. It is clear that the matrix with rows $v_m, \ldots, v_1, u_m, \ldots, u_1$ is modular.

Now let W be the real part of the matrix $(Z, I)' Y^{-1} (\overline{Z}, I)$, then

$$(CZ + D)Y^{-1}(CZ + D)^* = (C, D)W(C, D)'$$

Using the previous notation the series (3) may now be written

$$\Sigma \prod_{k=1}^{m} W\{\mathfrak{n}_k\}^{s_{m-k}-s_{m-k+1}-\frac{1}{2}}$$

the sum is over all ascending chains, $\mathfrak{n}_1 \subset \ldots \subset \mathfrak{n}_m$, of primitive submodules of the module of integral vectors with the property that $\mathfrak{n}_m$ is orthogonal to itself.

Now let k be an algebraic number field as before. Let $W = (W_1, \ldots, W_n)$ be an n-tuple of matrices satisfying the same conditions as above; let $\mathfrak{m}$ be a module of rank 2m in z_{2m}; and let J be a non-degenerate skew-symmetric form with coefficients in k. We suppose, moreover, that $J_i \overline{W}_i^{-1} \overline{J}_i' = W_i$, J_i denoting the conjugates of J, and that $\mathfrak{m} J = \mathfrak{m}^{-1}$. $\mathfrak{m}^{-1}$ is defined in Section 5. Then define

$$\chi(W, \mathfrak{m}, s) = \Sigma \prod_{k=1}^{m} W\{\mathfrak{n}_k\}^{s_{m-k}-s_{m-k+1}-\frac{1}{2}} \tag{3'}$$

the sum is over all ascending chains, $\mathfrak{n}_1 \subset \ldots \subset \mathfrak{n}_m$, of primitive submodules of M such that $\mathfrak{n}_m$ is orthogonal to itself with respect to J. Let

$$\Phi(W, \mathfrak{m}, s) = \prod_{i>j} a(\tfrac{1}{2} + s_i - s_j) a(\tfrac{1}{2} + s_i + s_j) \prod_i a(\tfrac{1}{2} + s_i) \chi(W, \mathfrak{m}, s) \tag{4'}$$

THEOREM 2. (i) The series (3') converges if $\mathrm{Re}(s_{k+1} - s_k) > \frac{1}{2}$, $k = 1, \ldots, m-1$ and $\mathrm{Re}(s_1) > \frac{1}{2}$.

(ii) $\Phi(W, \mathfrak{m}, s)$ may be analytically continued to an entire symmetric function of s.

(iii) $\Phi(W, \mathfrak{m}, \pm s_1, \ldots, \pm s_m) = \Phi(W, \mathfrak{m}, s_1, \ldots, s_m)$.

The discussion of Section 2 and pp. 58-77 of [5] should provide the reader with the necessary facts about Hecke's theta-formula and its relation to Dirichlet series. It leads immediately to a proof of Theorem 1 when m = 2. For other values of m the theorem is proved by induction in Section 4. Section 3 contains a preliminary discussion of the series (1'). In Section 5 another functional equation for $\Psi(T, \mathfrak{m}, s)$ is proved and Theorem 2 is proved in Section 6. In Section 7 the relation of $\Psi(T, \mathfrak{m}, s)$ to some Dirichlet series investigated by Koecher is discussed and in Section 8 a result of Klingen on the convergence of Eisenstein series is derived.

2. Let $T = (T_1, \ldots, T_n)$ be as above and consider the series

$$\Theta(T, \mathfrak{n}_1, \ldots, \mathfrak{n}_m) = \Sigma_{\alpha} e^{-\pi c \Sigma_{k=1}^{n} (\alpha_k T_k \bar{\alpha}'_k)} \tag{5}$$

$\mathfrak{n}_1, \ldots, \mathfrak{n}_m$, are m ideals in k; $c = (\prod_i \Delta \mathfrak{n}_i)^{\frac{-1}{mn}}$ with $\Delta\mathfrak{n}_i = \Delta N^2 \mathfrak{n}_i$; the sum is over all vectors $\alpha = (a_1, \ldots, a_m)$ with a_i in $\mathfrak{n}_i$; and α_k is the kth conjugate of α. Let $\{a_{i1}, \ldots, a_{in}\}$ be a basis for $\mathfrak{n}_i$, then $\mathfrak{n}_i = \Sigma_j a_{ij} x_{ij}$ with integral x_{ij} and

$$\Sigma_{k=1}^{n} \alpha_k T_k \bar{\alpha}'_k = (x_{11}, \ldots, x_{1n}, x_{21}, \ldots, x_{mn})\, S\, (x_{11}, \ldots, x_{mn})'$$

where, denoting for the moment conjugates by superscripts and setting $T_k = (t_{ij}^k)$, S is the matrix

$$\begin{bmatrix} \begin{matrix} a_{11}^1 \cdots a_{11}^n \\ \vdots \qquad \vdots \\ a_{1n}^1 \cdots a_{1n}^n \end{matrix} & & \\ & \ddots & \\ & & \begin{matrix} a_{m1}^1 \cdots a_{m1}^1 \\ \vdots \qquad \vdots \\ a_{mn}^1 \cdots a_{mn}^n \end{matrix} \end{bmatrix} \begin{bmatrix} \begin{matrix} t_{11}^1 & & \\ & \ddots & \\ & & t_{11}^n \end{matrix} & \cdots & \begin{matrix} t_{1m}^1 & & \\ & \ddots & \\ & & t_{1m}^n \end{matrix} \\ \vdots & & \vdots \\ \begin{matrix} t_{m1}^1 & & \\ & \ddots & \\ & & t_{m1}^n \end{matrix} & \cdots & \begin{matrix} t_{mn}^1 & & \\ & \ddots & \\ & & t_{mm}^n \end{matrix} \end{bmatrix} \begin{bmatrix} \begin{matrix} \bar{a}_{11}^1 \cdots \bar{a}_{1n}^1 \\ \vdots \qquad \vdots \\ \bar{a}_{11}^n \cdots \bar{a}_{1n}^n \end{matrix} & & \\ & \ddots & \\ & & \begin{matrix} \bar{a}_{m1}^1 \cdots \bar{a}_{mn}^1 \\ \vdots \qquad \vdots \\ \bar{a}_{m1}^n \cdots \bar{a}_{mn}^n \end{matrix} \end{bmatrix}$$

The usual considerations show that

$$\Theta(T, \mathfrak{n}_1, \ldots, \mathfrak{n}_m) = \prod_k |T_k|^{-\frac{1}{2}} \Theta(\bar{T}^{-1}, \mathfrak{n}'_1, \ldots, \mathfrak{n}'_m) \tag{6}$$

where, if ϑ is the different, $\mathfrak{n}'_i = \vartheta^{-1} \mathfrak{n}_i^{-1}$. It is not difficult to show that

$$|\Theta(T, \mathfrak{n}_1, \ldots, \mathfrak{n}_m) - 1| \leq C e^{-\frac{1}{2}\|(\pi c S)^{-1}\|^{-1}} \|(\pi c S)^{-1}\|^{\frac{mn}{2}} \tag{7}$$

Let $\mathfrak{m}$ be a module in z_m of rank m and consider the series

$$\varphi(T, \mathfrak{m}, t) = \Sigma_{\{n_1\}} T\{\mathfrak{n}_1\}^{-t}$$

the sum is over all primitive submodules of $\mathfrak{m}$ of rank 1. If $\mathfrak{n}$ is any submodule of $\mathfrak{m}$ of rank 1 let z be the one-dimensional subspace of z_m generated by $\mathfrak{n}$ and set $\mathfrak{n}_1 = z \cap \mathfrak{m}$. $\mathfrak{n}_1$ is primitive and if $b = \{a \in k \mid a\mathfrak{n}_1 \in \mathfrak{n}\}$ then b is an integral ideal and $\mathfrak{n} = b\mathfrak{n}_1$. This representation of n as the product of an integral ideal and a primitive submodule is unique. Thus

$$\zeta(2t)\varphi(T, \mathfrak{m}, t) = \Sigma_n T\{\mathfrak{n}\}^{-t}$$

The sum is now over all submodules of m of rank 1 and $\zeta(2t)$ is the zeta function of k. It is known that $\mathfrak{m} = A\mathfrak{m}'$; A is some $m \times m$ matrix in k and $\mathfrak{m}' = \{\alpha = (a_1, \ldots, a_m) \mid a_1 \in \mathfrak{r}, a_2, \ldots, a_m \in \mathfrak{o}\}$, $\mathfrak{r}$ being some ideal in k. If $A'TA = (A_1'T_1A_1, \ldots, A_n'T_nA_n)$, A_i being the conjugates of A, then $\varphi(T, m, t) = \varphi(A'TA, \mathfrak{m}', t)$. Consequently it may be assumed that $\mathfrak{m} = \mathfrak{m}'$. It is also convenient to take $|T_i| = 1$, $1 = 1, \ldots, m$; then

$$\zeta(2t)\varphi(T, \mathfrak{m}, t) = \Sigma_{\{\mathfrak{r}_i\}}(N^2\mathfrak{r}_i)^{-t}\Sigma_\alpha \prod_{k=1}^m (a_k T_k \bar{a}_k')^{-t}$$

$\mathfrak{r}_i$ runs over a set of representatives of ideal classes; $\alpha = (a_1, \ldots, a_n)$ with $a_1 \in \mathfrak{r}\mathfrak{r}_i^{-1}$, $a_j \in \mathfrak{r}_i^{-1}$, $j = 2, \ldots, m$; $\alpha = 0$ is excluded from the sum and no two α differ by multiplication with a unit. For if α is of this form for some i then $\mathfrak{r}_i\alpha$ is a submodule of rank 1 in m. Conversely if $\mathfrak{n}$ is a submodule of rank 1 it has previously been observed that it may be written as $\mathfrak{b}\beta$ where β is a vector in z_m and $\mathfrak{b}$ is an ideal in k. If $\mathfrak{b}$ is in the class of $\mathfrak{r}_i$, let $\mathfrak{b} = \mathfrak{r}_i(a)$ and $\alpha = a\beta$, then $n = \mathfrak{r}_i\alpha$ and α is of the above form. Moreover α is uniquely determined up to multiplication by a unit.

Multiply by

$$(N\mathfrak{r})^{\frac{2t}{m}}\pi^{-nt}2^{-2r_2t}\Delta^t\Gamma(t)^{r_1}\Gamma(2t)^{r_2}$$

and apply the usual transformation to obtain

$$\Sigma_{\{\mathfrak{n}_i\}}\Sigma_a \int_{-\infty}^{\infty} dz_1 \cdots \int_{-\infty}^{\infty} dz_{r+1}\, e^{-\pi c \Sigma_{k=1}^{r+1} d_k (a_k T_k \bar{a}_k') e^{z_k}}\, e^{t\Sigma_{k=1}^{r+1} d_k z_k}$$

$r + 1 = r_1 + r_2$, $c_i = (\Delta(\mathfrak{n}\mathfrak{n}_i^{-1})\Delta(\mathfrak{n}_i^{-1})^{m-1})^{\frac{-1}{mn}}$ and $d_i = 1$, $1 \leq i \leq r_1$, $d_i = 2$, $r_1 < i \leq r + 1$. The familiar change of variables gives

$$(8)\quad \Sigma_{\{\mathfrak{n}_i\}} \frac{N}{w} \int_{-\infty}^{\infty} e^{ntv} dv \int_{-\frac{1}{2}}^{\frac{1}{2}} d\eta_1 \cdots \int_{-\frac{1}{2}}^{\frac{1}{2}} d\eta_r \{\Theta(e^v T \prod_{\ell=1}^{r} |\varepsilon_\ell|^{2\eta_\ell}, \mathfrak{n}\mathfrak{n}_i^{-1}, \mathfrak{n}_i^{-1}, \ldots, \mathfrak{n}_i^{-1}) - 1\}$$

w is the order of the group of roots of unity in k and $\{\varepsilon_1, \ldots, \varepsilon_r\}$ is a system of fundamental units; for the meaning of N the reader is referred to [5].

It is easy to conclude from the estimate (7) that, if $t = \sigma + i\tau$ and $\sigma > \frac{m}{2} \geq \frac{1}{2}$,

$$(9)\quad |\varphi(T, \mathfrak{m}, t)| \leq f(\sigma)^*$$

Breaking the region of integration into two parts and changing the variable of integration (8) becomes

$$\Sigma_{\{\mathfrak{n}_i\}} \frac{N}{w} \int_0^{\infty} e^{ntv} \int_{-\frac{1}{2}}^{\frac{1}{2}} d\eta_1 \cdots \int_{-\frac{1}{2}}^{\frac{1}{2}} d\eta_r \{\Theta(e^v T \prod_{\ell=1}^{r} |\varepsilon_\ell|^{2\eta_\ell}, \mathfrak{n}\mathfrak{n}_i^{-1}, \ldots, \mathfrak{n}_i^{-1}) - 1\}$$

$$+ \Sigma_{\{\mathfrak{n}_i\}} \frac{N}{w} \int_0^{\infty} e^{-ntv} \int_{-\frac{1}{2}}^{\frac{1}{2}} d\eta_1 \cdots \int_{-\frac{1}{2}}^{\frac{1}{2}} d\eta_r \{\Theta(e^{-v} T \prod_{\ell=1}^{r} |\varepsilon_\ell|^{2\eta_\ell}, \mathfrak{n}\mathfrak{n}_i^{-1}, \ldots, \mathfrak{n}_i^{-1}) - 1\}$$

Apply formula (6) to obtain

$$\Sigma_{\{\mathfrak{n}_i\}} \frac{N}{w} \int_0^{\infty} e^{ntv} \int_{-\frac{1}{2}}^{\frac{1}{2}} d\eta_1 \cdots \int_{-\frac{1}{2}}^{\frac{1}{2}} d\eta_r \{\Theta(e^v T \prod_{\ell=1}^{r} |\varepsilon_\ell|^{2\eta_\ell}, \mathfrak{n}\mathfrak{n}_i^{-1}, \mathfrak{n}_i^{-1}, \ldots, \mathfrak{n}_i^{-1}) - 1\} +$$

$$\Sigma_{\{\mathfrak{n}_i\}} \frac{N}{w} \int_0^{\infty} e^{n(\frac{m}{2} - t)v} \int_{-\frac{1}{2}}^{\frac{1}{2}} d\eta_1 \cdots \int_{-\frac{1}{2}}^{\frac{1}{2}} d\eta_r \{\Theta(e^v T^{-1} \prod_{\ell=1}^{r} |\varepsilon_\ell|^{2\eta_\ell}, \vartheta^{-1}\mathfrak{n}^{-1}\mathfrak{n}_i, \vartheta^{-1}\mathfrak{n}_i, \ldots \vartheta^{-1}\mathfrak{n}_i) - 1\}$$

$$- \frac{Nh}{wnt} - \frac{Nh}{wn(\frac{m}{2} - t)}$$

---

*Here and in the following $f(\sigma)$ is used to denote a function of the real part of a complex vector which majorizes a function of the complex vector. The function it denotes may vary from line to line.

h is the class number of the field.

It is now easy to prove Theorem 1 when $m = 1$ or 2. Indeed if $m = 1$, $z_1 = k$, $\mathfrak{m}$ is an ideal $\mathfrak{n}$, and $\Psi(T, \mathfrak{m}, s) = N^{2s}\mathfrak{n}\prod_{i=1}^{n} t_i^s$. Since, when $m = 2$, $\Psi(T, \mathfrak{m}, s_1, s_2)$ is homogeneous of degree $s_1 + s_2$ in T_i, $i = 1, \ldots, n$, it may be assumed that $|T_i| = 1$. It may also be assumed that $\mathfrak{m}$ has the form of the module $\mathfrak{m}'$ described above. Then

$$\Omega(T, \mathfrak{m}, s_1, s_2) = T\{\mathfrak{m}\}^{s_2 + \frac{1}{4}} \varphi(T, \mathfrak{m}, \tfrac{1}{2} + s_2 - s_1)$$

and the series (1') converges when $\mathrm{Re}(\frac{1}{2} + s_2 - s_1) > 1$ or $\mathrm{Re}(s_2 - s_1) > \frac{1}{2}$. Moreover, since $T\{\mathfrak{m}\} = N^2\mathfrak{n}$, $\Psi(T, \mathfrak{m}, s_1, s_2)$ is $((s_2 - s_1)^2 - \frac{1}{4})T\{\mathfrak{m}\}^{s_1+s_2}$ times the function represented by (10) when $t = \frac{1}{2} + s_2 - s_1$. Thus it is an entire function. Since $|T_i| = 1$

$$\overline{T}_i^{-1} = \begin{bmatrix} 0 & 1 \\ -1 & 0 \end{bmatrix} T_i \begin{bmatrix} 0 & -1 \\ 1 & 0 \end{bmatrix}$$

and

$$\Theta(e^{\sqrt{-1}}\overline{T}^{-1}\prod_{\ell=1}^{r}|\varepsilon_\ell|^{2\eta_\ell}, \vartheta^{-1}\mathfrak{n}^{-1}\mathfrak{n}_i, \vartheta^{-1}\mathfrak{n}_i) = \Theta(e^{\mathrm{v}}T\prod_{\ell=1}^{r}|\varepsilon_\ell|^{2\eta_\ell}, \vartheta^{-1}\mathfrak{n}_i, \vartheta^{-1}\mathfrak{n}^{-1}\mathfrak{n}_i)$$

However the function (10) does not depend on the representatives of the ideal classes chosen so $\mathfrak{n}_i$ could be replaced by $\vartheta\mathfrak{n}\mathfrak{n}_i^{-1}$. The result is the same as that obtained by interchanging s_1 and s_2; thus (10) is a symmetric function of s_1 and s_2 and so is $\Psi(T, \mathfrak{m}, s_1, s_2)$.

3. If $\mathfrak{n}_k$ and $\mathfrak{m}_k$ are two primitive submodules of $\mathfrak{m}$ of rank k such that $\mathfrak{n}_k^{\ k} = \mathfrak{m}_k^{\ k}$ then $\mathfrak{n}_k$ and $\mathfrak{m}_k$ lie in the same k-dimensional subspace of z_m and, thus, must be the same. Consequently the series for $\Omega(T, \mathfrak{m}, s)$ is majorized by, setting $s = \sigma + i\tau$,

$$T\{\mathfrak{m}\}^{\sigma_m + \frac{m-1}{4}} \prod_{k=1}^{m-1} \varphi(T^k, \mathfrak{m}^k, \tfrac{1}{2} + \sigma_{k+1} - \sigma_k)$$

So, when $\frac{1}{2} + \sigma_{k+1} - \sigma_k > \frac{1}{2}\binom{m}{k}$, $k = 1, \ldots, m-1$, the series converges and, using (9),

$$|\Psi(T, \mathfrak{m}, s)| \leq f(\sigma) \prod_{i \neq j} (|s_i - s_j| + 1) \tag{11}$$

This is not the region of convergence promised in part (i) of Theorem 1; however (i) will follow from (ii) and Landau's Theorem on Dirichlet series with positive coefficients.

Before proceeding with the proof of (ii) it will be convenient to describe certain useful arrangements of the series (1'). That series may be written

$$T\{\mathfrak{m}\}^{\frac{m+1}{4}} \Sigma_{\{\mathfrak{n}_k\}} \{\Sigma \prod_{j=1}^{k} T\{\mathfrak{n}_j\}^{s_j - s_{j+1} - \frac{1}{2}}\} \{\Sigma \prod_{j=k+1}^{m} T\{\mathfrak{n}_j\}^{s_j - s_{j+1} - \frac{1}{2}}\}$$

The outer sum is over all primitive submodules of rank k. The first inner sum is over all chains, $\mathfrak{n}_1 \subset \mathfrak{n}_2 \subset \ldots \subset \mathfrak{n}_k$, of primitive submodules ending at $\mathfrak{n}_k$; the second is over all chains, $\mathfrak{n}_k \subset \mathfrak{n}_{k+1} \subset \ldots \subset \mathfrak{n}_m$, beginning at $\mathfrak{n}_k$.

It was observed above that for each $\mathfrak{n}_k$ there is a submodule $\mathfrak{p}$ such that $\mathfrak{m} = \mathfrak{n}_k \oplus \mathfrak{p}$. Choose bases $\{\alpha_1, \ldots, \alpha_k\}$ and $\{\alpha_{k+1}, \ldots, \alpha_m\}$ for the subspaces of z_m generated by $\mathfrak{n}_k$ and $\mathfrak{p}$ respectively. Then

$$\mathfrak{n}' = \{a = (a_1, \ldots, a_k) | \Sigma a_i \alpha_i \in \mathfrak{n}_k\}$$

and

$$\mathfrak{p}' = \{\beta = (b_{k+1}, \ldots, b_m) | \Sigma b_i \alpha_i \in p\}$$

are finitely generated modules in z_k and z_{m-k}. To simplify calculations assume that $\mathfrak{n}' = \{(a_1, \ldots, a_k) | a_1 \in \mathfrak{b}, a_2, \ldots, a_k \in \mathfrak{o}\}$; $\mathfrak{b}$ is some ideal in k. Let B be the matrix $(\alpha_1' \ldots \alpha_k')'$ and A the matrix $(\alpha_1' \ldots \alpha_m')'$; then set $R = BT\overline{B}'$. It is convenient to omit any explicit reference to the components in such equations.

There is a one-to-one correspondence between chains $\mathfrak{n}_1 \subset \ldots \subset \mathfrak{n}_k$ ending at $\mathfrak{n}_k$ and chains $\mathfrak{n}_1' \subset \ldots \subset \mathfrak{n}_k'$ in $\mathfrak{n}'$. Morevoer $T\{\mathfrak{n}_j\} = R\{\mathfrak{n}_j'\}$. Consequently the first inner sum is

$$\Omega(R, \mathfrak{n}', s_1 - s_{k+1} - \frac{k+1}{4}, \ldots, s_k - s_{k+1} - \frac{k+1}{4})$$

There is also a one-to-one correspondence between chains $\mathfrak{n}_k \subset \mathfrak{n}_{k+1} \subset \ldots \subset \mathfrak{n}_m$, chains in $\mathfrak{g}$, and chains in $\mathfrak{q}_1 \subset \mathfrak{q}_2 \subset \ldots \subset \mathfrak{q}_{m-k}$ in $\mathfrak{g}'$. Introduce the n-tuple of matrices

$$S = \begin{bmatrix} AT\overline{A}'_{(1,\ldots,k,k+1;1,\ldots,k,k+1)} \ldots AT\overline{A}'_{(1,\ldots,k,k+1;1,\ldots,k,m)} \\ \vdots \qquad\qquad \vdots \\ AT\overline{A}'_{(1,\ldots,k,m;1,\ldots,k,k+1)} \ldots AT\overline{A}'_{(1,\ldots,k,m;1,\ldots,k,m)} \end{bmatrix}$$

If $H = (h_{ij})$ is any matrix $H_{(i_1,\ldots,i_\ell;j_1,\ldots,j_\ell)}$ is the determinant of the matrix $(h_{i_u j_v})$, $u, v = 1, \ldots, \ell$. Since

$$S_{(i_1,\ldots,i_\ell;j_1,\ldots,j_\ell)}$$

is equal to

$$(AT\overline{A}'_{(1,\ldots,k;1,\ldots,k)})^{\ell-1} \times AT\overline{A}'_{(1,\ldots,k,i_1,\ldots,i_\ell;1,\ldots,k,j_1,\ldots,j_\ell)}$$

it is not difficult to show that

$$(12) \qquad T\{\mathfrak{n}_{k+\ell}\} = N^2\mathfrak{b} \prod (AT\overline{A}'_{(1,\ldots,k;1,\ldots,k)})^{1-\ell} S\{\mathfrak{q}_\ell\}$$

The product is the product of the indicated subdeterminants of the components of $AT\overline{A}'$. Consequently the second inner sum with the factor $T\{\mathfrak{m}\}^{\frac{m+1}{4}}$ incorporated is the product of

$$T\{\mathfrak{m}\}^{\frac{k}{4}}(N^2\mathfrak{g})^{s_{k+1}-\frac{m-k-1}{4}}\prod(AT\bar{A}'_{(1,\ldots,k;1,\ldots,k)})^{-s_{k+2}\cdots-s_m-\frac{m-k-1}{4}}$$

and

$$\Omega(S,\ \mathfrak{g}',\ s_{k+1},\ \ldots,\ s_m)$$

However

$$N^2\mathfrak{b}\prod AT\bar{A}'_{(1,\ldots,k;1,\ldots,k)} = T\{\mathfrak{n}_k\} \tag{13}$$

and the factor $T\{\mathfrak{n}_k\}^{-s_{k+2}-\cdots-s_m-\frac{m-k-1}{4}}$ may be absorbed into the first sum.

The result is

$$T\{\mathfrak{m}\}^{\frac{k}{4}}\Sigma_{\{\mathfrak{n}_k\}}(N^2\mathfrak{b})^{s_{k+1}+\cdots+s_m}\Omega(S,\ \mathfrak{g}',\ s_{k+1},\ \ldots,\ s_m)\Omega(R,\ \mathfrak{n}',\ r)$$

with

$$r = (s_1 - s_{k+1} - \cdots - s_m - \frac{m}{4},\ \ldots,\ s_k - s_{k+1} - \cdots - s_m - \frac{m}{4})$$

There is a corresponding representation of Ψ:

$$\Psi(T,\mathfrak{m},s) = \gamma_k(s)T\{\mathfrak{m}\}^{\frac{k}{4}}\Sigma_{\mathfrak{n}_k}(N^2\mathfrak{b})^{s_{k+1}+\cdots+s_m}\Psi(S,\mathfrak{g}',s_{k+1},\ldots,s_m)\Psi(R,\mathfrak{n}'',r) \tag{14}$$

with

$$\gamma_k(s) = \prod_{i>k,\ j\le k} a(\tfrac{1}{2} + s_i - s_j)$$

The series (14) converges if $\sigma_{k+1} - \sigma_k \geq b$ (b is a suitable positive constant), $k = 1, \ldots, m-1$. Assume that parts (ii) and (iii) of the theorem are true for k and $m-k$. Then the series is symmetric in the first k and last $m-k$ coordinates of s. Thus, if for some permutation π of $\{1, \ldots, m\}$ which leaves the sets $\{1, \ldots, k\}$ and $\{k+1, \ldots, m\}$ invariant $\sigma_{\pi(k+1)} - \sigma_{\pi(k)} \geq b$, $k = 1, \ldots, m-1$, the series will converge and the estimate (11) will be valid. It will now be shown that the series converges in the region defined by

(15) $$\sigma_i - \sigma_j \geq c_{m-k}(b) + c_k(b);\ i > k,\ j \leq k$$

$c_{m-k}(b)$ and $c_k(b)$ are constants obtained from the following lemma.

LEMMA. _If_ $\gamma = (\gamma_1, \ldots, \gamma_m)$ _is an_ m-_tuple of real numbers and_ b _is a positive constant there are_ m-_tuples_ γ', γ'' _such that_

(i) $\gamma = \frac{1}{2}(\gamma' + \gamma'')$.

(ii) $|\gamma_i - \gamma_i'| \leq c_m(b)$ _and_ $|\gamma_i - \gamma_i''| \leq c_m(b)$.

(iii) _There are permutations_ π' _and_ π'' _so that_

$$\gamma'_{\pi'(k+1)} - \gamma'_{\pi'(k)} \geq b,\ \gamma''_{\pi''(k+1)} - \gamma''_{\pi''(k)} \geq b,\quad k = 1, \ldots, m-1$$

$c_m(b)$ _is a constant depending only on_ m _and_ b.

Suppose the lemma has been proven for $1, \ldots, m-1$. It may be supposed that $c_1(b) \leq c_2(b) \leq \cdots \leq c_{m-1}(b)$ and that $\gamma_1 \geq \gamma_2 \geq \cdots \geq \gamma_m$. If $\gamma_1 - \gamma_m \geq (m-1)(2c_{m-1}(b) + b)$ then for some k, $\gamma_k - \gamma_{k+1} \geq 2c_{m-1}(b) + b$. Apply the lemma to the vectors $(\gamma_1, \ldots, \gamma_k)$ and $(\gamma_{k+1}, \ldots, \gamma_m)$ to obtain $\gamma_1', \ldots, \gamma_m', \gamma_1'', \ldots, \gamma_m''$. These m-tuples satisfy the conditions of the lemma if $c_m(b) \geq c_{m-1}(b)$. If $\gamma_1 - \gamma_m < (m-1)(2c_{m-1}(b) + b)$ set $a = 2(m-1)c_{m-1}(b) + mb$ and

$$\gamma_1' = \gamma_1 + (m-1)a,\ \gamma_2' = \gamma_2 + (m-2)a,\ \ldots,\ \gamma_m' = \gamma_m$$
$$\gamma_1'' = \gamma_1 - (m-1)a,\ \gamma_2'' = \gamma_2 - (m-2)a,\ \ldots,\ \gamma_m'' = \gamma_m$$

Then

$$\gamma_k' - \gamma_{k+1}' = a + \gamma_k - \gamma_{k+1} \geq a - (m-1)(2c_{m-1}(b) + b) = b$$
$$\gamma_{k+1}'' - \gamma_k'' = a + \gamma_{k+1} - \gamma_k \geq a - (m-1)(2c_{m-1}(b) + b) = b$$

This proves the lemma if $c_m(b) = (m-1)a$.

If $s = (s_1, \ldots, s_m) = (\sigma_1 + i\tau_1, \ldots, \sigma_m + i\tau_m)$ is in the region defined by

(13) apply the lemma to $(\sigma_1, \ldots, \sigma_k)$ and $(\sigma_{k+1}, \ldots, \sigma_m)$ to obtain $\sigma'_1, \ldots, \sigma'_m, \sigma''_1, \ldots, \sigma''_m$. Set

$$s' = (\sigma'_1 + i\tau_1, \ldots, \sigma'_m + i\tau_m)$$

$$s'' = (\sigma''_1 + i\tau_1, \ldots, \sigma''_m + i\tau_m)$$

$$s(t) = ts' + (1-t)s''$$

Then the series (14) may be written

$$T\{\mathfrak{m}\}^{\frac{k}{4}} \Sigma_{\{\mathfrak{n}_k\}} \frac{e^{-\frac{1}{4}}}{2\pi i} \int_{1-i\infty}^{1+i\infty} dt + \int_{i\infty}^{-i\infty} dt\, e^{t^2} \gamma_k(s(t))(N^2 \mathfrak{b})^{s_{k+1}(t)+\ldots+s_m(t)}$$

(16)

$$\cdot\ \Psi(S, \mathfrak{g}', s_{k+1}(t), \ldots, s_m(t))\Psi(R, \mathfrak{n}', r(t))$$

Because of the assumed validity of (iii) the integrals converge. Inverting the order of integration and summation gives a series with a convergent majorant of the form

$$|e^{t^2}| \prod_{i\neq j}(|s_i(t) - s_j(t)| + 1)\Sigma_{\{\mathfrak{n}_k\}} f(\mathfrak{n}_k, \sigma)$$

σ is the real part of t. Consequently (14) converges and is equal to

$$T\{\mathfrak{m}\}^{\frac{k}{4}} \frac{e^{-\frac{1}{4}}}{2\pi i} \int_{1-i\infty}^{1+i\infty} dt + \int_{i\infty}^{-i\infty} dt\, e^{t^2} \Psi(T, \mathfrak{m}, s(t))$$

So $\Psi(T, \mathfrak{m}, s)$ is defined; moreover

$$|\Psi(T, \mathfrak{m}, s)| \leq c \int_{-\infty}^{\infty} e^{-t^2} \{\prod_{i\neq j}(|s_i(it) - s_j(it)| + 1) f(\sigma')$$

$$+ \prod_{i\neq j}(|s_i(1+it) - s_j(1 + it)| + 1) f(\sigma'')\} dt$$

$$\leq f(\sigma) \prod_{i\neq j}(|s_i - s_j| + 1)$$

σ is the real part of s.

4. The theorem will now be proved by induction. It is sufficient to show that, for each $N > 0$, $\Psi(T, \mathfrak{m}, s)$ may be continued analytically to the region: $\Sigma_{i=1}^{m-2} |\sigma_i| < N$, σ_{m-1} and σ_m arbitrary. The diagram represents a decomposition of this region into four overlapping parts. The region I lies in the region defined

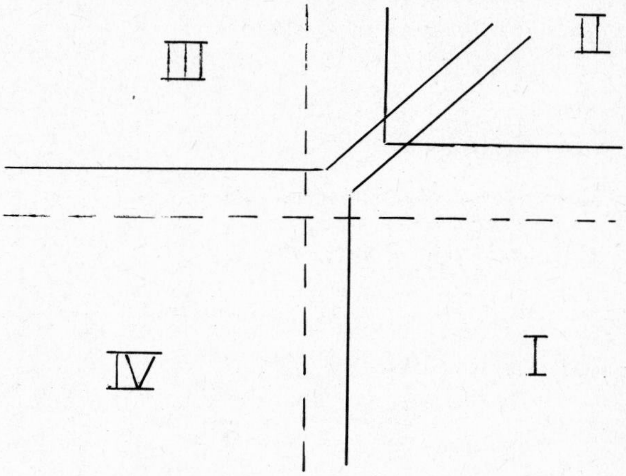

by (15) when $k = m - 1$. The region II lies in the region defined by (15) when $k = m - 2$. Moreover when $k = m - 1$ or $m - 2$ the assumption of Section 3 is part of the induction hypothesis. Consequently $\Psi(T, \mathfrak{m}, s)$ may be continued analytically to the regions I and II. Moreover it will be symmetric there in s_1 and s_2; consequently it may be extended to III. The inequality (11) will be valid in these regions.

To extend Ψ to the region IV let

$$\xi_1 = s_{m-1} + s_m$$
$$\xi_2 = s_{m-1} - s_m$$

Then, taking c large enough, the formula

$$\Psi(T, \mathfrak{m}, s) = \frac{e^{-\xi_2^2}}{2\pi i} \int_{c-i\infty}^{c+i\infty} d\zeta + \int_{-c+i\infty}^{-c-i\infty} d\zeta \, \frac{e^{\zeta^2} \Psi\left(T, \mathfrak{m}, s_1, \ldots, s_{m-2}, \frac{\xi_1 + \zeta}{2}, \frac{\xi_1 - \zeta}{2}\right)}{\zeta - \xi_2}$$

effects the desired continuation to IV. Moreover the inequality (11) is easily shown to remain valid.

5. If $\mathfrak{m}$ is a module of rank m in z_m let $\mathfrak{m}^{-1} = \{\beta \mid \alpha\beta' \in \mathfrak{v}$ for all $\alpha \in \mathfrak{m}\}$. As an essentially simple consequence of the definition

$$(17) \qquad \Psi(T, \mathfrak{m}, s) = \Psi(\overline{T}^{-1}, \mathfrak{m}^{-1}, -s)$$

Indeed, to establish this it is sufficient to show that

$$(18) \qquad \Omega(T, \mathfrak{m}, s_1, \ldots, s_m) = \Omega(\overline{T}^{-1}, \mathfrak{m}^{-1}, -s_m, \ldots, -s_1)$$

in the common region of convergence for the two series. If $\mathfrak{n}_k$ is a submodule of rank k let $\mathfrak{q}_{m-k} = \{\beta \in \mathfrak{m}^{-1} \mid \alpha\beta' = 0$ for all $\alpha \in \mathfrak{n}_k\}$. $\mathfrak{q}_{m-k}$ is primitive and corresponding to the chain $\mathfrak{n}_1 \subset \ldots \subset \mathfrak{n}_m$ is the chain $\mathfrak{q}_1 \subset \ldots \subset \mathfrak{q}_m = \mathfrak{m}^{-1}$. To prove (18) it is sufficient to show

$$(19) \quad T\{\mathfrak{m}\}^{\frac{m+1}{4}} \pi_{k=1}^{m} T\{\mathfrak{n}_k\}^{s_k - s_{k+1} - \frac{1}{2}} = \overline{T}^{-1}\{\mathfrak{m}^{-1}\}^{\frac{m+1}{4}} \pi_{k=0}^{m-1} \overline{T}^{-1}\{\mathfrak{q}_{m-k}\}^{s_k - s_{k+1} - \frac{1}{2}}$$

Of course, $s_0 = 0$. Replacing T by $AT\overline{A}'$ if necessary, it may be assumed that $\mathfrak{n}_k = \{\alpha = (a_1, \ldots, a_k, 0, \ldots, 0) \mid a_i \in \mathfrak{v}_i\}$, $k = 1, \ldots, m$; $\mathfrak{v}_i$, $i = 1, \ldots, m$, being some ideal in k. Then

$$\mathfrak{q}_{m-k} = \{\beta = (0, \ldots, 0, b_{k+1}, \ldots, b_m) \mid b_i \in \mathfrak{v}_i^{-1}\}, \ k = 0, \ldots, m-1$$

Since both sides are homogeneous of degree $\Sigma_i s_i$ in T_j it may be assumed that $|T_j| = 1$, $j = 1, \ldots, n$. Then

$$T\{\mathfrak{n}_k\} = N^2(\mathfrak{v}_1 \cdots \mathfrak{v}_k) \prod T_{(1, \ldots, k; 1, \ldots, k)}$$

and

$$\overline{T}^{-1}\{\mathfrak{q}_{m-k}\} = N^2(\mathfrak{v}_m^{-1} \cdots \mathfrak{v}_{k+1}^{-1}) \prod \overline{T}^{-1}_{(k+1, \ldots, m; k+1, \ldots, m)}$$

$$= N^2(\mathfrak{v}_m^{-1} \cdots \mathfrak{v}_{k+1}^{-1}) \prod T_{(1, \ldots, k; 1, \ldots, k)}$$

The product is over the indicated subdeterminants of the components of T or $\overline{T}^{-1}$; there is no convenient place for the subscripts. Thus the left side of (19) is

$$(N^2\mathfrak{n}_1)^{s_1+\frac{1-m}{4}}(N^2\mathfrak{n}_2)^{s_1+\frac{3-m}{4}}\dots(N^2\mathfrak{n}_m)^{s_m+\frac{m-1}{4}}\prod_{k=1}^{m}\prod(T_{(1,\dots,k;1,\dots,k)})^{s_k-s_{k+1}}$$

and the right side is

$$(N^2\mathfrak{n}_m^{-1})^{-s_m+\frac{1-m}{4}}\dots(N^2\mathfrak{n}_1^{-1})^{-s_1+\frac{m-1}{4}}\prod_{k=1}^{m}\prod(T_{(1,\dots,k;1,\dots,k)})^{s_k-s_{k+1}-\frac{1}{2}}$$

which establishes (19).

6. The proof of Theorem 2 will now be given omitting, however, that part of the analysis which is merely a repetition of the above. As in the proof of Theorem 1, the series for $\chi(W, \mathfrak{m}, s)$ is majorized by

$$\prod_{k=1}^{m} \varphi(W^k, \mathfrak{m}^k, \tfrac{1}{2} + s_{m-k+1} - s_{m-k})$$

with $s_0 = 0$. Consequently it converges for

(20) $$\mathrm{Re}(s_{k+1} - s_k) > b, \quad k = 1, \ldots, m-1, \ \mathrm{Re}(s_1) > b$$

b is some positive constant. The series for $\Phi(W, \mathfrak{m}, s)$ may be written

$$\prod_{i>j} a(\tfrac{1}{2}+s_i-s_j) \prod_i a(\tfrac{1}{2}+s_i) \Sigma_{\{n_m\}} \Sigma \prod_{i>j} a(\tfrac{1}{2}+s_i-s_j) \prod_{k=1}^{m} W\{\mathfrak{n}_k\}^{s_{m-k}-s_{m-k+1}-\frac{1}{2}}$$

The outer sum is over all primitve submodules which are orthogonal to themselves with respect to J; the inner sum is over all chains $\mathfrak{n}_1 \subset \ldots \subset \mathfrak{n}_m$ of primitive submodules which end at $\mathfrak{n}_m$. For each $\mathfrak{n}_m$ choose a basis $\{\alpha_1, \ldots, \alpha_m\}$ for the subspace of z_{2m} spanned by $\mathfrak{n}_m$. Set $A = (\alpha_1' \ldots \alpha_m')'$ and let $V = AW\bar{A}'$; then the inner sum is

$$\Psi(V, \mathfrak{n}, -s_m - \frac{m+1}{4}, \ldots, -s_1 - \frac{m+1}{4})$$

with $\mathfrak{n} = \{(a_1, \ldots, a_m) | \Sigma a_i \alpha_i \in \mathfrak{n}_m\}$. Then

(21) $$\Phi(W, \mathfrak{m}, s) = \prod_{i>j} a(\tfrac{1}{2}+s_i+s_j) \prod_i a(\tfrac{1}{2}+s_i) \Sigma \Psi(V, \mathfrak{n}, -s_m - \frac{m+1}{4}, \ldots, -s_1 - \frac{m+1}{4})$$

Using the lemma and techniques of Section 3 it can be shown that the series (21) converges for $\mathrm{Re}(s_i) > b_i$, $i = 1, \ldots, m$, and represents a symmetric function of $(s_1, \ldots, s_m)$.

To continue the function to negative values of the arguments the arrangement

(22) $$\prod_{i>j} a(\tfrac{1}{2}+s_i+s_j) a(\tfrac{1}{2}+s_i-s_j) \prod_{i=2}^{m} a(\tfrac{1}{2}+s_i) \Sigma\Sigma a(\tfrac{1}{2}+s_1) \prod_{k=1}^{m} W\{\mathfrak{n}_k\}^{s_{m-k}-s_{m-k+1}-\frac{1}{2}}$$

is used. The outer sum is over all chains $\mathfrak{n}_1 \subset \ldots \subset \mathfrak{n}_{m-1}$ such that $\mathfrak{n}_{m-1}$ is orthogonal to itself; the inner sum is over all primitive submodules $\mathfrak{n}_m$ such that $\mathfrak{n}_{m-1} \subset \mathfrak{n}_m \subset \mathfrak{n}^{\perp}_{m-1}$. $\mathfrak{n}^{\perp}_{m-1}$ is the orthogonal complement of $\mathfrak{n}_{m-1}$ with respect to J.

Let $\{\beta_1, \ldots, \beta_{m+1}\}$ be a basis for the subspace of z_{2m} generated by $\mathfrak{n}_{m-1}$. Set $B = (\beta'_1 \ldots \beta'_{m+1})'$ and $U = BW\overline{B}'$. Let

$$\mathfrak{p} = \{(b_1, \ldots, b_{m+1}) | \Sigma b_i \beta_i \in \mathfrak{n}^{\perp}_{m-1}\}$$

and

$$\mathfrak{q} = \{(b_1, \ldots, b_{m+1}) \Sigma b_i \beta_i \in \mathfrak{n}_{m-1}\}$$

Now in the argument preceding (12) replace $\mathfrak{n}_k$ by $\mathfrak{q}$, $\mathfrak{m}$ by $\mathfrak{p}$ and T by U. We conclude that there is a module $\mathfrak{q}_2$ of rank 2 in z_2, an n-tuple S of 2×2 matrices, an ideal $\mathfrak{b}$, and a one-to-one correspondence between the primitive submodules $\mathfrak{n}_m$ and primitive submodules, $\mathfrak{q}_1$, of rank 1 in $\mathfrak{q}_2$ such that

$$W\{\mathfrak{n}_m\} = N^2 \mathfrak{b}\, S\{\mathfrak{q}_1\}$$

Consequently the inner sum is

$$\prod_{k=1}^{m-1} W\{\mathfrak{n}_k\}^{s_{m-k} - s_{m-k+1} - \frac{1}{2}} (N^2 \mathfrak{b})^{-s_1 - \frac{1}{2}} \Psi(S, \mathfrak{q}_2, -s_1 - \tfrac{1}{4}, -\tfrac{1}{4})$$

which equals

$$\prod_{k=1}^{m-1} W\{\mathfrak{n}_k\}^{s_{m-k} - s_{m-k+1} - \frac{1}{2}} (N^2 \mathfrak{b})^{-s_1 - \frac{1}{2}} \Psi(S, \mathfrak{q}_2, -\tfrac{1}{4}, -s_1 - \tfrac{1}{4})$$

or

$$\prod_{k=1}^{m-1} W\{\mathfrak{n}_k\}^{s_{m-k} - s_{m-k+1} - \frac{1}{2}} (N^2 \mathfrak{b})^{-s_1 - \frac{1}{2}} S\{\mathfrak{q}_2\}^{-s_1} \Psi(S, \mathfrak{q}_2, s_1 - \tfrac{1}{4}, -\tfrac{1}{4})$$

However, by formulae (12) and (13)

$$S\{\mathfrak{q}^2\} = (N^2 \mathfrak{b})^{-2} W\{\mathfrak{n}^{\perp}_{m-1}\} W\{\mathfrak{n}_{m-1}\}$$

By the proof of formula (19)

$$W\{\mathfrak{n}^{\perp}_{m-1}\} = W\{\mathfrak{m}\}\overline{W}^{-1}\{\mathfrak{q}_{m-1}\}$$

$\mathfrak{q}_{m-1}$ is the orthogonal complement of $\mathfrak{n}^{\perp}_{m-1}$ in $\mathfrak{m}^{-1}$. Since $\mathfrak{m}^{-1} = \mathfrak{m}J$, $\mathfrak{q}_{m-1} = \mathfrak{n}_{m-1}J$. Moreover $W\{\mathfrak{m}\} = \prod_i |W_i| N^2(\mathfrak{m}^{2m})$; $(\mathfrak{m}^{-1})^{2m} = |J|\mathfrak{m}^{2m}$ so that $N^2(\mathfrak{m}^{2m}) = N^{-1}(|J|)$; and $J_i\overline{W}_i^{-1}\overline{J}_i^{-1} = W_i$ so that $N^2(|J|) = \prod_i |W_i|^2$. Consequently $W\{\mathfrak{m}\} = 1$. Finally

$$W\{\mathfrak{n}^{\perp}_{m-1}\} = \overline{W}^{-1}\{\mathfrak{n}_{m-1}J\} = W\{\mathfrak{n}_{m-1}\}$$

Thus the inner sum in (22) equals

$$\prod_{k=1}^{m-1} W\{\mathfrak{n}_k\}^{s_{m-k}-s_{m-k+1}-\frac{1}{2}} W\{\mathfrak{n}_{m-1}\}^{-s_1-s_2-\frac{1}{2}} (N^2\mathfrak{f})^{s_1-\frac{1}{2}} \Psi(S, \mathfrak{q}_2, s_1-\tfrac{1}{4}, -\tfrac{1}{4})$$

So it is an entire function of s_1 which is invariant when s_1 changes sign. Using previous methods it may be concluded that $\Phi(T, \mathfrak{m}, s)$ may be continued to the region: $\mathrm{Re}(s_i) > b_i$, $i = 2, \ldots, m$, s_1 arbitrary. It may then be continued to any domain obtained from this one by permuting the variables. The continuation to the entire m-dimensional space is then effected by Cauchy's integral formula.

7. Koecher [4] establishes, at least when k is the rational field, a functional equation for the series

(23) $$\zeta_k(T, \mathfrak{m}, t) = \Sigma_{\{n\}} T\{\mathfrak{n}\}^{-t}$$

where the sum is taken over all submodules of rank k of a given module $\mathfrak{m}$, of rank m, contained in z_m. It will be shown in this section that for special values of s the function $\Psi(T, \mathfrak{m}, s)$ reduces, apart from a factor depending only on t, to $\zeta_k(T, \mathfrak{m}, t)$ and that the functional equation for ζ_k is a special case of the functional equations for Ψ. The factor however has too many zeros and it is apparently not possible to deduce Koecher's results on the poles of ζ_k from the fact that Ψ is entire. These may be established by separate arguments similar to those above.

The series in (23) may be reduced to a sum over primitive submodules. If $\mathfrak{n}$ is a submodule of rank k then $\mathfrak{n}$ may be uniquely represented as $\mathfrak{a}\mathfrak{n}_k$; $\mathfrak{n}_k$ is a primitive submodule and $\mathfrak{a}$ is an integral right ideal in the ring of endomorphisms of $\mathfrak{n}_k$. Using the theory of algebras, as presented in [1], it is not difficult to show that (23) equals

(24) $$\zeta(2t) \ldots \zeta(2t - (k-1)\Sigma_{\{\mathfrak{n}_k\}} T\{\mathfrak{n}_k\}^{-t}$$

the sum now being over all primitive submodules of rank k, $\zeta(\cdot)$ is the zeta-function of the given field k.

Using formula (14) of Section 3 it may be shown by induction that

(25) $$\Psi(T, \mathfrak{m}, t - \frac{m-1}{4}, t - \frac{m-3}{4}, \ldots, t + \frac{m-1}{4}) = \gamma_m T\{\mathfrak{m}\}^t$$

with

$$\gamma_m = (\frac{Nh}{wn})^{m-1} (\frac{2\cdot 1}{4})^{m-2} (\frac{3\cdot 2}{4})^{m-3} \ldots \frac{(m-1)(m-2)}{4} \frac{m!}{2^{m-1}} b(1)^{m-1} b(\tfrac{3}{2})^{m-2} \ldots b(\frac{m-1}{2})$$

$$b(t) = \pi^{-nt} 2^{-2r_2 t} \Delta^t \Gamma(t)^{r_1} \Gamma(2t)^{r_2} \zeta(2t)$$

Indeed, by the induction hypothesis and formula (14) with $k = 1$

$$\Psi(T,\mathfrak{m}, s_1, t-\frac{m-3}{4}, \ldots, t+\frac{m-1}{4}) = \prod_{i=2}^{m} a(\tfrac{1}{2}+s_i-s_1)\gamma_{m-1} T\{\mathfrak{m}\}^{t+\frac{1}{2}} \Sigma_{\{n_1\}} T\{\mathfrak{n}_1\}^{s_1-t-\frac{m+1}{4}}$$

Setting $s_1 - t - \frac{m+1}{4} = -\frac{m}{2}$ or $s_1 = t - \frac{m-1}{4}$ and applying formula (10), (25) follows.

As a consequence

$$\Psi(T,\mathfrak{m}, s_1, \ldots, s_k, -\frac{m-1}{4}, -\frac{m-1}{4}+\tfrac{1}{2}, \ldots, -\frac{m-1}{4}+\frac{m-k-1}{2})$$

is equal to

$$\gamma_k(s)\gamma_{m-k} T\{\mathfrak{m}\}^{\frac{k}{4}} \Sigma_{\{\mathfrak{n}_k\}} (N^2\mathfrak{f})^{s_{k+1}+\ldots+s_m} S(\mathfrak{f}')^{\frac{-k}{4}} \Psi(R,\mathfrak{n}', s_1+(m-k)\frac{k}{4}-\frac{m}{4}, \ldots)$$

or

$$\gamma_k(s)\gamma_{m-k}\Sigma_{\{\mathfrak{n}_k\}} \Psi(R, \mathfrak{n}', s_1 - \frac{m-k}{4}, \ldots, s_k - \frac{m-k}{4})$$

Now let $s_k = -t + \frac{m-1}{4}, \ldots, s_1 = -t + \frac{m-1}{4} - \frac{k-1}{2}$ and apply (25) again to obtain

$$\gamma_k\gamma_{m-k}\gamma_k(s)\Sigma_{\{\mathfrak{n}_k\}} T\{\mathfrak{n}_k\}^{-t} \tag{26}$$

Let

$$a'(t) = \pi^{-nt} 2^{-2r_2 t} \Delta^t \Gamma(t)^{r_1} \Gamma(2t)^{r_2}$$

$$\psi(t) = t(t-\tfrac{1}{2})a'(t)\zeta(2t)$$

and

$$\Psi_k(T,\mathfrak{m}, t) = \prod_{j=0}^{k-1}(t-\tfrac{j}{2})(t-\tfrac{m-j}{2})a'(t-\tfrac{j}{2})\zeta_k(T,\mathfrak{m}, t)$$

Finally, let $\psi_k(t)$ be the function obtained by multiplying together all terms of the matrix

$$\begin{bmatrix} \psi(t-\frac{k}{2}) & \psi(t-\frac{k+1}{2}) & \cdots\cdots\cdots & \psi(t-\frac{m-2}{2}) \\ \psi(t-\frac{k-1}{2}) & & & \vdots \\ \vdots & & & \vdots \\ \psi(t-\frac{1}{2}) & \cdots\cdots\cdots & & \psi(t-\frac{m-k-1}{2}) \end{bmatrix}$$

Then, if $k < m$, (26) equals

$$\gamma_k \gamma_{m-k} \psi_k(t) \Psi_k(T, \mathfrak{m}, t)$$

Since $\psi(t) = \psi(\frac{1}{2} - t)$ replacing t by $\frac{m}{2} - t$ in the above matrix gives the same result as reflecting it in its centre. Consequently

$$\psi_k(t) = \psi_k(\frac{m}{2} - t)$$

Making use of the functional equations for ψ, we see that

$$\gamma_k \gamma_{m-k} \psi_k(t) \Psi_k(T, \mathfrak{m}, t)$$

is equal to

$$\Psi(T, \mathfrak{m}, -t + \frac{m-1}{4} - \frac{k-1}{2}, \ldots, -t + \frac{m-1}{4}, -\frac{m-1}{4}, \ldots, -\frac{m-1}{4} + \frac{m-k-1}{2})$$

or

$$\Psi(\overline{T}^{-1}, \mathfrak{m}^{-1}, t - \frac{m-1}{4}, \ldots, t - \frac{m-1}{4} + \frac{k-1}{2}, \frac{m-1}{4} - \frac{m-k-1}{2}, \ldots, \frac{m-1}{4})$$

This is the same as

$$T\{\mathfrak{m}\}^{-\frac{k}{2}} \Psi(\overline{T}^{-1}, \mathfrak{m}^{-1}, t - \frac{m}{2} + \frac{m-1}{4} - \frac{k-1}{2}, \ldots, t - \frac{m}{2} + \frac{m-1}{4}, -\frac{m-1}{4}, \ldots, -\frac{m-1}{4} + \frac{m-k-1}{2})$$

which equals

$$\gamma_k \gamma_{m-k} \psi_k(\frac{m}{2} - t) T\{\mathfrak{m}\}^{-\frac{k}{2}} \Psi_k(\overline{T}^{-1}, \mathfrak{m}^{-1}, \frac{m}{2} - t)$$

So

$$\Psi_k(T, \mathfrak{m}, t) = T\{\mathfrak{m}\}^{-\frac{k}{2}} \Psi_k(\overline{T}^{-1}, \mathfrak{m}^{-1}, \frac{m}{2} - t)$$

This is the functional equation of Koecher.

Suppose for the moment that k is the rational field. According to equation (3.17) of [4] $\Psi_k(T, \mathfrak{m}, t)$ is zero at the numbers common to $\{0, \frac{1}{2}, \ldots, \frac{k-1}{2}\}$ and $\{\frac{m-k+1}{2}, \ldots, \frac{m}{2}\}$. However if $k = m$, T is the identity matrix, and m is the lattice of integral vectors, then

$$\Psi_k(T, \mathfrak{m}, t) = \prod_{j=0}^{m-1} \Psi(t - \frac{j}{2})$$

But, as is well known, $\Psi(t)$ does not vanish for real values of t.

In view of this it seems worthwhile to sketch a proof of the

PROPOSITION. $\Psi_k(T, \mathfrak{m}, t)$ <u>is an entire function.</u>

It is only necessary to establish this for $k < m$ since $\Psi_m(T, \mathfrak{m}, t)$ may be expressed in terms of the zeta-function of k and the proposition follows from known properties of this zeta function. The proof is by induction. For $k = 1$ the proposition is a consequence of the discussion in Section 2. Suppose it is true for $k - 1$. Set

$$\alpha(t) = t(t - \frac{k}{2})a'(t)\zeta(2t)$$

$$\beta(t) = \prod_{j=0}^{k-2} (t - \frac{j}{2})(t - \frac{m-j-1}{2})a'(t - \frac{j}{2})\zeta(2t - j)$$

$$\gamma(t) = t(t - \frac{m-k+1}{2})a'(t)\Psi(2t)$$

and consider the series

$$\alpha(s_1)\beta(s_k)\gamma(s_1 + s_k - \frac{k-1}{2})\Sigma_{\mathfrak{n}_1 \subset \mathfrak{n}_k} T\{\mathfrak{n}_1\}^{-s_1} T\{\mathfrak{n}_k\}^{-s_k} \tag{27}$$

The sum is over all chains of primitive submodules of ranks 1 and k. Call the function defined by (27) $\varphi(T, \mathfrak{m}, s_1, s_k)$. To establish the proposition it will be shown that φ is an entire function of s_1 and s_k and that

$$\varphi(T, \mathfrak{m}, 0, t) = \frac{Nhk}{2wn} \Psi_k(T, \mathfrak{m}, t)$$

If σ_1 and σ_k are the real parts of s_1 and s_k the diagram represents a decomposition of the (s_1, s_k) space. $\varphi(T, \mathfrak{m}, s_1, s_k)$ is clearly analytic in the region I.

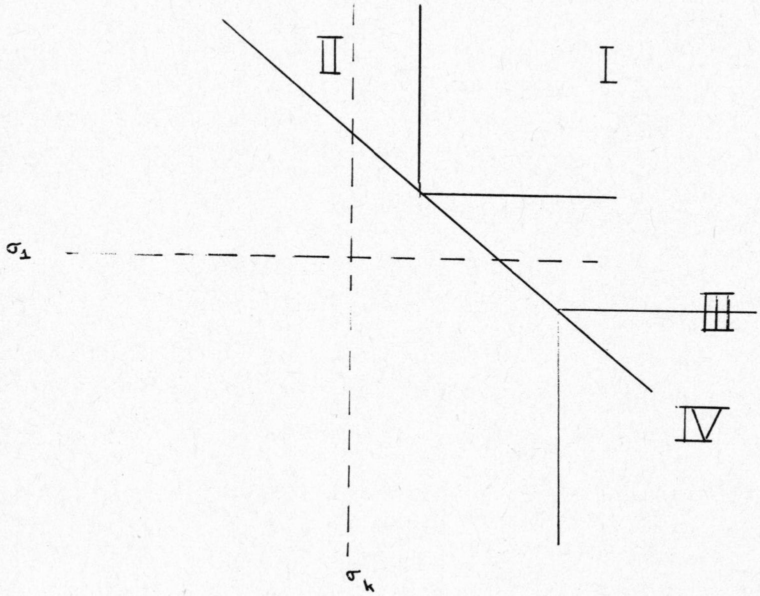

As before the continuation into the regions II, III, IV is effected by suitable arrangements of the series (27). Moreover in the regions I, II, III, IV, φ will have only polynomial growth on vertical lines. The proof of this is omitted; the analysis required is the same as above. Consequently Cauchy's integral formula may be applied to effect the continuation to the entire (s_1, s_k) space.

Since, in the notation of Section 5,

$$\Sigma_{\{\mathfrak{n}_k\}} \overline{T}^{-1}\{\mathfrak{n}_k\}^{-t} = T\{\mathfrak{m}\}^t \Sigma_{\{\mathfrak{n}_{m-k}\}} T\{\mathfrak{n}_{m-k}\}^{-t}$$

the functional equation for $\Psi_k(T, \mathfrak{m}, t)$ yields the equality of

(28a) $$\prod_{j=0}^{k-1} (t - \frac{j}{2})(t - \frac{m-j}{2}) a'(t - \frac{j}{2}) \zeta(2t - j) \Sigma_{\{\mathfrak{n}_k\}} T\{\mathfrak{n}_k\}^{-t}$$

and

$$(28b)\ T\{\mathfrak{m}\}^{\frac{m-k}{2}-t}\prod_{j=0}^{k-1}(t-\tfrac{j}{2})(t-\tfrac{m-j}{2})a'(\tfrac{m-j}{2}-t)\zeta(m-j-2t)\Sigma_{\{\mathfrak{n}_{m-k}\}}T\{\mathfrak{n}_{m-k}\}^{t-\frac{m}{2}}$$

in the sense that the functions represented by these series are equal. For brevity some equalities in the proof of the proposition have been written in this manner.

The first arrangement of the series is, in the notation of the argument preceding formula (12),

$$(29)\qquad \beta(s_k)\gamma(s_1+s_k-\tfrac{k-1}{2})\Sigma_{\{\mathfrak{n}_k\}}T\{\mathfrak{n}_k\}^{-s_k}\Psi_1(R,\mathfrak{n}',s_1)$$

which, as a consequence of (28), equals

$$(30)\quad \alpha(\tfrac{k}{2}-s_1)\beta(s_k)\gamma(s_1+s_k-\tfrac{k-1}{2})\Sigma_{\mathfrak{n}_{k-1}\subset\mathfrak{n}_k}T\{\mathfrak{n}_{k-1}\}^{s_1-\frac{k}{2}}\,T\{\mathfrak{n}_k\}^{\frac{k-1}{2}-s_1-s_k}$$

(30) converges in that part of the region II which is sufficiently far to the left of the σ_k-axis. (29) converges wherever (27) or (30) converge. Arguments similar to those of Section 3 show that it converges in all of I and II. In particular, if in (29), s_1 is set equal to zero the result is $\frac{Nhk}{2wn}\Psi_k(T,\mathfrak{m},s_k)$.

The second arrangement is, in the notation of Section 3 but with $\mathfrak{n}_1$ representing $\mathfrak{n}_k$,

$$(31)\qquad \alpha(s_1)\gamma(s_1+s_k-\tfrac{k-1}{2})\Sigma_{\{\mathfrak{n}_1\}}T\{\mathfrak{n}_1\}^{-s_1}(N^2\mathfrak{b})^{-s_k}\Psi_{k-1}(S,\mathfrak{p}',s_k)$$

Using (28) and simplifying (31) becomes the product of

$$(32a)\qquad \alpha(s_1)\beta(\tfrac{m-1}{2}-s_k)\gamma(s_1+s_k-\tfrac{k-1}{2})T\{\mathfrak{m}\}^{\frac{m-k}{2}-s_k}$$

and

$$(32b)\qquad \Sigma T\{\mathfrak{n}_1\}^{\frac{k-1}{2}-s_1-s_k}\,T\{\mathfrak{n}_{m-k+1}\}^{s_k-\frac{m-1}{2}}$$

The sum is over all chains $n_1\subset n_{m-k+1}$. This series converges in that part of the region III which is sufficiently far below the σ_1-axis. Consequently (31) converges in the regions I and III.

Replacing $\mathfrak{n}_k$ by $\mathfrak{n}_{m-k+1}$ in the definition of R and $\mathfrak{n}'$, write (32) as

$$\alpha(s_1)\beta(\frac{m-1}{2} - s_k)T\{\mathfrak{m}\}^{\frac{m-k}{2} - s_k}$$

times

$$\Sigma_{\{\mathfrak{n}_{m-k+1}\}} T\{\mathfrak{n}_{m-k+1}\}^{s_k - \frac{m-1}{2}} \Psi_1(R, \mathfrak{n}', s_1 + s_k - \frac{k-1}{2})$$

This is similar to the series (29) and by the same argument may be shown to converge in IV.

It should be remarked that if this sequence of rearrangements is carried one step further the functional equation is obtained.

The proposition implies that $\zeta_k(T, \mathfrak{m}, t)$ is a meromorphic function with at most a simple pole at $t = \frac{m}{2}, \ldots, \frac{m-k+1}{2}$. Some information about the residues may be obtained from the equations

$$\gamma_k \gamma_{m-k} \psi_k(\tfrac{j}{2}) \Psi_k(T, \mathfrak{m}, \tfrac{j}{2}) = \gamma_j \gamma_{m-j} \psi_j(\tfrac{k}{2}) \Psi_j(T, \mathfrak{m}, \tfrac{k}{2})$$

if $1 \leq j, k < m$ and

$$\gamma_k \gamma_{m-k} \psi_k(\tfrac{m}{2}) \Psi_k(T, \mathfrak{m}, \tfrac{m}{2}) = \gamma_m T\{\mathfrak{m}\}^{-\frac{k}{2}}$$

if $1 \leq k < m$ and $j = m$. To prove it observe that the left side is the value of Ψ at

$$(-\frac{j}{2} + \frac{m-1}{4} - \frac{k-1}{2}, \ldots, -\frac{j}{2} + \frac{m-1}{4}, -\frac{m-1}{4}, \ldots, -\frac{m-1}{4} + \frac{m-k-1}{2})$$

and the right side is the value of Ψ at

$$(-\frac{k}{2} + \frac{m-1}{4} - \frac{j-1}{2}, \ldots, -\frac{k}{2} + \frac{m-1}{4}, -\frac{m-1}{4}, \ldots, -\frac{m-1}{4} + \frac{m-j-1}{2})$$

But the second vector is obtained by permuting the coordinates of the first.

8. Let $Z_j = X_j + iY_j$, $Y_j > 0$, be n $m \times m$ matrices in the generalized upper half-plane and let k be a totally-real field of degree n. If $\mathfrak{n}_m$ is a module of rank m in z_{2m}, let $a_1, \ldots, a_m$ be a basis for the vector space generated by $\mathfrak{n}_m$ and let $\mathfrak{m}_m$ be the module generated over $\mathfrak{o}$ by $a_1, \ldots, a_m$. Then $\mathfrak{n}^m = \mathfrak{a}\mathfrak{m}^m$, where $\mathfrak{a}$ is some ideal in k whose class depends only on $\mathfrak{n}_m$. Let $A = (a'_1 \ldots a'_m)'$ and set

$$\lambda(Z_1, \ldots, Z_n, \mathfrak{n}_m) = \mathrm{N}\mathfrak{a} \prod_i |A_i(Z_i, I)|$$

A_i, $i = 1, \ldots, n$, are the conjugates of A. $\lambda(Z_1, \ldots, Z_n; \mathfrak{n}_m)$ does not depend on the basis chosen. Then the Eisenstein series are defined by

$$\varphi_g(Z_1, \ldots, Z_n; j) = \Sigma_{\{\mathfrak{n}_m\}} \lambda(Z_1, \ldots, Z_n; \mathfrak{n}_m)^{-g} \tag{33}$$

The sum is over those primitive submodules of rank m of the module of integral vectors in z_{2m} which are orthogonal to themselves with respect to the skew-symmetric form $\Sigma x_i y_{m+i} - y_i x_{m+i}$ and such that the ideal $\mathfrak{a}$ is in the class j. g is an even integer. It will now be shown that the series converges absolutely if $g > m + 1$ (cf. [3]).

Let W_i be the real part of the matrix $(Z_i, I)' Y_i^{-1} (\overline{Z}_i, I)$ and W the n-tuple $(W_1, \ldots, W_n)$. It follows from the discussion in Section 1 that

$$|\lambda(Z_1, \ldots, Z_n; \mathfrak{n}_m)|^2 = \prod_i |Y_i| W\{\mathfrak{n}_m\} \tag{34}$$

In the formula (21) set $s_m = t - \frac{1}{2}$, $s_{m-1} = t - 1, \ldots, s_1 = t - \frac{m}{2}$ and obtain, by formula (25),

$$\gamma_m \prod_{i>j} a(\tfrac{1}{2} + s_i + s_j) \prod_i a(\tfrac{1}{2} + s_i) \Sigma W\{\mathfrak{n}_m\}^{-t} \tag{35}$$

For $V\{\mathfrak{n}\} = W\{\mathfrak{n}_m\}$. The sum is over all primitive submodules of rank m which are orthogonal to themselves. Since (35) is an entire function the series converges to the right of the first real zero of the coefficient

$$\prod_{i>j} a(\tfrac{1}{2} + s_i + s_j) \prod_i a(\tfrac{1}{2} + s_i)$$

That is, where $s_i > \frac{1}{2}$, $i = 1, \ldots, m$, or $t > \frac{m+1}{2}$. It follows from (34) that (33) converges absolutely if $g > m + 1$.

REFERENCES

1. Deuring, M., Algebren, Chelsea, New York (1948).

2. Hardy, G. H. and Riesz, M., The General Theory of Dirichlet Series, Cambridge (1952).

3. Klingen, H., Eisensteinreihen zur Hilbertschen Modulgruppe n-ten Grades, Nach. der Akademie der Wissenschaften, Göttingen (1960).

4. Koecher, M., Uber Dirichlet Reihen mit Funktionalgleichungen, Journal für die reine und angewandte Mathematik, 192 (1953).

5. Landau, E., Einfuhrung in die ... Theorie der Algebraischen Zahlen ..., Chelsea, New York (1949).

6. Selberg, A., Harmonic Analysis and Discontinuous Groups ..., Report of International Colloquium on Zeta-Functions, Bombay (1956).

Appendix II

ADÈLE GROUPS

The principal theorem in the text, Theorem 7.7 is so formulated that it is impossible to understand its statement without knowing its proof as well, and that is technically complicated. In an attempt to remedy the situation, whose disadvantages are manifest, I shall reformulate the theorem in this appendix.

The first, obvious point is that it should be formulated adelicly, for a reductive algebraic group over a number field F. $\mathbb{A}$ will be the adèle ring of F. The typical function space which one has to understand in applications of the trace formula is of the following sort. Suppose Z is the centre of G and Z_0 a closed subgroup of $Z(\mathbb{A})$ for which $Z_0Z(F)$ is also closed and $Z_0Z(F)\backslash Z(\mathbb{A})$ is compact. Let ξ be a character of Z_0 trivial on $Z_0 \cap Z(F)$, which for the moment we take to be unitary, in order to postpone the explanation that would otherwise be necessary. Let $\mathcal{L} = \mathcal{L}(\xi)$ be the space of measurable functions φ on $G(F)\backslash G(\mathbb{A})$ satisfying

(i) $$\varphi(zg) = \xi(z)\varphi(g)$$

(ii) $$\int_{Z_0G(F)\backslash G(\mathbb{A})} |\varphi(g)|^2 dg < \infty$$

$\mathcal{L}$ is clearly a Hilbert space and, of course, $G(\mathbb{A})$ acts by right translations. The decompositions of $\mathcal{L}$ that we seek are to respect the action of $G(\mathbb{A})$. An obvious decomposition is

(1) $$\mathcal{L}(\xi) = \oplus_{\zeta} \mathcal{L}(\zeta)$$

where ζ runs over all extensions of ξ to $Z(F)\backslash Z(\mathbb{A})$. It seems therefore that we might as well take $Z_0 = Z(\mathbb{A})$.

However, this will not do for the induction which lies at the heart of the study of Eisenstein series. It is even necessary to drop the assumption that $Z_0 Z(F)\backslash Z(\mathbb{A})$ is compact but it is still demanded that ξ be unitary. In any case the set of all homomorphisms of $Z_0 Z(F)\backslash Z(\mathbb{A})$ into $\mathbb{R}^+$ is a finite-dimensional vector space $X(\mathbb{R})$ over $\mathbb{R}$. Multiplication by the scalar r takes χ to $z \longrightarrow \chi(z)^r$. The map that associates to $\chi \otimes c$ the character $z \longrightarrow \chi(z)^c$ extends to an injection of $X(\mathbb{C})$ into the set of characters of $Z_0 Z(F)\backslash Z(\mathbb{A})$. Thus the set D of extensions ζ of ξ to $Z_0 Z(F)\backslash Z(\mathbb{A})$ is a complex manifold, each component being an affine space. The component containing ζ is

$$\{\zeta\chi \mid \chi \in X(\mathbb{C})\}$$

The set D_0 of unitary characters in a component, a real subspace of the same dimension, is defined by $\operatorname{Re} \zeta = 0$, if

$$\operatorname{Re} \zeta = |\zeta|$$

The character $|\zeta|$ may be uniquely extended to a homomorphism ν of $G(\mathbb{A})$ into $\mathbb{R}^+$. We can define $\mathcal{L}(\zeta)$ by substituting for the condition (ii), the following:

(ii)'
$$\int_{Z(\mathbb{A})G(F)\backslash G(\mathbb{A})} \nu^{-2}(g)\,|\varphi(g)|^2 dg < \infty$$

Since we may uniquely extend elements of $X(\mathbb{R})$ to $G(\mathbb{A})$, we may also regard the elements of $X(\mathbb{C})$ as characters of $G(\mathbb{A})$. The map $\varphi \longrightarrow \varphi' = \chi\varphi$, that is,

$$\varphi'(g) = \chi(g)\varphi(g)$$

is an isomorphism of $\mathcal{L}(\zeta)$ with $\mathcal{L}(\zeta\chi)$. This enables us to regard the spaces $\mathcal{L}(\zeta)$ as an analytic bundle over D, the holomorphic sections locally on $\zeta\chi(\mathbb{C})$ being of the form

$$\chi(g)\{\Sigma_{i=1}^n a_i(\zeta\chi)\varphi_i(g)\}$$

with φ_i in $\mathcal{L}(\zeta)$ and a_i holomorphic with values in $\mathbb{C}$.

If φ lies in $\mathcal{L}(\xi)$ and is smooth with support, that is, compact modulo $Z_0G(F)$ and ζ lies in D set

$$\Phi(g;\zeta) = \int_{Z_0Z(F)\backslash Z(\mathbb{A})} \varphi(zg)\zeta^{-1}(z)dz$$

Then, if we take the dual Haar measure on D_0,

$$\varphi(g) = \int_{D_0} \Phi(g;\zeta)|d\zeta| \tag{2}$$

Indeed if $\chi \in X(\mathbb{R})$ is given then

$$\varphi(g) = \int_{\mathrm{Re}\,\zeta=\chi} \Phi(g,\zeta)|d\zeta|$$

There are various ways to define $|d\zeta|$ on $\mathrm{Re}\,\zeta = \chi$. The simplest is by transport of structure from D_0 to

$$D_0\chi = \{\zeta \mid \mathrm{Re}\,\zeta = \chi\}$$

The most intuitive is to define $|d\zeta|$ in terms of affine coordinates on the components. From (2) one easily deduces the direct integral decomposition

$$\mathcal{L}(\xi) = \int_{D_0}^{\oplus} \mathcal{L}(\zeta)|d\zeta| \tag{3}$$

A cusp form in $\mathcal{L}$ is defined by the condition that whenever N is the unipotent radical of a parabolic subgroup P over F different from G itself then

$$\int_{N(F)\backslash N(\mathbb{A})} \varphi(ng)dn = 0$$

for almost all g. It is sufficient to impose this condition for those P containing a given minimal P_0. We consider henceforth only such P and these we divide into classes $\{P\}$ of associate parabolic subgroups. The class $\{G\}$ consists of G alone. The space of cusp forms on $\mathcal{L}(\xi)$ will be denoted by $\mathcal{L}(\{G\}, \xi)$. For cusp forms the direct integral (3) becomes

$$\mathcal{L}(\{G\}, \xi) = \int_{D_0}^{\oplus} \mathcal{L}(\{G\}, \zeta)\,|d\zeta|$$

If $Z_0 Z(F)\backslash Z(\mathbb{A})$ is compact then $\mathcal{L}(\{G\}, \xi)$ decomposes into a direct sum of invariant, irreducible subspaces, and any irreducible representation of $G(\mathbb{A})$ occurs in $\mathcal{L}(\{G\}, \xi)$ with finite, perhaps zero, multiplicity. This is in particular so when ξ is replaced by ζ in D. Moreover the decomposition of $\mathcal{L}(\{G\}, \zeta)$ and $\mathcal{L}(\{G\}, \zeta\chi)$, $\chi \in X(\mathbb{C})$, are parallel.

Suppose P is a parabolic subgroup of G with Levi factor M. It is understood that P and M are defined over F and that P contains P_0. Since Z is contained in the centre of M, $\mathcal{L}(\{M\}, \xi)$ is defined as a space of functions on $M(\mathbb{A})$ and $M(\mathbb{A})$ acts on it. The representation

$$\operatorname{Ind}(G(\mathbb{A}), M(\mathbb{A}), \mathcal{L}(\{M\}, \xi))$$

is really a representation of $G(\mathbb{A})$ induced from the representation of $P(\mathbb{A})$ obtained from the homomorphism $P(\mathbb{A}) \longrightarrow M(\mathbb{A})$ and the action of $M(\mathbb{A})$ on $\mathcal{L}(\{M\}, \xi)$. It acts on the space of functions φ on $N(\mathbb{A})\backslash G(\mathbb{A})$ satisfying

(i) For all $g \in G(\mathbb{A})$:

$$\varphi(ng) \in \mathcal{L}(\{M\}, \xi)$$

(ii)

$$\int_{Z_0 N(\mathbb{A})P(F)\backslash G(\mathbb{A})} |\varphi(mg)|^2 dg < \infty$$

We denote this space of functions by $\mathcal{E}(P, \xi)$.

Let $D(M)$ and $D_0(M)$ be the analogues of D and D_0 when G is replaced by M. We may also define $\operatorname{Ind}(G(\mathbb{A}), M(\mathbb{A}), \mathcal{L}(\{M\}, \zeta))$ for $\zeta \in D(M)$. The induced representation is unitary if $\operatorname{Re}\zeta = \delta$, where δ is defined by the condition that $\delta^2(m)$ is the absolute value of the determinant of the restriction of $\mathrm{Ad}\, m$ to $\mathfrak{n}$, the Lie algebra of N. It is easily seen that

$$\mathrm{Ind}(G(\mathbb{A}),\ M(\mathbb{A}),\ \mathcal{L}(\{M\},\ \xi)) = \int_{D_0(M)}^{\oplus} \mathrm{Ind}(G(\mathbb{A}),\ M(\mathbb{A}),\ \mathcal{L}(\{M\},\ \zeta\delta))\,|d\zeta|$$

Thus if φ is a well-behaved function in $\mathcal{E}(P,\ \xi)$ and

$$\Phi(g,\ \zeta) = \int_{Z_0 Z_M(F)\backslash Z_M(\mathbb{A})} \varphi(ag)\zeta^{-1}(a)\delta^{-1}(a)da$$

then

$$\varphi(g) = \int_{D_0(M)} \Phi(\zeta,\ g)\,|d\zeta|$$

We cannot easily describe what, at least for the purpose immediately at hand, a well-behaved function in $\mathcal{E}(P,\ \xi)$ is without stepping slightly outside the categories introduced above. $X_M(\mathbb{R})$ is defined in the same way as $X(\mathbb{R})$ except that M replaces G. Set

$$M^0 = \{m \in M(\mathbb{A})\,|\,\chi(m) = 1 \ \text{ for all } \ \chi \in X(\mathbb{R})\}$$

M^0 contains $M(F)$ and the definitions made for $M(F)\backslash M(\mathbb{A})$ could also have been made for $M(F)\backslash M^o$. Fix a maximal compact subgroup of $\pi_v G(F_v) \subseteq G(A)$, where the product is taken over all infinite places. Let $\mathcal{E}_0(P,\ \xi)$ be the space of continuous functions φ in $\mathcal{E}(P,\ \xi)$ with the following properties:

(i) φ is K_∞-finite.

(ii) φ is invariant under a compact open subgroup of $G(\mathbb{A}_f)$.

(iii) For all $g \in G(\mathbb{A})$ the support of $m \longrightarrow \varphi(mg)$, a function on $M(\mathbb{A})$, is compact modulo M^o.

(iv) There is an invariant subspace V of the space of cusp forms on M^o transforming according to ξ which is the sum of finitely many irreducible subspaces, and for all $g \in G$ the function $m \longrightarrow \varphi(mg)$, now regarded as a function on M^o, lies in V.

The functions φ in $\mathcal{E}_0(P, \xi)$ will serve us well. In particular

$$\varphi^{\wedge}(g) = \Sigma_{P(F)\backslash G(F)}\varphi(\gamma g)$$

is a function in $\mathcal{L}(\xi)$. If φ_1 lies in $\mathcal{E}_0(P_1, \xi)$ and φ_2 lies in $\mathcal{E}_0(P_2, \xi)$ then $\hat{\varphi}_1$ and $\hat{\varphi}_2$ are orthogonal if P_1 and P_2 are not associate. If $\{P\}$ is a class of associate parabolic subgroups we let $\mathcal{L}(\{P\}, \xi)$ be the closure of the linear span of the functions $\varphi^{\wedge}$, with $\varphi \in \mathcal{E}_0(P, \xi)$ and $P \in \{P\}$. It is proved quite early in the theory (cf. Lemma 4.6) that

$$\mathcal{L}(\xi) = \oplus_{\{P\}} \mathcal{L}(\{P\}, \xi) \tag{4}$$

Abstractly seen, the main problem of the theory of Eisenstein series is to analyze the space $\mathcal{L}(\xi)$ or the spaces $\mathcal{L}(\{P\}, \xi)$ in terms of the cusp forms on the various M. This analysis is carried out – in principle – in the text. However, one can be satisfied with a more perspicuous statement if one is content to analyze $\mathcal{L}(\xi)$ in terms of the representations occurring discretely in the spaces of automorphic forms on the groups M.

It is clear that

$$\mathcal{L}(\{P\}, \xi) = \int_{D_0}^{\oplus} \mathcal{L}(\{P\}, \zeta)\,|d\zeta|$$

Let $\mathcal{L}(G, \{P\}, \zeta)$ be the closure of the sum of the irreducible invariant subspaces of $\mathcal{L}(\{P\}, \zeta)$ and let

$$\mathcal{L}(\{G\}, \{P\}, \xi) = \mathcal{L}(G, \{P\}, \xi) = \int_{D_0}^{\oplus} \mathcal{L}(G, \{P\}, \zeta)\,|d\zeta|$$

We write $\{P\} \succ \{P_1\}$ if there is a $P \in \{P\}$ and a $P_1 \in \{P_1\}$ with $P \supseteq P_1$. We shall construct a finer decomposition

$$(5)\qquad \mathcal{L}(\{P_1\},\ \xi) = \oplus_{\{P\}\succ\{P_1\}} \mathcal{L}(\{P\},\ \{P_1\},\ \xi)$$

If $P \in \{P\}$ let $\mathfrak{F} = \mathfrak{F}(\{P_1\})$ be the set of classes of associate parabolic subgroups $P_1(M)$ of M of the form

$$P_1(M) = M \cap P_1$$

with $P_1 \in \{P_1\}$ and $P_1 \subseteq P$. The space $\mathcal{L}(\{P\},\ \{P_1\},\ \xi)$ will be isomorphic to a subspace of

$$(6)\qquad \oplus_{P\in\{P\}} \oplus_{\mathfrak{F}} \operatorname{Ind}(G(\mathbb{A}),\ M(\mathbb{A}),\ \mathcal{L}(M,\ \{P_1(M)\},\ \xi))$$

which may also be written as

$$(7)\qquad \oplus_{P\in\{P\}} \oplus_{\mathfrak{F}} \int_{D_0(M)}^{\oplus} \operatorname{Ind}(G(\mathbb{A}),\ M(\mathbb{A}),\ \mathcal{L}(M,\ \{P_1(M)\},\ \zeta\delta)\,|d\zeta|$$

To describe these subspaces, we need the Eisenstein series.

The induced representations occurring in (6) act on a space $\mathcal{E}(P,\ \{P_1(M)\},\ \xi)$ of functions φ on $N(\mathbb{A})P(F)\backslash G(\mathbb{A})$ that satisfy the condition: for all $g \in G(\mathbb{A})$ the function $m \longrightarrow \varphi(mg)$ lies in $\mathcal{L}(M,\ \{P_1(M)\},\ \xi)$. We may also introduce $\mathcal{E}_0(P,\ \{P_1(M)\},\ \xi)$ in much the same manner as we introduced $\mathcal{E}_0(P,\ \xi)$. The induced representations in (7) act on spaces $\mathcal{E}(P,\ \{P_1(M)\},\ \zeta\delta)$ and the spaces $\mathcal{E}_0(P,\ \{P_1(M)\},\ \zeta)$, just as above, form a holomorphic vector bundle over $D_0(M)$.

If L is the lattice of rational characters of M over F then $X(\mathbb{R})$ may be imbedded in $L \otimes \mathbb{R}$, and the positive Weyl chamber in $X(\mathbb{R})$ with respect to P is well-defined. We write $\chi_1 > \chi_2$ if $\chi_1\chi_2^{-1}$ lies on it. If Φ lies in $\mathcal{E}_0(P,\ \{P_1(M)\},\ \zeta\delta)$ and $\operatorname{Re}\zeta > \delta$ the series

$$E(g,\ \Phi) = \Sigma_{P(F)\backslash G(F)} \Phi(\gamma g)$$

converges. For each g it may be analytically continued to a meromorphic function on the whole vector bundle, which will of course be linear on the fibres. It is an important part of the Corollary to Lemma 7.6 that none of its singular hyperplanes – the singularities all lie along hyperplanes – meet the set $\mathrm{Re}\,\zeta = 0$. If

$$\varphi = \int_{D_0(M)} \Phi(\zeta)\,|d\zeta|$$

with $\Phi(\zeta)$ in $\mathcal{E}_0(P, \{P_1(M)\}, \zeta\delta)$, lies in $\mathcal{E}_0(P, \{P_1(M)\}, \xi)$ then

$$T\varphi(g) = \ell.\mathrm{i.m.}\int E(g, \Phi(\zeta))\,|d\zeta|$$

exists, the limit being taken over an increasing exhaustive family of compact subsets of $D_0(M)$. The linear transformation $\varphi \longrightarrow T\varphi$ extends to a continuous linear transformation from $\mathcal{E}(P, \{P_1(M)\}, \xi)$ to $\mathcal{L}(\xi)$. By additivity we define it on

$$\oplus_{P\in\{P\}} \oplus_{\mathfrak{p}} \mathcal{E}(P, \{P_1(M)\}, \xi)$$

Then T commutes with the action of $G(\mathbb{A})$ and its image is, by definition, $\mathcal{L}(\{P\}, \{P_1\}, \xi)$. It has still to be explained how, apart from a constant factor, T is the composition of an orthogonal projection and an isometric imbedding. The functional equations now begin to play a role.

Suppose P and P' lie in $\{P\}$. If $\Phi = \oplus \Phi_{\mathfrak{p}}$ lies in

$$\oplus_{\mathfrak{p}} \mathcal{E}_0(P, \{P_1(M)\}, \xi)$$

we set

$$E(g, \Phi) = \Sigma_{\mathfrak{p}} E(g, \Phi_{\mathfrak{p}})$$

If $\mathrm{Re}\,\zeta > \delta$ consider

$$\int_{N'(F)\backslash N'(\mathbb{A})} E(ng, \Phi)dn$$

Since, as a function,

$$\Phi(g) = \Sigma_{\mathfrak{p}} \Phi_{\mathfrak{p}}(g)$$

this integral is equal to

$$\Sigma_{w \in N'(F)\backslash G(F)/P(F)} \int_{w^{-1}P(F)w \cap N'(F)\backslash N^1(\mathbb{A})} \Phi(w^{-1}ng)dn$$

We are only interested in those w for which

$$wMw^{-1} = M'$$

Then the integral equals

$$\Phi'(g) = \int_{wN(\mathbb{A})w^{-1} \cap N'(\mathbb{A})\backslash N'(\mathbb{A})} \Phi(w^{-1}ng)dn$$

and

$$\Phi \longrightarrow N(w)\Phi = \Phi'$$

is a linear transformation

$$\oplus_{\mathfrak{p}} \mathcal{E}_0(P, \{P_1(M)\}, \zeta\delta) \longrightarrow \oplus_{\mathfrak{p}'} \mathcal{E}_0(P', \{P_1'(M)\}, \zeta^{w^{-1}} \delta')$$

It is easy to turn

$$\mathrm{Hom}^{G(\mathbb{A})}(\oplus_{\mathfrak{p}} \mathcal{E}_0(P, \{P_1(M)\}, \zeta\delta), \oplus_{\mathfrak{p}'} \mathcal{E}_0(P', \{P_1'(M)\}, \zeta^{w^{-1}} \delta'))$$

into a holomorphic bundle on $D(M)$. $N(w)$ can be extended to a meromorphic section of it. Observe that $N(mw) = N(w)$ if $m \in M(F)$. The important functional equations are:

(i) If $w_2Mw_2^{-1} = M'$ and $w_1M'w_1^{-1} = M''$ then

$$N(w_1)N(w_2) = N(w_1w_2)$$

(ii) For any w

$$E(g,\ N(w)\Phi) = E(g,\ \Phi)$$

They are consequences of the rather turbid Lemma 7.4, immediate once its meaning is understood.

There is in addition a more elementary functional equation. We easily define a natural sesquilinear pairing

$$\{\oplus_{\mathfrak{F}}\ \mathcal{E}_0(P,\ \{P_1(M)\},\ \zeta\delta)\} \times \{\oplus_{\mathfrak{F}}\ \mathcal{E}_0(P,\ \{P_1(M)\},\ \bar{\zeta}^{-1}\delta)\} \longrightarrow \mathbb{C}$$

If K is a compact subgroup of $G(\mathbb{A})$ and $G(\mathbb{A})$ is a finite disjoint union

$$\bigcup N(\mathbb{A})M(\mathbb{A})g_i K$$

there are constants c_i so that

$$\int_{G(\mathbb{A})} f(g)dg = \Sigma_i c_i \int_{N(\mathbb{A})} dn \int_{M(\mathbb{A})} dm \int_K dk\ f(nmg_ik)$$

The pairing is

$$\langle\psi_1,\ \psi_2\rangle = \Sigma_i c_i \int_{Z_M(\mathbb{A})\backslash M(\mathbb{A})} dm \int_K dk\ \psi_1(mg_ik)\bar{\psi}_2(mg_ik)$$

According to Lemma 7.5 the adjoint $N^*(w)$ of

$$N(w) : \oplus\ \mathcal{E}_0(P,\ \{P_1(M)\},\ \zeta\delta) \longrightarrow \oplus\ \mathcal{E}_0(P',\ \{P_1'(M)\},\ \zeta^{w^{-1}}\delta)$$

is

$$N(w^{-1}) : \oplus\ \mathcal{E}_0(P',\ \{P_1'(M)\},\ \bar{\zeta}^{-w^{-1}}\delta) \longrightarrow \oplus\ \mathcal{E}_0(P,\ \{P_1(M)\},\ \bar{\zeta}^{-1}\delta)$$

The functional equations

$$N(w^{-1})N(w) = N(w)N(w^{-1}) = I$$

then imply that $N(w)$ is an isomorphism and an isometry when ζ is unitary.

The functional equations for the Eisenstein series imply that if

$$\varphi = \oplus \varphi_P$$

then $T\varphi(g)$, which is given by,

$$\ell.i.m. \Sigma_{P \in \{P\}} \int E(g, \Phi_P(\zeta)) |d\zeta|$$

is also equal to

$$\ell.i.m. \Sigma_{P \in \{P\}} \int E(g, \frac{1}{\omega} \Sigma N(w) \Phi_P(\zeta^w)) |d\zeta|$$

Here the sum is over all w such that, for some $P' \in \{P\}$, $wMw^{-1} = M'$ taken modulo $M(F)$, and ω is the number of terms in the sum. It is implicit that we have fixed a Levi factor of each P in $\{P\}$. The linear transformation

$$\oplus \Phi(\zeta) \longrightarrow \oplus \{\frac{1}{\omega} \Sigma N(w) \Phi_P(\zeta^w)\}$$

is the orthogonal projection U of the space (5) onto the closed, $G(\mathbb{A})$-invariant subspace defined by the equations

$$\Phi_{P'}(\zeta^{w-1}) = N(w) \Phi_P(\zeta)$$

whenever $wMw^{-1} = M'$. It is clear that $T = TU$. If $\oplus \Phi_P(\zeta)$ lies in the range of U then (Lemma 7.6)

$$\|T\varphi\|^2 = \omega \|\varphi\|^2$$

The main results of the text summarized, I would like to draw attention to a couple of questions that it seems worthwhile to pursue. The first, which I mention only in passing, is to extend the decompositions (4) and (5) to other function spaces, especially those needed for the study of cohomology groups (cf. [6]). The second involves a closer study of the operators $N(w)$. They have already led to many interesting, largely unsolved problems in the theory of automorphic forms and group representations ([4], [5]).

Suppose V is an irreducible invariant subspace of

$$\Sigma_{\mathfrak{X}} \mathcal{L}(M, \{P_1(M)\}, \zeta_0\delta)$$

If $\zeta = \zeta_0\chi$ lies in the same component as ζ_0 we may define

$$V_\zeta = \{\chi(m)\varphi(m) \mid \varphi \in V\}$$

as well as the spaces $\mathcal{E}(P, V_\zeta)$ on which the induced representations

$$\mathrm{Ind}(G(\mathbb{A}), M(\mathbb{A}), \rho_\zeta)$$

act. Here ρ_ζ is the representation of $M(\mathbb{A})$ on V_ζ. We may also introduce $\mathcal{E}_0(P, V_\zeta)$.

There are two ways of regarding the functions Φ in $\mathcal{E}(P, V_\zeta)$. Φ may be considered a function on $N(\mathbb{A})P(F)\backslash G(\mathbb{A})$ for which the function

$$m \longrightarrow \Phi(mg)$$

is for all g an element $F(g)$ of V. We may on the other hand emphasize F, from which Φ may be recovered; it is a function on $N(\mathbb{A})\backslash G(\mathbb{A})$ with values in V_ζ and

$$F(mg) = \rho_\zeta(m)F(g)$$

for all m and g.

If $wMw^{-1} = M'$ and $\zeta' = \zeta^{w^{-1}}$ we can introduce a space $V'_{\zeta'}$ and a representation $\rho'_{\zeta'}$ of $M'(\mathbb{A})$ on it in two different ways. Either $V_{\zeta'}$ is V_ζ and

$$\rho'_{\zeta'}(m') = \frac{\delta'(m')}{\delta(m)}\,\rho_\zeta(m) \qquad\qquad m = w^{-1}m'w$$

or

$$V_{\zeta'} = \{\varphi' \mid \varphi'(m') = \frac{\delta'(m')}{\delta(m)} \varphi(m)\}$$

and $\rho_{\zeta'}$ acts by right translations. With the second definition $V'_{\zeta'}$ is clearly a subspace of $\mathcal{L}(\zeta'\delta')$. Since $N(w)$ is easily seen to take $\mathcal{E}_0(P, V_\zeta)$ to $\mathcal{E}_0(P', V'_{\zeta'})$ we conclude that $V'_{\zeta'}$ lies in

$$\oplus_{\mathfrak{F}'} \mathcal{L}(M', \{P_1(M')\}, \zeta'\delta')$$

In terms of F and F', and the first definition of $V'_{\zeta'}$, we have

$$F'(g) = \int_{wN(\mathbb{A})w^{-1} \cap N'(\mathbb{A}) \backslash N'(\mathbb{A})} F(w^{-1}ng)dn$$

The integrals are now vector-valued. It is this definition of $N(w)$, which now takes F to F', that we prefer to work with. Of course the formula above is only valid for $\mathrm{Re}\, \zeta > \delta$. We write V as a tensor product over the places of F

$$V = \otimes V_v$$

Then $N(w)$ too becomes a product of local operators $N_v(w) : F_v \longrightarrow F'_v$ with

$$F'_v(g) = \int_{wN(F_v)w^{-1} \cap N'(F_v) \backslash N'(F_v)} F_v(w^{-1}ng)dn \qquad g \in G(F_v)$$

Suppose, in order to describe the second problem, that the L-functions and ε-factors intimated in [4] have been defined for all irreducible representations of $M(F_v)$ and all relevant representations of the associate group $M^\vee$ of M. Using the notions of [4] we see that $M^\vee$ acts on $\mathfrak{n}^\vee \cap w^{-1}\mathfrak{n}'^\vee w \backslash w^{-1}\mathfrak{n}'^\vee w$. Here $\mathfrak{n}^\vee$, $\mathfrak{n}'^\vee$ lie in the Lie algebra of the associate group $G^\vee$ and w is obtained from the isomorphism of the Weyl groups of G and $\overset{\circ}{G}{}^\vee$. Denote the above representation of the group $M^\vee$ by $r(w)$ and, in order to make room for a subscript, denote ρ_ζ by $\rho(\zeta)$. The calculations of [2], [3], and [5] suggest the introduction of a normalized intertwining operator $R_v(w)$ by the equation

$$N_v(w) = \frac{L(0, \rho_v(\zeta), \tilde{r}(w))}{\varepsilon(0, \rho_v(\zeta), r(w), \psi_v) L(1, \rho_v(\zeta), \tilde{r}(w))} R_v(w)$$

$\tilde{r}(w)$ is contragredient to $r(w)$. Exploiting the anticipated functional equation we obtain the global formula

$$N(w) = \otimes_v N_v(w) = \frac{L(0, \rho(\zeta), \tilde{r}(w))}{L(0, \rho(\zeta), r(w))} \otimes_v R_v(w)$$

If $s(w)$ is the representation of $M^\vee$ on $w^{-1} n'^\vee w$ then

$$r(w) - \tilde{r}(w) = s(w) - s(1)$$

and

$$\frac{L(0, \rho(\zeta), \tilde{r}(w))}{L(0, \rho(\zeta), r(w))} = \frac{L(0, \rho(\zeta), s(1))}{L(0, \rho(\zeta), s(w))}$$

If $w_2 M w_2^{-1} = M'$ and $w_1 M' w_1^{-1} = M''$ then $s'(1)$ composed with $m \longrightarrow w_2 m w_2^{-1}$ is $s(w_2)$ and $s'(w_1)$ composed with the same homomorphism is $s(w_1 w_2)$. Consequently the quotient of the two L-functions is multiplicative in w.

We are led to the following questions:

Is it possible to analytically continue the operators $R_v(w)$, which are at first defined for $\mathrm{Re}\, \zeta_v > 0$ to meromorphic functions on an entire component of the local analogue of $D(M)$? Is $R_v(w)$ then unitary on $D_0(M)$? Is the functional equati[on]

$$R_v(w_1) R_v(w_2) = R_v(w_1 w_2)$$

satisfied?

If r is archimedean, the L-functions and ε-factors can be defined ([7]). It is very likely that, in this case, answers to the above questions are contained in the work of Knapp-Stein [1]; but I have not tried to check this.

REFERENCES

1. A. Knapp and E. Stein, Singular integrals and the principal series III, PNAS, vol. 71 (1974).

2. K. F. Lai, On the Tamagawa number of quasi-split groups, Thesis, Yale University (1974).

3. ———, On the Tamagawa number of quasi-split groups, BAMS, vol. 82 (1976).

4. R. P. Langlands, Problems in the theory of automorphic forms in Lectures on Modern Analysis and Applications III, Springer-Verlag (1970).

5. ———, Euler Products, Yale University Press (1971).

6. ———, Modular forms and ℓ-adic representations, in Modular Functions of one Variable III, Springer-Verlag (1972).

7. ———, On the classification of irreducible representations of real algebraic groups, Institute for Advanced Study (1973).

Appendix III

EXAMPLES FOR §7

It might be a help to the reader who resolves to force his way through the jungle of Paragraph §7 to know the sources, apart from the author's expository inadequacy, of the complexity of the notation and the proofs. A number of unexpected and unwanted complications must be taken into account, and it may be asked whether they can really, under sufficiently unfavorable circumstances, arise or whether it was simply not within the author's power to eliminate them from consideration. Unfortunately they do arise, and they cannot be ignored unless a procedure radically different from that of the text be found.

I cannot comment on all the complexities, for a good deal of time has elapsed since the text was written, and I myself now have difficulty finding my way through it. But some of them were sufficiently vexing to imprint themselves indelibly on my memory, and these I shall try to explain.

Some of the notational elaborateness is of course purely formal, a result of the generality, and this part it is best to remove at once by fixing our attention on some special cases, in which the essential mathematics is nonetheless retained.

We take G to be the set of real points in a simply-connected Chevalley group and Γ to be the set of integral points. Fix a percuspidal subgroup P; then all other percuspidal subgroups are conjugate to it with respect to Γ. We take V and W to be the space of constant functions so that $\mathcal{E}(V, W)$ too consists of constant functions. The corresponding Eisenstein series we parametrize by λ in the dual of the Lie algebra $\mathfrak{a}$, rather than by an element in $\mathfrak{a}$ itself, as in the text. When writing it I was too strongly influenced by the then prevalent fashion of identifying $\mathfrak{a}$ with its dual.

We take Φ to be identically 1 and write $E(g, \lambda)$ instead of $E(g, \Phi, H)$. The constant term of $E(g, \lambda)$, that is

$$\int_{\Gamma \cap N \backslash N} E(ng, \lambda) dn$$

is then

$$\Sigma_{s \in \Omega} M(s, \lambda) e^{s\lambda(H(g)) + \rho(H(g))}$$

where $M(s, \lambda)$ is now a scalar which if G is $SL(2)$ can be easily computed. Lemma 6.1 then shows that it is in general equal to

$$\prod_{\substack{\alpha > 0 \\ s\alpha < 0}} \frac{\xi(\lambda(H_\alpha))}{\xi(1+\lambda(H_\alpha))}$$

Here

$$\xi(z) = \pi^{-\frac{z}{2}} \Gamma(\tfrac{z}{2}) \zeta(z)$$

and H_α is the coroot defined by

$$\lambda(H_\alpha) = 2\frac{(\lambda, \alpha)}{(\alpha, \alpha)}$$

The space $\mathcal{L}(\{P\}, \{V\}, W)$ is isomorphic to the space obtained by completing the space of complex-valued functions of λ holomorphic in the tube over a large ball and decaying sufficiently rapidly at infinity. The inner product is

(1) $$\frac{1}{(2\pi)^q} \int_{\mathrm{Re}\, \lambda = \lambda_0} \Sigma_{s \in \Omega} M(s, \lambda) \Phi(\lambda) \overline{\Psi(-s\bar{\lambda})} \, |d\lambda|$$

Here λ_0 must satisfy

$$\langle \lambda_0, \alpha \rangle > \langle \rho, \alpha \rangle$$

for all positive roots α. The integer q is the rank of G.

On the space $\mathcal{L}$ of functions on the set $\mathrm{Re}\, \lambda = 0$ square-integrable with respect to the measure

$$|\Omega| \cdot \frac{d\lambda}{(2\pi)^q}$$

we introduce the operator

$$Q : \Phi(\lambda) \longrightarrow \frac{1}{|\Omega|} \Sigma_s M(s^{-1}, s\lambda)\Phi(s\lambda)$$

Since

$$M(s, t\lambda)M(t, \lambda) = M(st, \lambda)$$

the operator Q is a projection. Its range consists of the functions satisfying

$$\Phi(s\lambda) = M(s, \lambda)\Phi(\lambda)$$

for all s and λ. Since

$$\overline{M(s, \lambda)} = M(s^{-1}, -s\bar{\lambda})$$

we infer also that Q is self-adjoint. The inner product of $Q\Phi$ and Ψ is given by (1).

If λ_0 were 0 we would infer that $\mathcal{L}(\{P\}, \{V\}, W)$ was isomorphic to the quotient of $\mathcal{L}$ by the kernel of Q or to the range of Q. This is the kind of concrete realization of the space $\mathcal{L}(\{P\}, \{V\}, W)$ which the theory of Eisenstein series seeks to give. If the functions $M(s, \lambda)$ had no poles in the region defined by

(2) $$\text{Re}\,\langle \lambda, \alpha \rangle \geq 0$$

for all positive α we could, because of the Cauchy integral theorem, replace λ_0 by 0. However, they do have poles. But we can deform the contour of integration in (1) to $\text{Re}\,\lambda = 0$ if the zeros of $\Phi(\lambda)$ compensate for the poles of the functions $M(s, \lambda)$. Therefore, the subspace of $\mathcal{L}(\{P\}, \{V\}, W)$ generated by such functions is isomorphic to the quotient of $\mathcal{L}$ by the kernel of Q and the inner product of the projection of the elements of $\mathcal{L}(\{P\}, \{V\}, W)$ represented by $\Phi(\lambda)$ and $\Psi(\lambda)$ on this subspace is given by (1) with λ_0 replaced by 0.

The inner product of the projections on the orthogonal complement of the subspace will be given by the residues which enter when we deform $\operatorname{Re} \lambda = \lambda_0$ to $\operatorname{Re} \lambda = 0$. This will be a sum of integrals of roughly the same type as (1), but over hyperplanes of dimension q-1. The procedure of §7 is to treat them in the same way, and then to proceed by induction until there is nothing left. The procedure is carried out fully for two simple examples in [1].

A number of difficulties can enter at the later stages which do not appear at first. The functions $M(s, \lambda)$ remain bounded as $\operatorname{Im} \lambda \longrightarrow \infty$ in the region defined by (2) so that the application of the residue theorem is clearly justified. However, at least in the general case when the functions $M(s, \lambda)$ are not explicitly known, it was necessary to deform the contour into regions in which, so far as I could see, the behaviour of the relevant functions as $\operatorname{Im} \lambda \longrightarrow \infty$ was no longer easy to understand. Some substitute for estimates was necessary. It is provided by unpleasant lemmas, such as Lemma 7.1, and the spectral theory of the operator A introduced in §6. The idea is, if I may use a one-dimensional diagram to illustrate it, to deform the contour as indicated and then to show

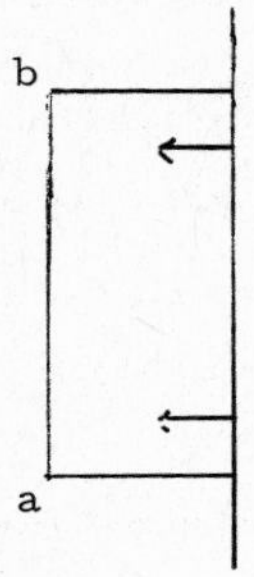

that at least on the range of an idempotent in the spectral decomposition of A associated to a finite interval only the interval [a, b] of the deformed contour matters. Of course for a given idempotent the interval [a, b] has to be taken sufficiently large. For function fields, this sort of problem would not arise.

At the first stage the functions $M(s, \lambda)$ have simple poles so that the residues which appear do not involve the derivatives of $\Phi(\lambda)$ or $\Psi(\lambda)$. At later stages this may not be so, and the elaborate discussion of notation with which §7 is prefaced is not to be evaded. The first – and only – example of such behaviour that I know is provided by the exceptional group of type G_2.

The root diagram for G_2 is:

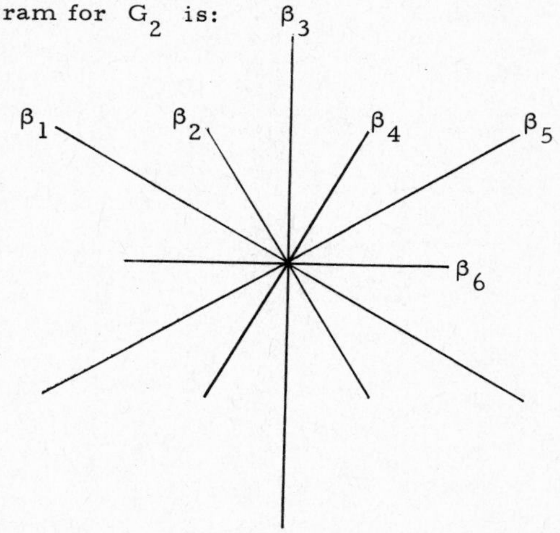

We take as coordinates of λ the numbers $z_1 = \lambda(H_{\beta_1})$, $z_2 = \lambda(H_{\beta_6})$ and the measure $|d\lambda|$ is then $dy_1 dy_2$. Since the poles of the functions $M(s, \lambda)$ all lie on hyperplanes defined by real equations we can represent the process of deforming the contour and the singular hyperplanes met thereby by a diagram in the real plane. The singularities that are met all lie on the hyperplanes s_i defined by

$$\lambda(H_{\beta_i}) = 1 \qquad\qquad 1 \leq i \leq 6$$

As can be seen in the diagram, if we move the contour along the dotted line indicated we may pick up residues at the points $\lambda_1, \ldots, \lambda_6$.

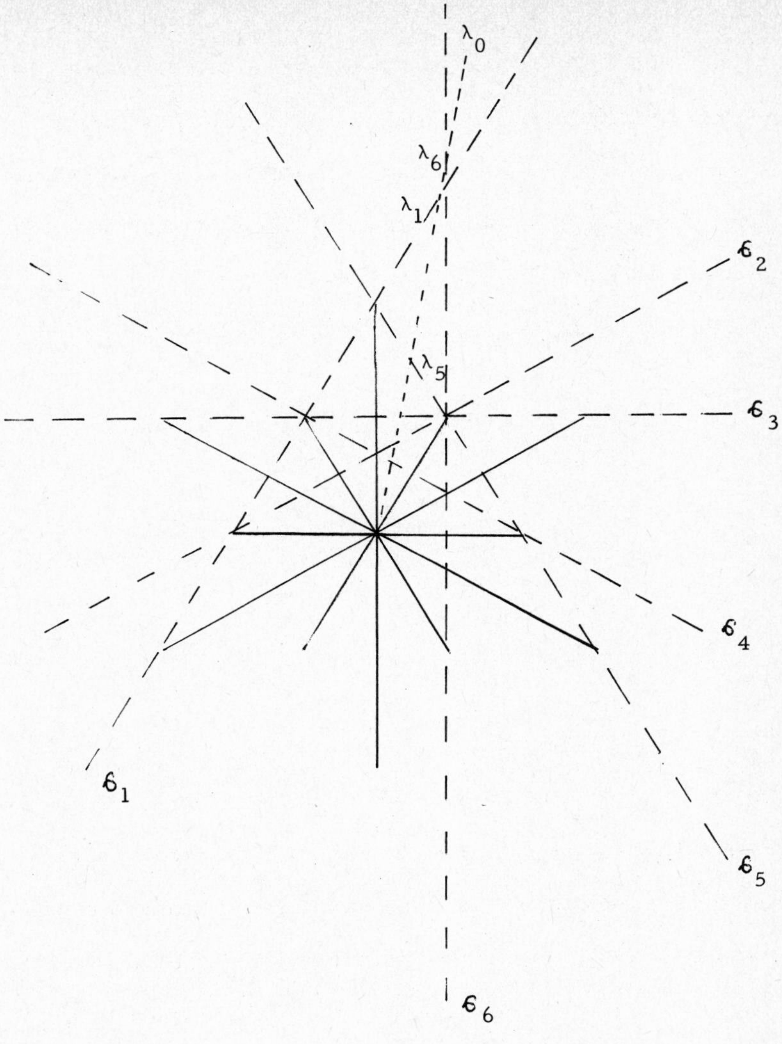

In order to write out the resulting residual integrals explicitly as in §7 we have to list the elements of $\Omega(\mathfrak{C}_i, \mathfrak{C}_j)$, and then tabulate the residues of $M(s, \lambda)$ on $\mathfrak{C}_i$ for each s in $\Omega(\mathfrak{C}_i, \mathfrak{C}_j)$. We first list the elements of the Weyl group, together with the positive roots that they send to negative roots. Let ρ_i be the reflection defined by β_i and $\sigma(\theta)$ the rotation through the angle θ.

(3)

	$\{\beta > 0 \mid \sigma\beta < 0\}$
1	
ρ_1	β_1
ρ_2	$\beta_1, \beta_2, \beta_3$
ρ_3	$\beta_1, \beta_2, \beta_3, \beta_4, \beta_5$
ρ_4	$\beta_2, \beta_3, \beta_4, \beta_5, \beta_6$
ρ_5	$\beta_4, \beta_5, \beta_6$
ρ_6	β_6
$\sigma(\frac{\pi}{3})$	β_1, β_2
$\sigma(\frac{2\pi}{3})$	$\beta_1, \beta_2, \beta_3, \beta_4$
$\sigma(\pi)$	$\beta_1, \beta_2, \beta_3, \beta_4, \beta_5, \beta_6$
$\sigma(\frac{4\pi}{3})$	$\beta_3, \beta_4, \beta_5, \beta_6$
$\sigma(\frac{5\pi}{3})$	β_5, β_6

Since an element of the Weyl group takes long roots to long roots and short roots to short roots, the set $\Omega(\mathfrak{G}_i, \mathfrak{G}_j)$ is empty unless i and j are both even or both odd. This allows us to consider the two sets $\{\mathfrak{G}_1, \mathfrak{G}_3, \mathfrak{G}_5\}$ and $\{\mathfrak{G}_2, \mathfrak{G}_4, \mathfrak{G}_6\}$ separately. We tabulate below the sets $\Omega(\mathfrak{G}_i, \mathfrak{G}_j)$, together with another more convenient labelling of the elements in them. The second labelling refers only to their action on $\mathfrak{G}_i$.

(4)

$\mathfrak{S}_i \backslash \mathfrak{S}_j$	$\mathfrak{S}_1$	$\mathfrak{S}_3$	$\mathfrak{S}_5$
$\mathfrak{S}_1$	$\rho_1 = \rho_+$ $\sigma(\pi) = \rho_-$	$\rho_2 = \sigma_+$ $\sigma(\frac{2\pi}{3}) = \sigma_-$	$\rho_3 = \tau_+$ $\sigma(\frac{\pi}{3}) = \tau_-$
$\mathfrak{S}_3$	$\rho_2 = \rho_+\rho_+\sigma_+^{-1} = \rho_-\rho_+\sigma_-^{-1}$ $\sigma(\frac{4\pi}{3}) = \rho_-\rho_+\sigma_+^{-1} = \rho_+\rho_+\sigma_-^{-1}$	$\rho_3 = \sigma_+\rho_+\sigma_+^{-1} = \sigma_-\rho_+\sigma_-^{-1}$ $\sigma(\pi) = \sigma_-\rho_+\sigma_+^{-1} = \sigma_+\rho_+\sigma_-^{-1}$	$\rho_4 = \tau_+\rho_+\sigma_+^{-1} = \tau_-\rho_+\sigma_-^{-1}$ $\sigma(\frac{2\pi}{3}) = \tau_-\rho_+\sigma_+^{-1} = \tau_+\rho_+\sigma_-^{-1}$
$\mathfrak{S}_5$	$\rho_3 = \rho_+\rho_+\tau_+^{-1} = \rho_-\rho_+\tau_-^{-1}$ $\sigma(\frac{5\pi}{3}) = \rho_+\rho_+\tau_-^{-1} = \rho_-\rho_+\tau_+^{-1}$	$\rho_4 = \sigma_+\rho_+\tau_+^{-1} = \sigma_-\rho_+\tau_-^{-1}$ $\sigma(\frac{4\pi}{3}) = \sigma_-\rho_+\tau_+^{-1} = \sigma_+\rho_+\tau_-^{-1}$	$\rho_5 = \tau_+\rho_+\tau_+^{-1} = \tau_-\rho_+\tau_-^{-1}$ $\sigma(\pi) = \tau_-\rho_+\tau_+^{-1} = \tau_+\rho_+\tau_-^{-1}$

(5)

$\mathfrak{S}_i \backslash \mathfrak{S}_j$	$\mathfrak{S}_2$	$\mathfrak{S}_4$	$\mathfrak{S}_6$
$\mathfrak{S}_1$	$\rho_2 = \rho_+\tau_+\rho_+^{-1} = \rho_-\tau_+\rho_-^{-1}$ $\sigma(\pi) = \rho_-\tau_+\rho_+^{-1} = \rho_+\tau_+\rho_-^{-1}$	$\rho_3 = \sigma_+\tau_+\rho_+^{-1} = \sigma_-\tau_+\rho_-^{-1}$ $\sigma(\frac{2\pi}{3}) = \sigma_-\tau_+\rho_+^{-1} = \sigma_+\tau_+\rho_-^{-1}$	$\rho_4 = \tau_+\tau_+\rho_+^{-1} = \tau_-\tau_+\rho_-^{-1}$ $\sigma(\frac{\pi}{3}) = \tau_-\tau_+\rho_+^{-1} = \tau_+\tau_+\rho_-^{-1}$
$\mathfrak{S}_4$	$\rho_3 = \rho_+\tau_+\sigma_+^{-1} = \rho_-\tau_+\sigma_-^{-1}$ $\sigma(\frac{4\pi}{3}) = \rho_+\tau_+\sigma_-^{-1} = \rho_-\tau_+\sigma_+^{-1}$	$\rho_4 = \sigma_+\tau_+\sigma_+^{-1} = \sigma_-\tau_+\sigma_-^{-1}$ $\sigma(\pi) = \sigma_-\tau_+\sigma_+^{-1} = \sigma_+\tau_+\sigma_-^{-1}$	$\rho_5 = \tau_+\tau_+\sigma_+^{-1} = \tau_-\tau_+\sigma_-^{-1}$ $\sigma(\frac{2\pi}{3}) = \tau_-\tau_+\sigma_+^{-1} = \tau_+\tau_+\sigma_-^{-1}$
$\mathfrak{S}_6$	$\rho_4 = \rho_+$ $\sigma(\frac{5\pi}{3}) = \rho_-$	$\rho_5 = \sigma_+$ $\sigma(\frac{4\pi}{3}) = \sigma_-$	$\rho_6 = \tau_+$ $\sigma(\pi) = \tau_-$

Of course the ρ_+, ρ_-, etc., which appear in the two tables are distinct, but there is no point in encumbering the notation with primes or superscripts.

We have next to choose a coordinate on each of the $\mathfrak{S}_i$ and calculate the residues of $M(s, \lambda)$, $s \in \Omega(\mathfrak{S}_i, \mathfrak{S}_j)$, with respect to it. The coordinate will be denoted z and will be the restriction of the coordinate on the total λ-space indicated in the table below.

$\mathcal{E}_1$	$\mathcal{E}_2$	$\mathcal{E}_3$	$\mathcal{E}_4$	$\mathcal{E}_5$	$\mathcal{E}_6$
$\frac{3}{2}+z_2$	$\frac{1}{2}-z_1$	$\frac{3}{2}-\lambda(H_{\beta_2})$	$\frac{1}{2}-\lambda(H_{\beta_5})$	$\frac{3}{2}-z_2$	$\frac{1}{2}+z_1$

To calculate the residue we have to choose near $\mathcal{E}_i$ as coordinates $\lambda(H_{\beta_i})$ and $\pm\lambda(H_{\beta_j})$ where $z = a_i \pm \lambda(H_{\beta_j})$ and express the other coordinates $\lambda(H_{\beta_k})$ in terms of them.

Principal coordinates   Other coordinates

1) $\lambda(H_{\beta_1})$, $\lambda(H_{\beta_6})$ $H\beta_2 = 3H_{\beta_1} + H_{\beta_6}$ $H_{\beta_3} = 2H_{\beta_1} + H_{\beta_6}$

          $H_{\beta_4} = 3H_{\beta_1} + 2H_{\beta_6}$ $H_{\beta_5} = H_{\beta_1} + H_{\beta_6}$

2) $\lambda(H_{\beta_2})$, $-\lambda(H_{\beta_1})$ $H_{\beta_3} = H_{\beta_2} - H_{\beta_1}$ $H_{\beta_4} = 2H_{\beta_2} - 3H_{\beta_1}$

          $H_{\beta_5} = H_{\beta_2} - 2H_{\beta_1}$ $H\beta_6 = H_{\beta_2} - 3H_{\beta_1}$

3) $\lambda(H_{\beta_3})$, $-\lambda(H_{\beta_2})$ $H_{\beta_1} = H_{\beta_2} - H_{\beta_3}$ $H_{\beta_4} = 3H_{\beta_3} - H_{\beta_2}$

          $H_{\beta_5} = 2H_{\beta_3} - H_{\beta_2}$ $H_{\beta_6} = 3H_{\beta_3} - 2H_{\beta_2}$

4) $\lambda(H_{\beta_4})$, $-\lambda(H_{\beta_5})$ $H_{\beta_1} = H_{\beta_4} - 2H_{\beta_5}$ $H_{\beta_2} = 2H_{\beta_4} - 3H_{\beta_5}$

          $H_{\beta_3} = H_{\beta_4} - H_{\beta_5}$ $H_{\beta_6} = 3H_{\beta_5} - H_{\beta_4}$

5) $\lambda(H_{\beta_5})$, $-\lambda(H_{\beta_6})$ $H_{\beta_1} = H_{\beta_5} - H_{\beta_6}$ $H_{\beta_2} = 3H_{\beta_5} - 2H_{\beta_6}$

          $H_{\beta_3} = 2H_{\beta_5} - H_{\beta_6}$ $H_{\beta_4} = 3H_{\beta_5} - H_{\beta_6}$

6) $\lambda(H_{\beta_6})$, $\lambda(H_{\beta_1})$ $H_{\beta_2} = 3H_{\beta_1} + H_{\beta_6}$ $H_{\beta_3} = 2H_{\beta_1} + H_{\beta_6}$

          $H_{\beta_4} = 3H_{\beta_1} + 2H_{\beta_6}$ $H_{\beta_5} = H_{\beta_1} + H_{\beta_6}$

In table (6) the residues $n(\sigma, z)$ or $n(\sigma, \lambda)$, $\lambda = \lambda(z)$, for the elements of table (4) are given and in table (7) those for the elements of table (5). To obtain them one uses the formula for $M(s, \lambda)$, the table (3), and the relations (5). To make sure that there is no ambiguity I observe that, for example, the entry in the third row and third column of (6) is $n(\tau_+\rho_+\sigma_+^{-1}, z)$ and corresponds to the third row and third column of (4). The residue of $\frac{\xi(2)}{\xi(1+2)}$ at $z = 1$ is $\frac{1}{\xi(2)}$. Thus, for example, the residue of $M(\sigma(\pi), \lambda)$ on $\mathfrak{C}_1$ is

$$\frac{1}{\xi(2)} \frac{\xi(z-\frac{3}{2})}{\xi(z-\frac{1}{2})} \frac{\xi(z-\frac{1}{2})}{\xi(z+\frac{1}{2})} \frac{\xi(z+\frac{1}{2})}{\xi(z+\frac{3}{2})} \frac{\xi(z+\frac{3}{2})}{\xi(z+\frac{5}{2})} \frac{\xi(2z)}{\xi(1+2z)} = \frac{1}{\xi(2)} \frac{\xi(z-\frac{3}{2})\xi(2z)}{\xi(z+\frac{5}{2})\xi(1+2z)}$$

To save space the factor $\frac{1}{\xi(2)}$, which should appear before all entries, is omitted and $\xi(az+b)$ is written $(az+b)$

(6)

1	$\frac{(z+\frac{1}{2})}{(z+\frac{5}{2})}$	$\frac{(z-\frac{1}{2})}{(z+\frac{5}{2})} \frac{(2z)}{(2z+1)}$
$\frac{(z-\frac{3}{2})}{(z+\frac{5}{2})} \frac{(2z)}{(1+2z)}$	$\frac{(z+\frac{1}{2})}{(z+\frac{5}{2})} \frac{(2z)}{(2z+1)}$	$\frac{(z+\frac{3}{2})}{(z+\frac{5}{2})}$
$\frac{(\frac{1}{2}-z)}{(\frac{5}{2}-z)}$	$\frac{(\frac{1}{2}-z)}{(\frac{5}{2}-z)} \frac{(z+\frac{1}{2})}{(z+\frac{5}{2})}$	$\frac{(\frac{3}{2}-z)}{(\frac{5}{2}-z)} \frac{(z+\frac{1}{2})}{(z+\frac{5}{2})} \frac{(2z)}{(2z+1)}$
$\frac{(z+\frac{1}{2})}{(z+\frac{5}{2})} \frac{(2z)}{(2z+1)}$	$\frac{(\frac{1}{2}-z)}{(\frac{5}{2}-z)} \frac{(z+\frac{1}{2})}{(z+\frac{5}{2})} \frac{(2z)}{(2z+1)}$	$\frac{(\frac{1}{2}-z)}{(\frac{5}{2}-z)} \frac{(z+\frac{3}{2})}{(z+\frac{5}{2})}$
$\frac{(z-\frac{1}{2})}{(z+\frac{5}{2})} \frac{(2z)}{(2z+1)}$	$\frac{(z+\frac{1}{2})}{(z+\frac{5}{2})} \frac{(\frac{3}{2}-z)}{(\frac{5}{2}-z)} \frac{(2z)}{(2z+1)}$	$\frac{(z+\frac{3}{2})}{(z+\frac{5}{2})} \frac{(\frac{3}{2}-z)}{(\frac{5}{2}-z)}$
$\frac{(\frac{3}{2}-z)}{(\frac{5}{2}-z)}$	$\frac{(z+\frac{1}{2})}{(z+\frac{5}{2})} \frac{(\frac{3}{2}-z)}{(\frac{5}{2}-z)}$	$\frac{(z-\frac{1}{2})}{(z+\frac{5}{2})} \frac{(\frac{3}{2}-z)}{(\frac{5}{2}-z)} \frac{(2z)}{(2z+1)}$

(7)

$\frac{(\frac{1}{2}-z)}{(\frac{3}{2}-z)}\frac{(\frac{1}{2}+z)}{(\frac{3}{2}+z)}$	$\frac{(\frac{1}{2}-z)}{(\frac{3}{2}-z)}\frac{(\frac{1}{2}+z)}{(\frac{3}{2}+z)}\frac{(\frac{1}{2}+3z)}{(\frac{3}{2}+3z)}\frac{(2z)}{(2z+1)}$	$\frac{(\frac{1}{2}+z)}{(\frac{3}{2}+z)}\frac{(-\frac{1}{2}+3z)}{(\frac{3}{2}+3z)}\frac{(2z)}{(2z+1)}$
$\frac{(\frac{1}{2}-z)}{(\frac{3}{2}-z)}\frac{(\frac{1}{2}+z)}{(\frac{3}{2}+z)}\frac{(-\frac{1}{2}+3z)}{(\frac{3}{2}+3z)}\frac{(2z)}{(2z+1)}$	$\frac{(\frac{1}{2}-z)}{(\frac{3}{2}-z)}\frac{(\frac{1}{2}+z)}{(\frac{3}{2}+z)}\frac{(\frac{1}{2}+3z)}{(\frac{3}{2}+3z)}$	$\frac{(\frac{1}{2}-z)}{(\frac{3}{2}-z)}$
$\frac{(2z)}{(2z+1)}\frac{(3z+\frac{1}{2})}{(3z+\frac{3}{2})}\frac{(z+\frac{1}{2})}{(z+\frac{3}{2})}\frac{(\frac{1}{2}-z)}{(\frac{3}{2}-z)}$	$\frac{(3z+\frac{1}{2})}{(3z+\frac{3}{2})}\frac{(z+\frac{1}{2})}{(z+\frac{3}{2})}\frac{(\frac{1}{2}-z)}{(\frac{3}{2}-z)}\frac{(\frac{1}{2}-3z)}{(\frac{3}{2}-3z)}$	$\frac{(\frac{1}{2}-z)}{(\frac{3}{2}-z)}\frac{(\frac{1}{2}-3z)}{(\frac{3}{2}-3z)}$
$\frac{(z+\frac{1}{2})}{(z+\frac{3}{2})}\frac{(\frac{1}{2}-z)}{(\frac{3}{2}-z)}\frac{(\frac{1}{2}-3z)}{(\frac{3}{2}-3z)}$	$\frac{(2z)}{(2z+1)}\frac{(3z+\frac{1}{2})}{(3z+\frac{3}{2})}\frac{(z+\frac{1}{2})}{(z+\frac{3}{2})}\frac{(\frac{1}{2}-z)}{(\frac{3}{2}-z)}\frac{(\frac{1}{2}-3z)}{(\frac{3}{2}-3z)}$	$\frac{(2z)}{(2z+1)}\frac{(3z+\frac{1}{2})}{(3z+\frac{3}{2})}\frac{(z+\frac{1}{2})}{(z+\frac{3}{2})}$
$\frac{(3z-\frac{1}{2})}{(3z+\frac{3}{2})}\frac{(2z)}{(2z+1)}\frac{(z+\frac{1}{2})}{(z+\frac{3}{2})}$	$\frac{(3z+\frac{1}{2})}{(3z+\frac{3}{2})}\frac{(z+\frac{1}{2})}{(z+\frac{3}{2})}$	1
$\frac{(z+\frac{1}{2})}{(z+\frac{3}{2})}$	$\frac{(2z)}{(2z+1)}\frac{(3z+\frac{1}{2})}{(3z+\frac{3}{2})}\frac{(z+\frac{1}{2})}{(z+\frac{3}{2})}$	$\frac{(z-\frac{1}{2})}{(z+\frac{3}{2})}\frac{(3z-\frac{1}{2})}{(3z+\frac{3}{2})}\frac{(2z)}{(2z+1)}$

The difference between (1) and the analogous integral with $\lambda_0 = 0$ is the sum of

(8) $$\Sigma_{i=1}^{3}\Sigma_{j=1}^{3}\Sigma_{\sigma\in\Omega(\mathcal{E}_{2i},\mathcal{E}_{2j})}\frac{1}{2\pi i}\int_{\operatorname{Re}\lambda=\lambda_{2i}} \overline{n(\sigma,\lambda)\Phi(\lambda)\Psi(-\sigma\bar{\lambda})}dz$$

and

(9) $$\Sigma_{i=1}^{3}\Sigma_{j=1}^{3}\Sigma_{\sigma\in\Omega(\mathcal{E}_{2i-1},\mathcal{E}_{2j-1})}\frac{1}{2\pi i}\int_{\operatorname{Re}\lambda=\lambda_{2i-1}} \overline{n(\sigma,\lambda)\Phi(\lambda)\Psi(-\sigma\bar{\lambda})}dz$$

Here $\lambda = \lambda(z)$. If we follow the procedure of §7, we deform the contours to $\operatorname{Re}\lambda = \lambda(0)$. The resulting expressions give the inner product of the projections on the one-dimensional spectrum. The residues which arise during the deformation when added together give the inner product of the projections on the spectrum of dimension 0. We shall see that the subspace corresponding to the discrete spectrum, that is, the spectrum of dimension 0, is of dimension two, consisting of the constant functions and another eigenspace of dimension one.

Before carrying out the deformation and computing the residues explicitly, we write out for the collections $\{\mathfrak{S}_1, \mathfrak{S}_3, \mathfrak{S}_5\}$ and $\{\mathfrak{S}_2, \mathfrak{S}_4, \mathfrak{S}_6\}$ the matrix M(H) figuring in Lemma 7.4, observing as a check upon tables (6) and (7) that they satisfy the conclusion of Lemma 7.4, that is, they are both or rank one. H is now $\lambda = \lambda(z)$ and the matrix elements are functions of z. The matrices are given in tables (10) and (11). Once again, to save space the factor $\frac{1}{\xi(2)}$ has been omitted from all entries and $\xi(az+b)$ is written simply $(az+b)$. In (10) the element s^o of the text is ρ_+; in (11) it is τ_+. Thus if $\lambda = \lambda(z)$ the entry in the box of (10) with row labelled σ_+ and column ρ_- is $\xi(2)n(\sigma_+\rho_+\rho_-^{-1}, \rho_-\rho_+\lambda)$. It should perhaps be stressed that if $\lambda = \lambda(z)$ then for all σ the coordinate of $-\sigma\lambda$ is $\pm z$.

Since none of the functions $n(\sigma, \lambda)$ has a singularity on $\mathrm{Re}\,\lambda = \lambda(0)$, we may deform each of the terms in (8) and (9) separately. Since there are eighteen terms in each of the two expressions, and since some of the residues arising are complicated, the computation will be lengthy. Nonetheless it is best to write it out completely, for one appreciates better the difficulties faced in §7 if one sees the procedure which was there described in an abstract form carried out in a specific case, which is after all relatively simple. Suppose that, near $z = 1$,

$$\xi(z) = \frac{1}{z-1} + a + b(z-1) + O((z-1)^2)$$

1) We begin by finding the residues for $\mathfrak{S}_1$. At λ_1

$$\frac{3}{2} < z < \frac{5}{2}$$

Let $R(\rho_+)$, $R(\rho_-)$, and so on, denote the residues arising from the corresponding terms of (8). Making use of (4) and (6) we obtain the following results. Observe

$$\{\mathcal{C}_1,\ \mathcal{C}_3,\ \mathcal{C}_5\}$$

(10)

	ρ_+	ρ_-	σ_+	σ_-	τ_+	τ_-
ρ_+	1	$\frac{(-z-\frac{3}{2})}{(\frac{5}{2}-z)}\frac{(-2z)}{(1-2z)}$	$\frac{(\frac{1}{2}-z)}{(\frac{5}{2}-z)}$	$\frac{(\frac{1}{2}-z)}{(\frac{5}{2}-z)}\frac{(-2z)}{(1-2z)}$	$\frac{(-z-\frac{1}{2})}{(-z+\frac{5}{2})}\frac{(-2z)}{(1-2z)}$	$\frac{(\frac{3}{2}-z)}{(\frac{5}{2}-z)}$
ρ_-	$\frac{(z-\frac{3}{2})}{(z+\frac{5}{2})}\frac{(2z)}{(2z+1)}$	1	$\frac{(z+\frac{1}{2})}{(z+\frac{5}{2})}\frac{(2z)}{(2z+1)}$	$\frac{(\frac{1}{2}+z)}{(\frac{5}{2}+z)}$	$\frac{(\frac{3}{2}+z)}{(\frac{5}{2}+z)}$	$\frac{(z-\frac{1}{2})}{(z+\frac{5}{2})}\frac{(2z)}{(1+2z)}$
σ_+	$\frac{(z+\frac{1}{2})}{(z+\frac{5}{2})}$	$\frac{(\frac{1}{2}-z)}{(\frac{5}{2}-z)}\frac{(-2z)}{(1-2z)}$	$\frac{(\frac{1}{2}-z)}{(\frac{5}{2}-z)}\frac{(\frac{1}{2}+z)}{(\frac{5}{2}+z)}$	$\frac{(\frac{1}{2}+z)}{(\frac{5}{2}+z)}\frac{(\frac{1}{2}-z)}{(\frac{5}{2}-z)}\frac{(-2z)}{(1-2z)}$	$\frac{(\frac{3}{2}+z)}{(\frac{5}{2}+z)}\frac{(\frac{1}{2}-z)}{(\frac{5}{2}-z)}\frac{(-2z)}{(1-2z)}$	$\frac{(\frac{1}{2}+z)}{(\frac{5}{2}+z)}\frac{(\frac{3}{2}-z)}{(\frac{5}{2}-z)}$
σ_-	$\frac{(z+\frac{1}{2})}{(z+\frac{5}{2})}\frac{(2z)}{(2z+1)}$	$\frac{(\frac{1}{2}-z)}{(\frac{5}{2}-z)}$	$\frac{(\frac{1}{2}-z)}{(\frac{5}{2}-z)}\frac{(\frac{1}{2}+z)}{(\frac{5}{2}+z)}\frac{(2z)}{(1+2z)}$	$\frac{(\frac{1}{2}+z)}{(\frac{5}{2}+2)}\frac{(\frac{1}{2}-z)}{(\frac{5}{2}-z)}$	$\frac{(\frac{1}{2}-z)}{(\frac{5}{2}-z)}\frac{(\frac{3}{2}+z)}{(\frac{5}{2}+z)}$	$\frac{(\frac{3}{2}-z)}{(\frac{5}{2}-z)}\frac{(\frac{1}{2}+z)}{(\frac{5}{2}+z)}\frac{(2z)}{(1+2z)}$
τ_+	$\frac{(z-\frac{1}{2})}{(z+\frac{5}{2})}\frac{(2z)}{(2z+1)}$	$\frac{(\frac{3}{2}-z)}{(\frac{5}{2}-z)}$	$\frac{(\frac{3}{2}-z)}{(\frac{5}{2}-z)}\frac{(z+\frac{1}{2})}{(z+\frac{5}{2})}\frac{(2z)}{(2z+1)}$	$\frac{(\frac{1}{2}+z)}{(\frac{5}{2}+z)}\frac{(\frac{3}{2}-z)}{(\frac{5}{2}-z)}$	$\frac{(\frac{3}{2}-z)}{(\frac{5}{2}-z)}\cdot\frac{(\frac{3}{2}+z)}{(\frac{5}{2}+z)}$	$\frac{(\frac{3}{2}-z)}{(\frac{5}{2}-z)}\frac{(z-\frac{1}{2})}{(\frac{5}{2}+z)}\frac{(2z)}{(1+2z)}$
τ_-	$\frac{(z+\frac{3}{2})}{(z+\frac{5}{2})}$	$\frac{(-z-\frac{1}{2})}{(\frac{5}{2}-z)}\frac{(-2z)}{(1-2z)}$	$\frac{(\frac{1}{2}-z)}{(\frac{5}{2}-z)}\frac{(\frac{3}{2}+z)}{(\frac{5}{2}+z)}$	$\frac{(\frac{3}{2}+z)}{(\frac{5}{2}+z)}\frac{(\frac{1}{2}-z)}{(\frac{5}{2}-z)}\frac{(-2z)}{(1-2z)}$	$\frac{(\frac{3}{2}+z)}{(\frac{5}{2}+z)}\frac{(-z-\frac{1}{2})}{(\frac{5}{2}-z)}\frac{(-2z)}{(1-2z)}$	$\frac{(\frac{3}{2}+z)}{(\frac{5}{2}+z)}\frac{(\frac{3}{2}-z)}{(\frac{5}{2}-z)}$

that the relevant singularities occur at the intersections of $\mathfrak{S}_1$ with some other $\mathfrak{S}_j$.

$$R(\rho_+) = 0$$

$$R(\rho_-) = -\frac{\xi(3)}{\xi(2)\xi^2(4)}\Phi(\beta_3)\overline{\Psi}(\beta_3) + \frac{1}{2\xi(2)\xi(3)}\Phi(\beta_2)\overline{\Psi}(\beta_2)$$

$$R(\sigma_+) = \frac{1}{\xi(2)\xi(3)}\Phi(\beta_2)\overline{\Psi}(\beta_2)$$

The term $R(\sigma_-)$ is more complicated because the poles of $n(\sigma_-, \lambda)$ are not simple. We let D_i be the differential operator

$$D_i\Phi(\lambda) = \frac{d}{dt}\Phi(\lambda + t\beta_i)\Big|_{t=0}$$

Then, as the conscientious reader will readily verify, $R(\sigma_-)$ is the sum of

$$\frac{1}{2\xi^2(2)\xi(3)}\{\Phi(\beta_2)D_6\overline{\Psi}(\beta_4) + D_4\Phi(\beta_2)\overline{\Psi}(\beta_4)\}$$

and

$$\left\{\frac{3a}{2\xi^2(2)\xi(3)} - \frac{\xi'(2)}{\xi(3)\xi^3(2)} - \frac{\xi'(3)}{2\xi^2(2)\xi^2(3)}\right\}\Phi(\beta_2)\overline{\Psi}(\beta_4)$$

Moreover $R(\tau_+)$ is the sum of

$$\frac{-1}{2\xi^2(2)\xi(3)}\{\Phi(\beta_2)D_2\overline{\Psi}(\beta_4) + D_4\Phi(\beta_2)\overline{\Psi}(\beta_4)\}$$

and

$$\frac{\xi(3)}{\xi(2)\xi^2(4)}\Phi(\beta_3)\overline{\Psi}(\beta_3) + \left\{\frac{-a}{2\xi^2(2)\xi(3)} + \frac{\xi'(3)}{2\xi^2(2)\xi^2(3)} + \frac{\xi'(2)}{\xi^3(2)\xi(3)}\right\}\Phi(\beta_2)\overline{\Psi}(\beta_4)$$

while

$$R(\tau_-) = 0$$

Adding these six terms together we see that the total residue from $\mathfrak{S}_1$ is

$\{\mathfrak{S}_2, \mathfrak{S}_4, \mathfrak{S}_6\}$

11)

	ρ_+	ρ_-	σ_+
ρ_+	$\frac{(\frac{1}{2}+z)}{(\frac{3}{2}+z)} \frac{(\frac{1}{2}-z)}{(\frac{3}{2}-z)}$	$\frac{(\frac{1}{2}-z)}{(\frac{3}{2}-z)} \frac{(\frac{1}{2}+z)}{(\frac{3}{2}+z)} \frac{(3z-\frac{1}{2})}{(3z+\frac{3}{2})} \frac{(2z)}{(1+2z)}$	$\frac{(2z)}{(1+2z)} \frac{(\frac{1}{2}+3z)}{(\frac{3}{2}+3z)} \frac{(\frac{1}{2}+z)}{(\frac{3}{2}+z)} \frac{(\frac{1}{2}-z)}{(\frac{3}{2}-z)}$
ρ_-	$\frac{(\frac{1}{2}+z)}{(\frac{3}{2}+z)} \frac{(\frac{1}{2}-z)}{(\frac{3}{2}-z)} \frac{(-\frac{1}{2}-3z)}{(\frac{3}{2}-3z)} \frac{(-2z)}{(1-2z)}$	$\frac{(\frac{1}{2}-z)}{(\frac{3}{2}-z)} \frac{(\frac{1}{2}+z)}{(\frac{3}{2}+z)}$	$\frac{(\frac{1}{2}+z)}{(\frac{3}{2}+z)} \frac{(\frac{1}{2}-z)}{(\frac{3}{2}-z)} \frac{(\frac{1}{2}-3z)}{(\frac{3}{2}-3z)}$
σ_+	$\frac{(\frac{1}{2}+z)}{(\frac{3}{2}+z)} \frac{(\frac{1}{2}-z)}{(\frac{3}{2}-z)} \frac{(-2z)}{(1-2z)} \frac{(\frac{1}{2}-3z)}{(\frac{3}{2}-3z)}$	$\frac{(\frac{1}{2}-z)}{(\frac{3}{2}-z)} \frac{(\frac{1}{2}+z)}{(\frac{3}{2}+z)} \frac{(3z+\frac{1}{2})}{(3z+\frac{3}{2})}$	$\frac{(\frac{1}{2}+3z)}{(\frac{3}{2}+3z)} \frac{(\frac{1}{2}+z)}{(\frac{3}{2}+z)} \frac{(\frac{1}{2}-z)}{(\frac{3}{2}-z)} \frac{(\frac{1}{2}-3z)}{(\frac{3}{2}-3z)}$
σ_-	$\frac{(\frac{1}{2}+z)}{(\frac{3}{2}+z)} \frac{(\frac{1}{2}-z)}{(\frac{3}{2}-z)} \frac{(\frac{1}{2}-3z)}{(\frac{3}{2}-3z)}$	$\frac{(\frac{1}{2}-z)}{(\frac{3}{2}-z)} \frac{(\frac{1}{2}+z)}{(\frac{3}{2}+z)} \frac{(2z)}{(1+2z)} \frac{(3z+\frac{1}{2})}{(3z+\frac{3}{2})}$	$\frac{(\frac{1}{2}+z)}{(\frac{3}{2}+z)} \frac{(\frac{1}{2}-z)}{(\frac{3}{2}-z)} \frac{(2z)}{(1+2z)} \frac{(\frac{1}{2}+3z)}{(\frac{3}{2}+3z)} \frac{(\frac{1}{2}-3z)}{(\frac{3}{2}-3z)}$
τ_+	$\frac{(\frac{1}{2}-z)}{(\frac{3}{2}-z)} \frac{(-2z)}{(1-2z)} \frac{(-\frac{1}{2}-3z)}{(\frac{3}{2}-3z)}$	$\frac{(\frac{1}{2}-z)}{(\frac{3}{2}-z)}$	$\frac{(\frac{1}{2}-z)}{(\frac{3}{2}-z)} \frac{(\frac{1}{2}-3z)}{(\frac{3}{2}-3z)}$
τ_-	$\frac{(\frac{1}{2}+z)}{(\frac{3}{2}+z)}$	$\frac{(\frac{1}{2}+z)}{(\frac{3}{2}+z)} \frac{(2z)}{(2z+1)} \frac{(3z-\frac{1}{2})}{(3z+\frac{3}{2})}$	$\frac{(\frac{1}{2}+z)}{(\frac{3}{2}+z)} \frac{(2z)}{(1+2z)} \frac{(\frac{1}{2}+3z)}{(\frac{3}{2}+3z)}$

$$\{\mathfrak{G}_2,\ \mathfrak{G}_4,\ \mathfrak{G}_6\}$$

σ_-	τ_+	τ_-	
$\frac{(\frac{1}{2}-z)}{(\frac{3}{2}-z)}\frac{(\frac{1}{2}+z)}{(\frac{3}{2}+z)}\frac{(\frac{1}{2}+3z)}{(\frac{3}{2}+3z)}$	$\frac{(z+\frac{1}{2})}{(z+\frac{3}{2})}\frac{(2z)}{(2z+1)}\frac{(3z-\frac{1}{2})}{(3z+\frac{3}{2})}$	$\frac{(\frac{1}{2}-z)}{(\frac{3}{2}-z)}$	ρ_+
$\frac{(-2z)}{(1-2z)}\frac{(\frac{1}{2}-3z)}{(\frac{3}{2}-3z)}\frac{(\frac{1}{2}-z)}{(\frac{3}{2}-z)}\frac{(\frac{1}{2}+z)}{(\frac{3}{2}+z)}$	$\frac{(z+\frac{1}{2})}{(z+\frac{3}{2})}$	$\frac{(\frac{1}{2}-z)}{(\frac{3}{2}-z)}\frac{(-2z)}{(1-2z)}\frac{(-\frac{1}{2}-3z)}{(\frac{3}{2}-3z)}$	ρ_-
$\frac{(\frac{1}{2}-z)}{(\frac{3}{2}-z)}\frac{(\frac{1}{2}+z)}{(\frac{3}{2}+z)}\frac{(-2z)}{(1-2z)}\frac{(\frac{1}{2}-3z)}{(\frac{3}{2}-3z)}\frac{(\frac{1}{2}+3z)}{(\frac{3}{2}+3z)}$	$\frac{(3z+\frac{1}{2})}{(3z+\frac{3}{2})}\frac{(z+\frac{1}{2})}{(z+\frac{3}{2})}$	$\frac{(\frac{1}{2}-z)}{(\frac{3}{2}-z)}\frac{(-2z)}{(1-2z)}\frac{(\frac{1}{2}-3z)}{(\frac{3}{2}-3z)}$	σ_+
$\frac{(\frac{1}{2}-3z)}{(\frac{3}{2}-3z)}\frac{(\frac{1}{2}-z)}{(\frac{3}{2}-z)}\frac{(\frac{1}{2}+z)}{(\frac{3}{2}+z)}\frac{(\frac{1}{2}+3z)}{(\frac{3}{2}+3z)}$	$\frac{(z+\frac{1}{2})}{(z+\frac{3}{2})}\frac{(2z)}{(2z+1)}\frac{(3z+\frac{1}{2})}{(3z+\frac{3}{2})}$	$\frac{(\frac{1}{2}-3z)}{(\frac{3}{2}-3z)}\frac{(\frac{1}{2}-z)}{(\frac{3}{2}-z)}$	σ_-
$\frac{(\frac{1}{2}-z)}{(\frac{3}{2}-z)}\frac{(-2z)}{(1-2z)}\frac{(\frac{1}{2}-3z)}{(\frac{3}{2}-3z)}$	1	$\frac{(-\frac{1}{2}-z)}{(\frac{3}{2}-z)}\frac{(-2z)}{(1-2z)}\frac{(-\frac{1}{2}-3z)}{(\frac{3}{2}-3z)}$	τ_+
$\frac{(\frac{1}{2}+z)}{(\frac{3}{2}+z)}\frac{(\frac{1}{2}+3z)}{(\frac{3}{2}+3z)}$	$\frac{(z-\frac{1}{2})}{(z+\frac{3}{2})}\frac{(2z)}{(1+2z)}\frac{(3z-\frac{1}{2})}{(3z+\frac{3}{2})}$	1	τ_-

(11)

$$\frac{3}{2\xi(2)\xi(3)}\Phi(\beta_2)\overline{\Psi}(\beta_2) + \frac{a}{\xi^2(2)\xi(3)}\Phi(\beta_2)\overline{\Psi}(\beta_4) - \frac{1}{2\xi^2(2)\xi(3)}\Phi(\beta_2)D_1\overline{\Psi}(\beta_4)$$

Since there is considerable cancellation involved in these calculations which cannot be predicted from general principles, the interested reader is advised to verify each step for himself.

2) The residues for $\mathfrak{C}_2$ are easier to find. The coordinate of the point λ_2 satisfies $\frac{1}{6} < z < \frac{1}{2}$.

$$R(\rho_+\tau_+\rho_+^{-1}) = 0$$

$$R(\rho_-\tau_+\rho_+^{-1}) = \frac{-\xi^2(\frac{1}{3})\xi(\frac{2}{3})}{3\xi^2(2)\xi^2(\frac{4}{3})\xi(\frac{5}{3})}\Phi\left(\frac{\beta_3}{3}\right)\overline{\Psi}\left(\frac{\beta_3}{3}\right)$$

$$R(\sigma_+\tau_+\rho_+^{-1}) = \frac{\xi^2(\frac{1}{3})\xi(\frac{2}{3})}{3\xi^2(2)\xi^2(\frac{4}{3})\xi(\frac{5}{3})}\Phi\left(\frac{\beta_3}{3}\right)\overline{\Psi}\left(\frac{\beta_3}{3}\right)$$

$$R(\sigma_-\tau_+\rho_+^{-1}) = \frac{\xi(\frac{1}{3})\xi(\frac{2}{3})}{3\xi^2(2)\xi(\frac{4}{3})\xi(\frac{5}{3})}\Phi\left(\frac{\beta_3}{3}\right)\overline{\Psi}\left(\frac{\beta_5}{3}\right)$$

$$R(\tau_+\tau_+\rho_+^{-1}) = \frac{-\xi(\frac{1}{3})\xi(\frac{2}{3})}{3\xi^2(2)\xi(\frac{4}{3})\xi(\frac{5}{3})}\Phi\left(\frac{\beta_3}{3}\right)\overline{\Psi}\left(\frac{\beta_5}{3}\right)$$

$$R(\tau_-\tau_+\rho_+^{-1}) = 0$$

The sum of these six terms is 0. It is clear from the diagram of the spaces $\mathfrak{C}_i$ that there are no residues for $\mathfrak{C}_3$ or $\mathfrak{C}_4$. The residues from $\mathfrak{C}_5$ and $\mathfrak{C}_6$ are however extremely complicated.

5) The coordinate of λ_5 satisfies $\frac{1}{2} < z < \frac{3}{2}$. Putting our head down and bashing on we obtain the following results for the residues. $R(\rho_+\rho_+\tau_+^{-1})$ is the sum of

$$\frac{-1}{2\xi^2(2)\xi(3)}\{\Phi(\beta_4)D_4\overline{\Psi}(\beta_2) + D_2\Phi(\beta_4)\overline{\Psi}(\beta_2)\}$$

and

$$\left\{\frac{-a}{2\xi^2(2)\xi(3)} + \frac{\xi'(3)}{2\xi^2(2)\xi^2(3)} + \frac{\xi'(2)}{\xi^3(2)\xi(3)}\right\} \Phi(\beta_4)\overline{\Psi}(\beta_2)$$

$R(\rho_-\rho_+\tau_+^{-1})$ is easier to find; it equals

$$\frac{-1}{\xi^2(2)} \Phi(\beta_4)\overline{\Psi}(-\beta_6)$$

Since $-\beta_6$ does not lie in the dual of the positive chamber, we infer from Lemma 7.5 that this term will be cancelled by another, for it cannot remain when all the residues are added together.

The other terms grow more complicated. $R(\sigma_+\rho_+\tau_+^{-1})$ is the sum of the following expressions.

$$\frac{-1}{2\xi^3(2)\xi(3)} \{D_2\Phi(\beta_4)D_6\overline{\Psi}(\beta_4) + \tfrac{1}{2}\Phi(\beta_4)D_6^2\overline{\Psi}(\beta_4) + \tfrac{1}{2}D_2^2\Phi(\beta_4)\overline{\Psi}(\beta_4)\}$$

$$\frac{1}{\xi(2)}\left\{\frac{1}{2}\left(\frac{\xi'(2)}{\xi^3(2)\xi(3)} + \frac{\xi'(3)}{\xi^2(2)\xi^2(3)}\right) - \frac{a}{\xi^2(2)\xi(3)}\right\}\left\{D_2\Phi(\beta_4)\overline{\Psi}(\beta_4) + \Phi(\beta_4)D_6\overline{\Psi}(\beta_4)\right\}$$

$$\frac{1}{\xi(2)}\left\{a\left(\frac{\xi'(2)}{\xi^3(2)\xi(3)} + \frac{\xi'(3)}{\xi^2(2)\xi^2(3)}\right) + \frac{1}{\xi^2(2)\xi(3)}\left(\frac{a^2}{2} - 3b\right)\right\} \Phi(\beta_4)\overline{\Psi}(\beta_4)$$

$$\frac{1}{\xi(2)}\left\{\frac{5}{4\xi(2)\xi(3)}\left(\frac{\xi''(2)}{\xi^2(2)} - \frac{2(\xi'(2))^2}{\xi^3(2)}\right) + \frac{1}{4\xi^2(2)}\left(\frac{\xi''(3)}{\xi^2(3)} - \frac{2(\xi'(3))^2}{\xi^3(3)}\right)\right\} \Phi(\beta_4)\overline{\Psi}(\beta_4)$$

$$\frac{-1}{\xi(2)}\left(\frac{\xi'(2)\xi'(3)}{\xi^3(2)\xi^2(3)} - \frac{2(\xi'(2))^2}{\xi^4(2)\xi(3)}\right)\Phi(\beta_4)\overline{\Psi}(\beta_4)$$

$R(\sigma_-\rho_+\tau_+^{-1})$ is not so bad; it is the sum of

$$\frac{1}{\xi^2(2)\xi(3)}\{\Phi(\beta_4)D_6\overline{\Psi}(\beta_2) - D_2\Phi(\beta_4)\overline{\Psi}(\beta_2)\}$$

and

$$\left\{\frac{\xi'(3)}{\xi^2(2)\xi^2(3)} - \frac{\xi'(2)}{\xi^3(2)\xi(3)}\right\} \Phi(\beta_4)\overline{\Psi}(\beta_2)$$

$R(\tau_+\rho_+\tau_+^{-1})$ is simply

$$\frac{-1}{\xi(2)\xi(3)}\Phi(\beta_4)\overline{\Psi}(\beta_6)$$

With $R(\tau_-\rho_+\tau_+^{-1})$ complications appear once again. It is the sum of the following terms.

$$\frac{1}{2\xi^3(2)\xi(3)}\{D_2\Phi(\beta_4)D_2\overline{\Psi}(\beta_4) + \frac{1}{2}D_2^2\Phi(\beta_4)\overline{\Psi}(\beta_4) + \frac{1}{2}\Phi(\beta_4)D_2^2\overline{\Psi}(\beta_4)\}$$

$$-\frac{1}{2\xi(2)}\left\{\frac{\xi'(2)}{\xi^3(2)\xi(3)} + \frac{\xi'(3)}{\xi^2(2)\xi^2(3)}\right\}\{D_2\Phi(\beta_4)\overline{\Psi}(\beta_4) + \Phi(\beta_4)D_2\overline{\Psi}(\beta_4)\}$$

$$\frac{1}{\xi(2)}\left\{\frac{1}{2}\left(\frac{\xi'(2)\xi'(3)}{\xi^3(2)\xi^2(3)} - \frac{2(\xi'(2))^2}{\xi^4(2)\xi(3)}\right) + \frac{1}{\xi^2(2)\xi(3)}\left(3b - \frac{3a^2}{2}\right)\right\}\Phi(\beta_4)\overline{\Psi}(\beta_4)$$

$$\frac{1}{2\xi(2)}\left\{\frac{-5}{2\xi(2)\xi(3)}\left(\frac{\xi''(2)}{\xi^2(2)} - \frac{2(\xi'(2))^2}{\xi^3(2)}\right) - \frac{1}{2\xi^2(2)}\left(\frac{\xi''(3)}{\xi^2(3)} - \frac{2(\xi'(3))^2}{\xi^3(3)}\right)\right\}\Phi(\beta_4)\overline{\Psi}(\beta_4)$$

We add up all the terms above and find that the total contribution from $\mathfrak{E}_5$ is the sum of the following six expressions:

$$\left\{\frac{a}{\xi(2)}\left(\frac{\xi'(2)}{\xi^3(2)\xi(3)} + \frac{\xi'(3)}{\xi^2(2)\xi^2(3)}\right) - \frac{a^2}{\xi^3(2)\xi(3)}\right\}\Phi(\beta_4)\overline{\Psi}(\beta_4)$$

$$-\frac{a}{\xi^3(2)\xi(3)}D_2\Phi(\beta_1)\overline{\Psi}(\beta_4) - \frac{a}{\xi^3(2)\xi(3)}\Phi(\beta_4)D_6\overline{\Psi}(\beta_4)$$

$$-\frac{1}{\xi^2(2)}\Phi(\beta_4)\overline{\Psi}(-\beta_6) - \frac{1}{\xi(2)\xi(3)}\Phi(\beta_4)\overline{\Psi}(\beta_6)$$

$$-\frac{1}{2\xi^2(2)\xi(3)}\Phi(\beta_4)D_1\overline{\Psi}(\beta_2) - \frac{3}{2\xi^2(2)\xi(3)}D_2\Phi(\beta_4)\overline{\Psi}(\beta_2)$$

$$\frac{1}{\xi^2(2)\xi(3)}\left(\frac{-a}{2} + \frac{3\xi'(3)}{2\xi(3)}\right)\Phi(\beta_4)\overline{\Psi}(\beta_2) + \frac{1}{4\xi^3(2)\xi(3)}\Phi(\beta_4)(D_2^2 - D_6^2)\overline{\Psi}(\beta_4)$$

$$\frac{1}{2\xi^3(2)\xi(3)}D_2\Phi(\beta_4)D_1\overline{\Psi}(\beta_4) - \frac{1}{2\xi(2)}\left(\frac{\xi'(2)}{\xi^3(2)\xi(3)} + \frac{\xi'(3)}{\xi^2(2)\xi^2(3)}\right)\Phi(\beta_4)D_1\overline{\Psi}(\beta_4)$$

The term involving $\bar{\Psi}(-\beta_6)$ has not yet disappeared.

The reader will be losing heart, for we still have $\mathfrak{S}_6$ to work through. He is urged to persist, for the final result is very simple. I do not know the reason.

(6) The coordinate of λ_6 is greater than $\frac{3}{2}$. It will be seen from the diagram of the spaces $\mathfrak{S}_i$ that we may pick up residues at three points, at the intersection of $\mathfrak{S}_6$ and $\mathfrak{S}_1$, at the common intersection of $\mathfrak{S}_6$, $\mathfrak{S}_5$, $\mathfrak{S}_3$, and $\mathfrak{S}_2$, and at the intersection of $\mathfrak{S}_6$ with $\mathfrak{S}_4$. The corresponding values of z are $\frac{3}{2}$, $\frac{1}{2}$, and $\frac{1}{6}$. The contribution $R(\rho_+)$ is the sum of the following terms:

$$\frac{1}{6\xi^3(2)\xi(3)}\{D_3\Phi(\beta_4)D_5\bar{\Psi}(\beta_4) + \frac{1}{2}D_3^2\Phi(\beta_4)\bar{\Psi}(\beta_4) + \Phi(\beta_4)D_5\bar{\Psi}(\beta_4)\}$$

$$\frac{1}{\xi(2)}\left\{\frac{a}{\xi^2(2)\xi(3)} - \frac{1}{2}\left(\frac{\xi'(3)}{\xi^2(2)\xi^2(3)} + \frac{\xi'(2)}{\xi^3(2)\xi(3)}\right)\right\}\left\{D_3\Phi(\beta_4)\bar{\Psi}(\beta_4) + \Phi(\beta_4)D_5\bar{\Psi}(\beta_4)\right\}$$

$$\frac{1}{\xi^3(2)\xi(3)}\left\{\left(\frac{11}{6}a^2 + \frac{7}{3}b\right) - a\left(\frac{3\xi'(3)}{\xi(3)} + \frac{3\xi'(2)}{\xi(2)}\right)\right\}\Phi(\beta_4)\bar{\Psi}(\beta_4)$$

$$\frac{1}{6\xi(2)}\left\{\frac{9\xi'(2)\xi'(3)}{\xi^3(2)\xi^2(3)} + \frac{2(\xi'(2))^2}{\xi^4(2)\xi(3)}\right\}\Phi(\beta_4)\bar{\Psi}(\beta_4)$$

$$\frac{-1}{12\xi^2(2)}\left\{\frac{9}{\xi(2)}\left(\frac{\xi''(3)}{\xi^2(3)} - \frac{2(\xi'(3))^2}{\xi^3(3)}\right) + \frac{5}{\xi(3)}\left(\frac{\xi''(2)}{\xi^2(2)} - \frac{2(\xi'(2))^2}{\xi^3(2)}\right)\right\}\Phi(\beta_4)\bar{\Psi}(\beta_4)$$

$$-\frac{\xi(\frac{1}{3})\xi(\frac{2}{3})}{3\xi^2(2)\xi(\frac{5}{3})\xi(\frac{4}{3})}\Phi(\frac{\beta_5}{3})\bar{\Psi}(\frac{\beta_3}{3})$$

The value of $R(\rho_-)$ is simply

$$\frac{1}{\xi^2(2)}\Phi(\beta_4)\bar{\Psi}(-\beta_6)$$

It cancels the term for $\mathfrak{S}_5$ which had troubled us.

The value of $R(\sigma_+)$ is

$$\frac{1}{\xi(2)\xi(3)}\Phi(\beta_4)\bar{\Psi}(\beta_6) + \frac{\xi(\frac{2}{3})}{3\xi^2(2)\xi(\frac{5}{3})}\Phi(\frac{\beta_5}{3})\bar{\Psi}(\frac{\beta_5}{3})$$

For $R(\sigma_-)$ we obtain the sum of three terms

$$\frac{1}{2\xi^2(2)\xi(3)}\{D_3\Phi(\beta_4)\bar{\Psi}(\beta_2) + \Phi(\beta_4)D_1\bar{\Psi}(\beta_2)\}$$

$$\left\{\frac{3a}{2\xi^2(2)\xi(3)} - \frac{3\xi'(3)}{2\xi^2(2)\xi^2(3)}\right\}\Phi(\beta_4)\bar{\Psi}(\beta_2)$$

$$\frac{\xi(\frac{1}{3})\xi(\frac{2}{3})}{3\xi(\frac{4}{3})\xi(\frac{5}{3})\xi^2(2)}\Phi(\frac{\beta_5}{3})\bar{\Psi}(\frac{\beta_3}{3})$$

$R(\tau_+)$ is of course zero, but $R(\tau_-)$ is the sum of the following nine terms. δ is now one-half the sum of the positive roots.

$$\frac{1}{\xi(2)\xi(6)}\Phi(\delta)\bar{\Psi}(\delta)$$

$$\frac{-1}{6\xi^3(2)\xi(3)}\{D_3\Phi(\beta_4)D_3\bar{\Psi}(\beta_4) + \frac{1}{2}D_3^2\Phi(\beta_4)\bar{\Psi}(\beta_4) + \frac{1}{2}\Phi(\beta_4)D_3^2\bar{\Psi}(\beta_4)\}$$

$$\frac{1}{2\xi(2)}\left\{\frac{\xi'(3)}{\xi^2(2)\xi^2(3)} + \frac{\xi'(2)}{\xi^3(2)\xi(3)}\right\}\left\{\Phi(\beta_4)D_3\bar{\Psi}(\beta_4) + D_3\Phi(\beta_4)\bar{\Psi}(\beta_4)\right\}$$

$$\frac{-2a}{3\xi^3(2)\xi(3)}\{\Phi(\beta_4)D_3\bar{\Psi}(\beta_4) + D_3\Phi(\beta_4)\bar{\Psi}(\beta_4)\}$$

$$\frac{2a}{\xi(2)}\left\{\frac{\xi'(3)}{\xi^2(2)\xi^2(3)} + \frac{\xi'(2)}{\xi^3(2)\xi(3)}\right\}\Phi(\beta_4)\bar{\Psi}(\beta_4)$$

$$\frac{-1}{6\xi(2)}\left\{\frac{9\xi'(2)\xi'(3)}{\xi^3(2)\xi^2(3)} + \frac{2(\xi'(2))^2}{\xi^4(2)\xi(3)}\right\}\Phi(\beta_4)\bar{\Psi}(\beta_4)$$

$$\frac{-1}{6\xi^3(2)\xi(3)}(a^2 - 14b)\Phi(\beta_4)\bar{\Psi}(\beta_4)$$

$$\frac{1}{6\xi(2)}\left\{\frac{5}{2\xi(2)\xi(3)}\left(\frac{\xi''(2)}{\xi^2(2)} - \frac{2(\xi'(2))^2}{\xi^3(2)}\right) + \frac{9}{2\xi^2(2)}\left(\frac{\xi''(3)}{\xi^2(3)} - \frac{2(\xi'(3))^2}{\xi^3(3)}\right)\right\}\Phi(\beta_4)\bar{\Psi}(\beta_4)$$

$$\frac{-\xi(\frac{1}{3})}{3\xi(\frac{4}{3})\xi(\frac{5}{3})\xi^2(2)}\Phi(\frac{\beta_5}{3})\bar{\Psi}(\frac{\beta_5}{3})$$

Adding the six contributions together we see that the total residue from

$\mathcal{E}_6$ is the sum of the following terms:

$$\frac{1}{\xi^2(2)}\Phi(\beta_4)\overline{\Psi}(-\beta_6) + \frac{1}{\xi(2)\xi(3)}\Phi(\beta_4)\overline{\Psi}(\beta_6) + \frac{1}{\xi(2)\xi(6)}\Phi(\rho)\overline{\Psi}(\rho)$$

$$\frac{1}{2\xi^2(2)\xi(3)}\{\Phi(\beta_4)D_1\overline{\Psi}(\beta_2) + D_3\Phi(\beta_4)\overline{\Psi}(\beta_2)\}$$

$$\frac{3}{2\xi^2(2)\xi(3)}\left(a - \frac{\xi'(3)}{\xi(3)}\right)\Phi(\beta_4)\overline{\Psi}(\beta_2) + \frac{1}{12\xi^3(2)\xi(2)}\Phi(\beta_4)(D_5^2 - D_3^2)\overline{\Psi}(\beta_4)$$

$$\frac{-1}{6\xi^3(2)\xi(3)}D_3\Phi(\beta_4)D_1\overline{\Psi}(\beta_4) + \frac{1}{2\xi(2)}\left(\frac{\xi'(3)}{\xi^2(2)\xi^2(3)} + \frac{\xi'(2)}{\xi^3(2)\xi(3)}\right)\Phi(\beta_4)D_1\overline{\Psi}(\beta_4)$$

$$\frac{a}{3\xi^3(2)\xi(3)}D_3\Phi(\beta_4)\overline{\Psi}(\beta_4) + \frac{a}{\xi^3(2)\xi(3)}\Phi(\beta_4)(D_5 - \tfrac{2}{3}D_3)\overline{\Psi}(\beta_4)$$

$$\frac{5a^2}{3\xi^3(2)\xi(3)}\Phi(\beta_4)\overline{\Psi}(\beta_4) - \frac{a}{\xi(2)}\left(\frac{\xi'(2)}{\xi^3(2)\xi(3)} + \frac{\xi'(3)}{\xi^2(2)\xi^2(3)}\right)\Phi(\beta_4)\overline{\Psi}(\beta_4)$$

Now all we have to do is add together the contributions from $\mathcal{E}_1, \ldots, \mathcal{E}_6$. The result may be expressed simply in matrix notation as:

$$\begin{bmatrix} \Psi(\rho) \\ \Psi(\beta_2) \\ \Psi(\beta_4) \\ D_1\Psi(\beta_4) \end{bmatrix}^* \begin{bmatrix} \frac{1}{\xi(2)\xi(6)} & 0 & 0 & 0 \\ 0 & \frac{3}{2\xi(2)\xi(3)} & \frac{a}{\xi^2(2)\xi(3)} & \frac{-1}{2\xi^2(2)\xi(3)} \\ 0 & \frac{a}{\xi^2(2)\xi(3)} & \frac{2a^2}{3\xi^3(2)\xi(3)} & \frac{-a}{3\xi^3(2)\xi(3)} \\ 0 & \frac{-1}{2\xi^2(2)\xi(3)} & \frac{-a}{3\xi^3(2)\xi(3)} & \frac{1}{6\xi^3(2)\xi(3)} \end{bmatrix} \begin{bmatrix} \Phi(\rho) \\ \Phi(\beta_2) \\ \Phi(\beta_4) \\ D_1\Phi(\beta_4) \end{bmatrix}$$

That the matrix turns out to be symmetric and positive-definite is a check on our calculations. Since it is of rank two, the discrete spectrum contains two points.

One of the associated subspaces is the space of constant functions. The constant term of the functions in the other space is not a sum of pure exponentials. The appearance of a second point in the discrete spectrum is a surprise. One wonders what its significance is.

In the example just discussed the functions $n(\sigma, \lambda)$ were analytic on the line $\mathrm{Re}\,\lambda = \lambda(0)$, and the corresponding residues of the Eisenstein series must be as well. This may not always be so, and one must be content with a weaker assertion, that of Lemma 7.6. This is seen already with the one-dimensional spectrum for the group of type A_3.

This is the group $SL(4)$. We may take as coordinates of λ, parameters z_1, z_2, z_3, z_4 with $\Sigma z_i = 0$. The elements of the Weyl group are permutations and

$$M(s, \lambda) = \prod_{\substack{i<j \\ s(i)>s(j)}} \frac{\xi(z_i - z_j)}{\xi(1+z_i - z_j)}$$

At the first stage the integration will be taken over the set $\mathrm{Re}\, z_i = z_i^o$ with $z_i^o - z_j^o > 1$ if $i<j$. Then it is moved to $\mathrm{Re}\, z_i = 0$. Residues are obtained on the hyperplanes $\mathcal{E}_{ij}$ defined by $z_i - z_j = 1$. These give the two-dimensional spectrum. In order to obtain the one-dimensional spectrum the integration has then to be moved to

$$\mathrm{Re}\, z_k = \frac{\delta_{ki}}{2} - \frac{\delta_{kj}}{2}$$

where δ_{ki} is Kronecker's delta. If $M(s, \lambda)$ has a singularity on $\mathcal{E}_{ij}$ then $s(i) > s(j)$. If $k < \ell$ and $s(k) > s(\ell)$ with $(ij) \neq (k\ell)$ then

$$\frac{\xi(z_k - z_\ell)}{\xi(1+z_k - z_\ell)}$$

is a factor of the residue. If $k \neq j$ then $1 + z_k - z_\ell \geq 1$ during the deformation and the zeros of the denominator play no role. If $k=j$ then $s(i)>s(\ell)$ and the residue contains the factor

$$\frac{\xi(z_i-z_\ell)}{\xi(1+z_i-z_\ell)}\ \frac{\xi(z_j-z_\ell)}{\xi(1+z_j-z_\ell)}$$

Since $z_i = 1+z_j$ on $\mathfrak{S}_{ij}$ the denominator is again harmless. The relevant singularities lie on the intersection of $\mathfrak{S}_{ij}$ with some $\mathfrak{S}_{i'j'}$.

Because we are interested in the one-dimensional spectrum and want to proceed as expeditiously as possible, we shall only write down those two-dimensional residues which in turn yield one-dimensional residues. We take $z_i^o - z_{i+1}^o > z_{i+1}^o - z_4^o$, $i = 1, 2$.

1) $i=1$, $j=4$. When we deform the two-dimensional integral on $\mathfrak{S}_{14}$ we pick up no residues. So this hyperplane may be ignored.

2) $i=1$, $j=3$. Because of our choice of z_i^o, the only singular hyperplane that we meet during the deformation is $\mathfrak{S}_{14}$. The intersection is $\mathfrak{S} = (\frac{2}{3}, 0, \frac{-1}{3}, \frac{-1}{3}) + (u, v, u, u)$ with $3u+v=0$. We obtain contributions from those s for which $s(4) < s(1)$ and $s(3) < s(1)$. For these we obtain the following results:

s		R(s)
$(1234) \to (3412)$	$(\frac{2}{3}, 0, \frac{-1}{3}, \frac{-1}{3}) + (u, v, u, u) \to (\frac{-1}{3}, \frac{-1}{3}, \frac{2}{3}, 0) + (u, u, u, v)$	$(23)(24)$
$\to (4312)$	$\to (\frac{-1}{3}, \frac{-1}{3}, \frac{2}{3}, 0) + (u, u, u, v)$	$-(23)(24)$
$\to (3421)$	$\to (\frac{-1}{3}, \frac{-1}{3}, 0, \frac{2}{3}) + (u, u, v, u)$	$(12)(23)(34)$
$\to (4321)$	$\to (\frac{-1}{3}, \frac{-1}{3}, 0, \frac{2}{3}) + (u, u, v, u)$	$-(12)(23)(34)$
$\to (3241)$	$\to (\frac{-1}{3}, 0, \frac{-1}{3}, \frac{2}{3}) + (u, v, u, u)$	$(12)(23)$
$\to (4231)$	$\to (\frac{-1}{3}, 0, \frac{-1}{3}, \frac{2}{3}) + (u, v, u, u)$	$-(12)(24)$
$\to (2341)$	$\to (0, \frac{-1}{3}, \frac{-1}{3}, \frac{2}{3}) + (v, u, u, u)$	(12)
$\to (2431)$	$\to (0, \frac{-1}{3}, \frac{-1}{3}, \frac{2}{3}) + (v, u, u, u)$	$-(12)$

The symbol $(k\ell)$ is an abbreviation for

$$\frac{\xi(z_k - z_\ell)}{\xi(\ +z_k - z_\ell)}$$

and we have omitted from all the $R(s)$ a common constant. But this is unimportant, for we see that the residues cancel in pairs and that $\mathfrak{S}_{13}$ contributes nothing to the one-dimensional spectrum.

3) $i = 1$, $j = 2$. There will be singularities at the intersections of $\mathfrak{S}_{12}$ with $\mathfrak{S}_{13}$ and $\mathfrak{S}_{14}$. Because of our choice of z_i^o, they are the only ones which affect our calculations.

$$\mathfrak{S}_{12} \cap \mathfrak{S}_{13} = (\frac{2}{3}, \frac{-1}{3}, \frac{-1}{3}, 0) + (u, u, u, v)$$

$$\mathfrak{S}_{12} \cap \mathfrak{S}_{14} = (\frac{2}{3}, \frac{-1}{3}, 0, \frac{-1}{3}) + (u, u, v, u)$$

If s contributes to the residue on the first intersection then $s(2) < s(1)$ and $s(3) < s(1)$. If s_0 is the interchange of (2) and (3) then ss_0 has the same effect on $\mathfrak{S}_{12} \cap \mathfrak{S}_{13}$, but the residues of $R(s)$ and $R(ss_0)$ are of opposite sign because

$$\frac{\xi(z_2 - z_3)}{\xi(1 + z_2 - z_3)}$$

is -1 when $z_2 = z_3$. Thus the contribution of the first intersection to the one-dimensional spectrum is 0.

If s contributes to the residue on the second intersection then $s(2) < s(1)$ and $s(4) < s(1)$. The possibilities are given below.

s	R(s)
(1234) → (2413)	(43)
→ (4213)	-(43)
→ (2431)	(13)(34)
→ (4231)	-(13)(34)
→ (2341)	(13)
→ (4321)	-(13)(23)(34)
→ (3241)	(13)(23)
→ (3421)	-(13)(23)

Since

$$z_1 - z_3 = -(z_3 - z_4)$$

on the intersection

$$\frac{\xi(z_2-z_3)}{\xi(1+z_2-z_3)}\,\frac{\xi(z_3-z_4)}{\xi(1+z_3-z_4)} = 1$$

Once again the cancellation is complete.

4) $i = 2$, $j = 4$. The poles occur at the intersection of $\mathcal{S}_{24}$ with $\mathcal{S}_{12}$, $\mathcal{S}_{13}$, and $\mathcal{S}_{14}$. These intersections are:

$$\mathcal{S}_{24} \cap \mathcal{S}_{12} = (1, 0, 0, -1) + (u, u, v, u)$$
$$\mathcal{S}_{24} \cap \mathcal{S}_{13} = \left(\frac{1}{2}, \frac{1}{2}, \frac{-1}{2}, \frac{-1}{2}\right) + (u, v, u, v)$$
$$\mathcal{S}_{24} \cap \mathcal{S}_{14} = \left(\frac{1}{3}, \frac{1}{3}, 0, \frac{-2}{3}\right) + (u, u, v, u)$$

We list in the three cases the relevant s and the corresponding residues.

a)	s		R(s)
(1234)	$\to$ (4213)	$(1,0,0,-1)+(u,u,v,u) \to (-1,0,1,0)+(u,u,u,v)$	(34)
	$\to$ (4231)	$\to (-1,0,0,1)+(u,u,v,u)$	(13)(34)
	$\to$ (4321)	$\to (-1,0,0,1)+(u,v,u,u)$	(13)(23)(34)
	$\to$ (3421)	$\to (0,1,0,-1)+(v,u,u,u)$	(13)(23)

We have omitted the common factor $\frac{1}{\xi(2)\xi(3)}$.

b)	s		R(s)
(1234)	$\to$ (3142)	$(\frac{1}{2},\frac{1}{2},\frac{-1}{2},\frac{-1}{2})+(u,v,u,v) \to (\frac{-1}{2},\frac{1}{2},\frac{-1}{2},\frac{1}{2})+(u,u,v,v)$	(23)
	$\to$ (3412)	$\to (\frac{-1}{2},\frac{-1}{2},\frac{1}{2},\frac{1}{2})+(u,v,u,v)$	(14)(23)
	$\to$ (3421)	$\to (\frac{-1}{2},\frac{-1}{2},\frac{1}{2},\frac{1}{2})+(u,v,v,u)$	(12)(14)(23)
	$\to$ (4312)	$\to (\frac{-1}{2},\frac{-1}{2},\frac{1}{2},\frac{1}{2})+(v,u,u,v)$	(14)(23)(34)
	$\to$ (4321)	$\to (\frac{-1}{2},\frac{-1}{2},\frac{1}{2},\frac{1}{2})+(v,u,v,u)$	(12)(14)(23)(34)
	$\to$ (4231)	$\to (\frac{-1}{2},\frac{1}{2},\frac{-1}{2},\frac{1}{2})+(v,v,u,u)$	(12)(14)(34)

We have omitted a common factor $\frac{1}{\xi^2(2)}$.

c) If s contributes to the residue for the third intersection then $s(4) < s(1)$ and $s(4) < s(2)$. If s_0 interchanges 1 and 2 and leaves 3 and 4 fixed, then ss_0 contributes as well. Since

$$R(s) = -R(ss_0)$$

the total contribution will be 0.

5) $i = 2$, $j = 3$. The relevant poles occur at the intersection of $\mathfrak{S}_{23}$ with $\mathfrak{S}_{12}$, $\mathfrak{S}_{13}$, $\mathfrak{S}_{14}$, and $\mathfrak{S}_{24}$. These intersections are as follows:

$$\mathcal{E}_{23} \cap \mathcal{E}_{12} = (1, 0, -1, 0) + (u, u, u, v)$$

$$\mathcal{E}_{23} \cap \mathcal{E}_{13} = (\frac{1}{3}, \frac{1}{3}, \frac{-2}{3}, 0) + (u, u, u, v)$$

$$\mathcal{E}_{23} \cap \mathcal{E}_{14} = (\frac{1}{2}, \frac{1}{2}, \frac{-1}{2}, \frac{-1}{2}) + (u, v, v, u)$$

$$\mathcal{E}_{23} \cap \mathcal{E}_{24} = (0, \frac{2}{3}, \frac{-1}{3}, \frac{-1}{3}) + (u, v, v, v)$$

Again we list the pertinent s and the corresponding R(s).

a) s R(s)

s		R(s)
(1234) → (3214)	$(1, 0, -1, 0) + (u, u, u, v) \to (-1, 0, 1, 0) + (u, u, u, v)$	1
→ (3241)	$\to (-1, 0, 0, 1) + (u, u, v, u)$	(14)
→ (3421)	$\to (-1, 0, 0, 1) + (u, v, u, u)$	(14)(24)
→ (4321)	$\to (0, -1, 0, 1) + (v, u, u, u)$	(14)(24)(34)

Again a common factor $\frac{1}{\xi(2)\xi(3)}$ has been omitted.

b) The same argument as above establishes that the total contribution from this intersection is 0.

c) s R(s)

s		R(s)
(1234) → (4132)	$(\frac{1}{2}, \frac{1}{2}, \frac{-1}{2}, \frac{-1}{2}) + (u, v, v, u) \to (\frac{-1}{2}, \frac{1}{2}, \frac{-1}{2}, \frac{1}{2}) + (u, u, v, v)$	(24)(34)
→ (4312)	$\to (\frac{-1}{2}, \frac{-1}{2}, \frac{1}{2}, \frac{1}{2}) + (u, v, u, v)$	(13)(24)(34)
→ (4321)	$\to (\frac{-1}{2}, \frac{-1}{2}, \frac{1}{2}, \frac{1}{2}) + (u, v, v, u)$	(12)(13)(24)(34)
→ (3412)	$\to (\frac{-1}{2}, \frac{-1}{2}, \frac{1}{2}, \frac{1}{2}) + (v, u, u, v)$	(13)(24)
→ (3421)	$\to (\frac{-1}{2}, \frac{-1}{2}, \frac{1}{2}, \frac{1}{2}) + (v, u, v, u)$	(12)(13)(24)
→ (3241)	$\to (\frac{-1}{2}, \frac{1}{2}, \frac{-1}{2}, \frac{1}{2}) + (v, v, u, u)$	(12)(13)

d) Here again the total contribution is 0.

6) $i = 3$, $j = 4$. The intersections with any of the other $\mathcal{E}_{i'j'}$ are now relevant.

These intersections are as follows:

$$\mathcal{S}_{34} \cap \mathcal{S}_{12} = (\tfrac{1}{2}, \tfrac{-1}{2}, \tfrac{1}{2}, \tfrac{-1}{2}) + (u, u, v, v)$$

$$\mathcal{S}_{34} \cap \mathcal{S}_{13} = (1, 0, 0, -1) + (u, v, u, u)$$

$$\mathcal{S}_{34} \cap \mathcal{S}_{14} = (\tfrac{1}{3}, 0, \tfrac{1}{3}, \tfrac{-2}{3}) + (u, v, u, u)$$

$$\mathcal{S}_{34} \cap \mathcal{S}_{23} = (0, 1, 0, -1) + (u, v, v, v)$$

$$\mathcal{S}_{34} \cap \mathcal{S}_{24} = (0, \tfrac{1}{3}, \tfrac{1}{3}, \tfrac{-2}{3}) + (u, v, v, v)$$

Again we take each possibility in order and list the pertinent s and the corresponding R(s).

a)

s		R(s)
$(1234) \to (2143)$	$(\tfrac{1}{2}, \tfrac{-1}{2}, \tfrac{1}{2}, \tfrac{-1}{2}) + (u, u, v, v) \to (\tfrac{-1}{2}, \tfrac{1}{2}, \tfrac{-1}{2}, \tfrac{1}{2}) + (u, u, v, v)$	1
$\to (2413)$	$\to (\tfrac{-1}{2}, \tfrac{-1}{2}, \tfrac{1}{2}, \tfrac{1}{2}) + (u, v, u, v)$	(14)
$\to (2431)$	$\to (\tfrac{-1}{2}, \tfrac{-1}{2}, \tfrac{1}{2}, \tfrac{1}{2}) + (u, v, v, u)$	(13)(14)
$\to (4213)$	$\to (\tfrac{-1}{2}, \tfrac{-1}{2}, \tfrac{1}{2}, \tfrac{1}{2}) + (v, u, u, v)$	(14)(24)
$\to (4231)$	$\to (\tfrac{-1}{2}, \tfrac{-1}{2}, \tfrac{1}{2}, \tfrac{1}{2}) + (v, u, v, u)$	(13)(14)(24)
$\to (4321)$	$\to (\tfrac{-1}{2}, \tfrac{1}{2}, \tfrac{-1}{2}, \tfrac{1}{2}) + (v, v, u, u)$	(13)(14)(23)(24)

A common factor $\frac{1}{\xi^2(2)}$ has been omitted.

b)

s		R(s)
$(1234) \to (4312)$	$(1, 0, 0, -1) + (u, v, u, u) \to (-1, 0, 1, 0) + (u, u, u, v)$	(23)(24)
$\to (4321)$	$\to (-1, 0, 0, 1) + (u, u, v, u)$	(12)(23)(24)
$\to (4231)$	$\to (-1, 0, 0, 1) + (u, v, u, u)$	(12)(24)
$\to (2431)$	$\to (0, -1, 0, 1) + (v, u, u, u)$	(12)

The common factor $\frac{1}{\xi(2)\xi(3)}$ has been omitted.

c) The total contribution is again 0.

d) s		R(s)
(1234) → (1432)	(0, 1, 0, -1) + (u, v, v, v) → (0, -1, 0, 1) + (u, v, v, v)	1
→ (4132)	→ (-1, 0, 0, 1) + (v, u, v, v)	(14)
→ (4312)	→ (-1, 0, 0, 1) + (v, v, u, v)	(13)(14)
→ (4321)	→ (-1, 0, 1, 0) + (v, v, v, u)	(12)(13)(14)

Again the common factor $\frac{1}{\xi(2)\xi(3)}$ has been omitted.

e) The total contribution is 0.

The one-dimensional spectrum is therefore determined by two collections of subspaces. The first collection is formed by

$$(0, 1, 0, -1) + (u, v, v, v)$$

$$(1, 0, 0, -1) + (v, u, v, v)$$

$$(1, 0, 0, -1) + (v, v, u, v)$$

$$(1, 0, -1, 0) + (v, v, v, u)$$

For any two $\mathcal{S}$ and $\mathcal{T}$ of these subspaces, the set $\Omega(\mathcal{S}, \mathcal{T})$ consists of a single element. The matrix M(H) figuring in Lemma 7.4 is given, apart from the factor $\frac{1}{\xi(2)\xi(3)}$, in Table (12). It is, as it must be, of rank one. However, it does have singularities at $u = v = 0$, that is, on the line over which we must finally integrate.

This is disconcerting at first, but, as shown in the text, presents no insurmountable problem. The constant term of the Eisenstein series, or system, associated to the line (1, 0, -1, 0) + (u, v, v, v) is apart from the factor

$$\frac{1}{\xi(2)\xi(3)}\, e^{\frac{3}{2}z_1+\frac{1}{2}z_2-\frac{1}{2}z_3-\frac{3}{2}z_4}$$

given by the sum of

(12)

	$(0,1,0,-1)+(u,v,v,v)$	$(1,0,0,-1)+(v,u,v,v)$	$(1,0,0,-1)+(v,v,u,v)$	$(1,0,-1,0)+(v,v,v,u)$
$(0,1,0,-1)+(u,v,v,v)$	1	$\frac{\xi(1+v-u)}{\xi(2+v-u)}$	$\frac{\xi(v-u)}{\xi(2+v-u)}$	$\frac{\xi(v-u-1)}{\xi(v-u+2)}$
$(1,0,0,-1)+(v,u,v,v)$	$\frac{\xi(1+u-v)}{\xi(2+u-v)}$	$\frac{\xi(1+v-u)}{\xi(2+v-u)}\frac{\xi(1+u-v)}{\xi(2+u-v)}$	$\frac{\xi(v-u)}{\xi(2+v-u)}\frac{\xi(1+u-v)}{\xi(2+u-v)}$	$\frac{\xi(v-u)}{\xi(2+v-u)}$
$(1,0,0,-1)+(v,v,u,v)$	$\frac{\xi(u-v)}{\xi(2+u-v)}$	$\frac{\xi(1+v-u)}{\xi(2+v-u)}\frac{\xi(u-v)}{\xi(2+u-v)}$	$\frac{\xi(1+v-u)}{\xi(2+v-u)}\frac{\xi(1+u-v)}{\xi(2+u-v)}$	$\frac{\xi(1+v-u)}{\xi(2+v-u)}$
$(1,0,-1,0)+(v,v,v,u)$	$\frac{\xi(u-v-1)}{\xi(2+u-v)}$	$\frac{\xi(u-v)}{\xi(2+u-v)}$	$\frac{\xi(1+u-v)}{\xi(2+u-v)}$	1

The entries are $M(ts^{o}s^{-1}, ss^{o}H)$. The rows are indexed by t, and the columns by s. The entry at the top of a given column is the result of applying ss^{o} to the entry at the top of the first column. The rows are labelled in a similar fashion.

$$e^{-z_2+z_4} e^{uz_1+vz_2+vz_3+vz_4} + \frac{\xi(u-v-1)}{\xi(u-v+2)} e^{-z_1+z_3} e^{vz_1+vz_2+vz_3+vz_4}$$

which has no poles on the line $\mathrm{Re}(u-v) = 0$ and

$$\frac{\xi(1+u-v)}{\xi(2+u-v)} e^{-z_1+z_4} \left\{ e^{vz_1+uz_2+vz_3+vz_4} + \frac{\xi(u-v)}{\xi(1+u-v)} e^{vz_1+vz_2+uz_3+vz_4} \right\}$$

Since the factor $\frac{\xi(u-v)}{\xi(1+u-v)}$ equals -1 at $u = v$, this term too has no poles on the line $\mathrm{Re}(u-v) = 0$. Thus the constant term, and hence the Eisenstein series itself, is analytic on that line. This is a simple illustration of the corollary to Lemma 7.6.

The second collection is formed by

$$(\tfrac{1}{2}, \tfrac{-1}{2}, \tfrac{1}{2}, \tfrac{-1}{2}) + (u, u, v, v)$$

$$(\tfrac{1}{2}, \tfrac{1}{2}, \tfrac{-1}{2}, \tfrac{-1}{2}) + (u, v, u, v)$$

$$(\tfrac{1}{2}, \tfrac{1}{2}, \tfrac{-1}{2}, \tfrac{-1}{2}) + (u, v, v, u)$$

The sets $\Omega(\mathfrak{a}, \mathfrak{a})$ now consist of two elements. The matrix of Lemma 7.4 is given in Table (13). It may be readily verified that it is of rank one.

Reference

1. R. P. Langlands, Eisenstein series, in Algebraic Groups and Discontinuous Subgroups, Amer. Math. Soc. (1966).

(13)

	$(\frac{1}{2}, \frac{-1}{2}, \frac{1}{2}, \frac{-1}{2})$ + (u, u, v, v)	$(\frac{1}{2}, \frac{-1}{2}, \frac{1}{2}, \frac{-1}{2})$ + (v, v, u, u)	$(\frac{1}{2}, \frac{1}{2}, \frac{-1}{2}, \frac{-1}{2})$ + (u, v, v, u)
$(\frac{1}{2}, \frac{-1}{2}, \frac{1}{2}, \frac{1}{2})$ + (u, u, v, v)	1	$\frac{\xi(v-u)}{\xi(2+v-u)} \frac{\xi(v-u-1)}{\xi(v-u+1)}$	$\frac{\xi(v-u)}{\xi(2+v-u)}$
$(\frac{1}{2}, \frac{-1}{2}, \frac{1}{2}, \frac{-1}{2})$ + (v, v, u, u)	$\frac{\xi(u-v)}{\xi(2+u-v)} \frac{\xi(u-v-1)}{\xi(u-v+1)}$	1	$\frac{\xi(u-v)}{\xi(2+u-v)}$
$(\frac{1}{2}, \frac{1}{2}, \frac{-1}{2}, \frac{-1}{2})$ + (u, v, v, u)	$\frac{\xi(u-v)}{\xi(2+u-v)}$	$\frac{\xi(v-u)}{\xi(2+v-u)}$	$\frac{\xi(v-u)}{\xi(2+v-u)} \frac{\xi(u-v)}{\xi(2+u-v)}$
$(\frac{1}{2}, \frac{1}{2}, \frac{-1}{2}, \frac{-1}{2})$ + (v, u, u, v)	$\frac{\xi(u-v)}{\xi(2+u-v)}$	$\frac{\xi(v-u)}{\xi(2+v-u)}$	$\frac{\xi(1+u-v)}{\xi(2+u-v)} \frac{\xi(1+v-u)}{\xi(2+v-u)}$
$(\frac{1}{2}, \frac{1}{2}, \frac{-1}{2}, \frac{-1}{2})$ + (u, v, u, v)	$\frac{\xi(1+u-v)}{\xi(2+u-v)}$	$\frac{\xi(v-u)}{\xi(1+v-u)} \frac{\xi(v-u)}{\xi(2+v-u)}$	$\frac{\xi(1+u-v)}{\xi(2+u-v)} \frac{\xi(v-u)}{\xi(2+v-u)}$
$(\frac{1}{2}, \frac{1}{2}, \frac{-1}{2}, \frac{-1}{2})$ + (v, u, v, u)	$\frac{\xi(u-v)}{\xi(1+u-v)} \frac{\xi(u-v)}{\xi(2+u-v)}$	$\frac{\xi(1+v-u)}{\xi(2+v-u)}$	$\frac{\xi(u-v)}{\xi(2+u-v)} \frac{\xi(1+v-u)}{\xi(2+v-u)}$

The principle according to which the entries are indexed is the same as in the preceding table.

$(\frac{1}{2}, \frac{1}{2}, \frac{-1}{2}, \frac{-1}{2})$ $+ (v, u, u, v)$	$(\frac{1}{2}, \frac{1}{2}, \frac{-1}{2}, \frac{-1}{2})$ $+ (u, v, u, v)$	$(\frac{1}{2}, \frac{1}{2}, \frac{-1}{2}, \frac{-1}{2})$ $+ (v, u, v, u)$	
$\frac{\xi(v-u)}{\xi(2+v-u)}$	$\frac{\xi(1+v-u)}{\xi(2+v-u)}$	$\frac{\xi(v-u)}{\xi(2+v-u)} \frac{\xi(v-u)}{\xi(1+v-u)}$	
$\frac{\xi(u-v)}{\xi(2+u-v)}$	$\frac{\xi(u-v)}{\xi(2+u-v)} \frac{\xi(u-v)}{\xi(1+u-v)}$	$\frac{\xi(1+u-v)}{\xi(2+u-v)}$	
$\frac{\xi(1+v-u)}{\xi(2+v-u)} \frac{\xi(1+u-v)}{\xi(2+u-v)}$	$\frac{\xi(u-v)}{\xi(2+u-v)} \frac{\xi(1+v-u)}{\xi(2+v-u)}$	$\frac{\xi(v-u)}{\xi(2+v-u)} \frac{\xi(1+u-v)}{\xi(2+u-v)}$	(13)
$\frac{\xi(u-v)}{\xi(2+u-v)} \frac{\xi(v-u)}{\xi(2+v-u)}$	$\frac{\xi(u-v)}{\xi(2+u-v)} \frac{\xi(1+v-u)}{\xi(2+v-u)}$	$\frac{\xi(v-u)}{\xi(2+v-u)} \frac{\xi(1+u-v)}{\xi(2+v-u)}$	
$\frac{\xi(v-u)}{\xi(2+v-u)} \frac{\xi(1+u-v)}{\xi(2+u-v)}$	$\frac{\xi(1+u-v)}{\xi(2+u-v)} \frac{\xi(1+v-u)}{\xi(2+v-u)}$	$\frac{\xi(v-u)}{\xi(2+v-u)} \frac{\xi(1+u-v)}{\xi(2+u-v)} \frac{\xi(v-u)}{\xi(1+v-u)}$	
$\frac{\xi(1+v-u)}{\xi(2+v-u)} \frac{\xi(u-v)}{\xi(2+u-v)}$	$\frac{\xi(u-v)}{\xi(2+u-v)} \frac{\xi(1+v-u)}{\xi(2+v-u)} \frac{\xi(u-v)}{\xi(1+u-v)}$	$\frac{\xi(1+v-u)}{\xi(2+v-u)} \frac{\xi(1+u-v)}{\xi(2+u-v)}$	

Appendix IV

THE SIMPLEST CASE

I have been requested to append an independent exposition of the methods employed in the text in the simplest case, that of a Fuchsian subgroup Γ of $G = PSL(2, \mathbb{R})$ with a single cusp, the Eisenstein series being taken to be invariant under right multiplication by elements of $K = PSO(2, \mathbb{R})$. The methods of the text when applied to $SL(2, \mathbb{R})$ are basically those of Selberg, with the inner product formula of §4 taking the place of what Harish-Chandra has called the Maass-Selberg relation. But this and a few other minor modifications do not affect the essence of the proof.

In order to be as brief as possible, I shall tailor the exposition to the needs of a competent analyst familiar with the first part of Lang's book and the geometry of fundamental domains. Moreover I shall use the Maass-Selberg relation as well as the inner product formula.

If

$$g = \begin{pmatrix} \alpha & 0 \\ 0 & \alpha^{-1} \end{pmatrix} \begin{pmatrix} 1 & x \\ 0 & 1 \end{pmatrix} k \qquad \alpha = \alpha(g) > 0$$

with k in K and λ is a complex number, set

$$F(g, \lambda) = \alpha^{\lambda+1}$$

If P is the group of upper-triangular matrices and the cusp is supposed to lie at infinity then the Eisenstein series

$$E(g, \lambda) = \Sigma_{\Gamma \cap P \backslash \Gamma} F(\gamma g, \lambda)$$

converges for $\operatorname{Re} \lambda > 1$. It is continuous as a function of g and λ and analytic as a function of λ in this region. It needs to be analytically continued.

If N is the group of matrices in P with eigenvalues 1 then

$$\int_{\Gamma \cap N \backslash N} E(ng, \lambda)dn \tag{1}$$

is easily evaluated. We take the measure of $\Gamma \cap N \backslash N$ to be 1 and write Γ as a union of double cosets

$$(\Gamma \cap N)\gamma(\Gamma \cap P)$$

The integral then becomes the sum over these double cosets of

$$\int_{(\Gamma \cap N) \cap \gamma^{-1}(\Gamma \cap P)\gamma \backslash N} F(\gamma ng, \lambda)dn$$

If γ lies in the trivial double coset this integral is equal to

$$F(g, \lambda)$$

Otherwise it is

$$\int_N F(\gamma ng, \lambda)dn$$

Writing

$$\gamma = \begin{pmatrix} 1 & x \\ 0 & 1 \end{pmatrix} \begin{pmatrix} a & 0 \\ 0 & a^{-1} \end{pmatrix} \begin{pmatrix} 0 & 1 \\ -1 & 0 \end{pmatrix} \begin{pmatrix} 1 & y \\ 0 & 1 \end{pmatrix} \qquad a > 0$$

we see that this integral equals

$$a^{\lambda+1} \left\{ \int_N F\left(\begin{pmatrix} 0 & 1 \\ -1 & 0 \end{pmatrix} n, \lambda \right) dn \right\} F(g, -\lambda)$$

and conclude that the integral (1) is equal to

$$F(g, \lambda) + M(\lambda)F(g, -\lambda)$$

where $M(\lambda)$ is analytic for $\operatorname{Re} \lambda > 1$. The analytic continuation of $E(g, \lambda)$ is bound up with that of $M(\lambda)$.

If ϕ is a smooth, compactly supported function on $N\backslash G/K$ we may write

$$\phi(g) = \frac{1}{2\pi}\int_{\mathrm{Re}\,\lambda=\lambda_0} \Phi(\lambda)F(g, \lambda)\,|d\lambda|$$

where $\Phi(\lambda)$ is an entire function. The function

$$\phi^\wedge(g) = \Sigma_{\Gamma\cap P\backslash\Gamma}\,\phi(\gamma g)$$

is smooth and compactly supported on $\Gamma\backslash G$ and in particular lies in $L^2(\Gamma\backslash G)$. It is given by

$$\phi^\wedge(g) = \frac{1}{2\pi}\int_{\mathrm{Re}\,\lambda=\lambda_0} \Phi(\lambda)E(g, \lambda)\,|d\lambda| \qquad \lambda_0 > 1 \tag{2}$$

If we have chosen the Haar measures properly we may calculate the inner product

$$(\phi^\wedge, \psi^\wedge) = \int_{\Gamma\backslash G} \phi^\wedge(g)\overline{\psi}^\wedge(g)dg$$

as follows. Substitute the formula (2) for $\phi^\wedge(g)$ and write out $\psi^\wedge(g)$ according to its definition. We obtain

$$\frac{1}{2\pi}\int_{\mathrm{Re}\,\lambda=\lambda_0} \Phi(\lambda)\left\{\int_{\Gamma\backslash G} E(g, \lambda)\Sigma_{\Gamma\cap P\backslash\Gamma}\overline{\psi}(\gamma g)dg\right\}|d\lambda|$$

The inner integral is equal to

$$\int_{\Gamma\cap P\backslash G} E(g, \lambda)\overline{\psi}(g)dg = \int_0^\infty \left\{a^{\lambda+1} + M(\lambda)a^{-\lambda+1}\right\} a^{-2}\overline{\psi}\left(\begin{pmatrix} a & 0 \\ 0 & a^{-1}\end{pmatrix}\right)\frac{da}{a}$$

By the Fourier inversion formula this integral is equal to

$$\overline{\Psi}(-\overline{\lambda}) + M(\lambda)\overline{\Psi}(\overline{\lambda})$$

We see that the inner product is given by

$$(3)\qquad \frac{1}{2\pi}\int_{\mathrm{Re}\,\lambda=\lambda_0} \{\Phi(\lambda)\overline{\Psi}(-\overline{\lambda}) + M(\lambda)\Phi(\lambda)\overline{\Psi}(\overline{\lambda})\}\,|d\lambda|$$

We can already deduce a great deal from the fact that (3) defines an inner product which is positive semi-definite. By approximation, we may extend the inner product to the space of functions analytic and bounded in some strip $|\mathrm{Re}\,\lambda| < 1+\varepsilon$, $\varepsilon > 0$, and decreasing to 0 at infinity faster than any polynomial. We denote it by $(\Phi(\cdot), \Psi(\cdot))$. We may form the completion with respect to this inner product and obtain a Hilbert space H.

If f is bounded and analytic in some strip $|\mathrm{Re}\,\lambda| < 1+\varepsilon$, $\varepsilon > 0$, and

$$f(-\lambda) = f(\lambda)$$

then

$$(f(\cdot)\Phi(\cdot),\ \Psi(\cdot)) = (\Phi(\cdot),\ f^*(\cdot)\Psi(\ ))$$

Here

$$f^*(\lambda) = \overline{f}(-\overline{\lambda})$$

Suppose

$$\sup_{|\mathrm{Re}\,\lambda|<1+\varepsilon} |f(\lambda)| < k$$

Then

$$g(\lambda) = \sqrt{k^2 - f^*(\lambda)f(\lambda)}$$

is analytic and bounded for $|\mathrm{Re}\,\lambda| < 1+\varepsilon$. Moreover

$$g(\lambda) = g(-\lambda)$$

and

$$g^*(\lambda) = g(\lambda)$$

Thus

$$((k^2 - f^*(\cdot)f(\cdot))\Phi(\cdot),\ \Phi(\cdot)) = (g(\cdot)\Phi(\cdot),\ g(\cdot)\Phi(\cdot))$$

We conclude that multiplication by f extends to a bounded linear operator on H with adjoint given by multiplication by f^*.

If $\mu > 1$ we may in particular take

$$f(\lambda) = \frac{1}{\mu - \lambda^2}$$

The associated operator is bounded and self-adjoint. Its range is clearly dense. We deduce that multiplication by λ^2 defines an unbounded self-adjoint operator A on H with

$$R(\mu,\ A) = \frac{1}{\mu - A}$$

being the operator defined by the given f.

If $\mathrm{Re}\,\mu > \lambda_0 > 1$ then

$$(R(\mu^2,\ A)\Phi(\cdot),\ \Psi(\cdot)) = \frac{1}{2\pi}\int_{\mathrm{Re}\,\lambda=\lambda_0} \frac{1}{\mu^2 - \lambda^2}\{\Phi(\lambda)\overline{\Psi}(-\overline{\lambda}) + M(\lambda)\Phi(\lambda)\overline{\Psi}(\overline{\lambda})\}\,|d\lambda|$$

This integral may be evaluated by moving the line of integration off to the right. We obtain the sum of

$$\frac{1}{2\pi}\{\Phi(\mu)\overline{\Psi}(-\overline{\mu}) + M(\mu)\Phi(\mu)\overline{\Psi}(\overline{\mu})\} \tag{4}$$

and, if λ_1 is very large,

$$\frac{1}{2\pi}\int_{\mathrm{Re}\,\lambda=\lambda_1} \frac{1}{\mu^2 - \lambda^2}\{\Phi(\lambda)\overline{\Psi}(-\overline{\lambda}) + M(\lambda)\Phi(\lambda)\overline{\Psi}(\overline{\lambda})\}\,|d\lambda|$$

The resolvent $R(\mu^2,\ A)$ is certainly analytic in the domain $\mathrm{Re}\,\mu > 0$, $\mu \notin (0,\ 1]$. We infer that the expression (4) is too. Taking

$$\Phi(\mu) = \Psi(\mu) = e^{\mu^2}$$

we even deduce that $M(\mu)$ is analytic in the same region.

We next continue the function $E(g, \lambda)$ into this region. Observe that if f is a continuous function on G with compact support and invariant under multiplication by elements of K from the left or the right then

$$r(f)F(g, \lambda) = \int_G F(gh, \lambda)f(h)dh$$

is equal to

$$\alpha_f(\lambda)F(g, \lambda)$$

Here $\alpha_f(\lambda)$ is an entire function of λ and for any given λ we may choose f so that $\alpha_f(\lambda) \neq 0$. We conclude immediately from the definition of $E(g, \lambda)$ that

$$r(f)E(g, \lambda) = \alpha_f(\lambda)E(g, \lambda) \qquad \operatorname{Re}\lambda > 1$$

If $\lambda \longrightarrow E(\cdot, \lambda)$ can be analytically continued when regarded as a function with values in the space of locally integrable functions on $\Gamma\backslash G$ this relation will persist and we may infer that the continuation yields in fact a continuous function of g and λ.

We now introduce two auxiliary functions. If

$$g = \begin{pmatrix} \alpha & 0 \\ 0 & \alpha^{-1} \end{pmatrix} \begin{pmatrix} 1 & x \\ 0 & 1 \end{pmatrix} k \qquad \alpha > 0$$

let

$$F'(g, \lambda) = \begin{cases} F(g, \lambda) & \alpha \leq 1 \\ 0 & \alpha > 1 \end{cases}$$

and let

$$F''(g, \lambda) = \begin{cases} F(g, \lambda) & \alpha \leq 1 \\ -M(\lambda)F(g, -\lambda) & \alpha \geq 1 \end{cases}$$

If $\text{Re}\,\lambda > 1$, $\text{Re}\,\mu > 1$ we may invoke an approximation argument and apply our inner product formula to the pairs

(i) $\phi(g) = F'(g, \lambda) \qquad \psi(g) = F'(g, \mu)$

(ii) $\phi(g) = F''(g, \lambda) \qquad \psi(g) = F''(g, \mu)$

For the first pair the Fourier transform of ϕ is

$$\Phi(z) = \frac{1}{\lambda - z}$$

Thus if

$$E'(g, \lambda) = \Sigma_{\Gamma \cap P \backslash \Gamma} F'(g, \lambda)$$

then

$$(E'(\cdot, \lambda), E'(\cdot, \mu))$$

is equal to

$$\frac{1}{2\pi i} \int_{\text{Re}\, z = \lambda_0} \frac{1}{(\lambda - z)(\bar{\mu} + z)} + \frac{M(z)}{(\lambda - z)(\bar{\mu} - z)} dz$$

We evaluate the integral by moving the vertical line of integration off to the right. The result is

$$\frac{1}{\lambda + \bar{\mu}} + \frac{M(\lambda)}{\bar{\mu} - \lambda} + \frac{M(\bar{\mu})}{\lambda - \bar{\mu}} = \omega(\lambda, \bar{\mu})$$

In general

$$(5) \qquad \left(\frac{\partial^n}{\partial \lambda^n} E'(\cdot, \lambda), \frac{\partial^n}{\partial \mu^n} E'(\cdot, \mu) \right) = \frac{\partial^{2n}}{\partial \lambda^n \partial \bar{\mu}^n} \omega(\lambda, \bar{\mu})$$

Thus if λ_1 is any point with $\text{Re}\,\lambda_1 > 1$

$$\Sigma_{n=0}^{\infty} \frac{1}{n!} |\lambda - \lambda_1|^n \left\| \frac{\partial^n}{\partial \lambda^n} E'(\cdot, \lambda) \right\|$$

converges in the largest circle about λ_1 which does not meet the real or imaginary axis. Since the formula (5) persists in any region in $\operatorname{Re}\lambda > 0$, $\operatorname{Re}\mu > 0$, $\lambda, \mu \notin (0, 1]$ to which the functions in it can be analytically continued we deduce by iteration that

$$\lambda \longrightarrow E'(\cdot, \lambda)$$

may be analytically continued as a function with values in $L^2(\Gamma\backslash G)$ to the region $\operatorname{Re}\lambda > 0$, $\lambda \notin (0, 1]$. Since

$$\Sigma_{\Gamma\cap P\backslash\Gamma}(F(\gamma g, \lambda) - F'(\gamma g, \lambda))$$

is clearly an analytic function of λ, $E(g, \lambda)$ can itself be continued to this region.

For the second pair the Fourier transform of ϕ is

$$\Phi(z) = \frac{1}{\lambda - z} - \frac{M(\lambda)}{\lambda + z}$$

The integrand occurring in the formula for

$$(E''(\cdot, \lambda), E''(\cdot, \mu))$$

where

$$E''(g, \lambda) = \Sigma_{\Gamma\cap P\backslash\Gamma} F''(g, \lambda)$$

will now be the sum of eight terms. They can each be easily evaluated by moving the line of integration to the left or right. Carrying out the evaluation and summing one obtains

$$(6) \qquad \frac{1}{\lambda+\bar{\mu}}\{1 - M(\lambda)\overline{M(\mu)}\} - \frac{1}{\lambda-\bar{\mu}}\{M(\lambda) - \overline{M(\mu)}\}$$

The formula just obtained remains valid for $\operatorname{Re}\lambda > 0$, $\operatorname{Re}\mu > 0$, $\lambda, \mu \notin (0, 1]$. Since (6) is positive when $\lambda = \mu$ we infer that $M(\lambda)$ is bounded in the neighbourhood of any point different from 0 on the imaginary axis. By this we mean that it is bounded in the intersection of a small disc about that point with the region in which $M(\lambda)$ has so far been defined. We shall deduce that $\|E''(\cdot, \lambda)\|$ is also bounded in such a neighbourhood.

Assuming this for the moment we return to (6) once again and conclude that

$$|M(\lambda)| \longrightarrow 1$$

as $\lambda \longrightarrow i\tau$, a point on the imaginary axis different from 0. Of course we are constrained to approach it from the right-hand side. Since

$$\overline{M(\lambda)} = M(\lambda)$$

we also have

$$\lim_{\sigma \downarrow 0} M^{-1}(\sigma - i\tau) - M(\sigma + i\tau) = 0$$

We define

(7) $$M(\lambda) = M^{-1}(-\lambda)$$

for $\operatorname{Re} \lambda < 0$, $\lambda \in [-1, 0)$ and infer from the reflection principle that $M(\lambda)$ can then be extended across the imaginary axis as well. It is defined and meromorphic outside the interval $[-1, 1]$ and satisfies the functional equation (7).

To complete the proof of the analytic continuation and the functional equation we need a lemma. Suppose $\lambda_1, \lambda_2, \ldots$ is a sequence of points and $\lambda_k \longrightarrow \lambda$. Suppose in addition that for each λ_k we are given a continuous function $E_k(g)$ on $\Gamma \backslash G$ with the following properties:

(i) There is a constant a and constants $c_k > 0$ so that

$$|E_k(g)| \leq c_k \alpha(g)^a$$

for $\alpha(g) \geq \varepsilon > 0$. Here ε is fixed.

(ii) $E_k(g)$ is orthogonal to all rapidly decreasing cusp forms.

(iii) If f is a continuous, compactly supported function in G bi-invariant under K then

$$r(f)E_k(g) = \alpha_f(\lambda_k)E_k(g)$$

(iv)

$$\int_{\Gamma\cap N\backslash N} E_k(ng)dn = A_k F(g, \lambda_k) + B_k F(g, -\lambda_k)$$

with A_k, B_k in $\mathbb{C}$.

Then if the sequences $\{A_k\}$, $\{B_k\}$ are bounded the inequalities of (i) are valid with a bounded sequence c_k. Moreover, if the sequences $\{A_k\}$, $\{B_k\}$ converge then the sequence $\{E_k(g)\}$ converges uniformly on compact sets.

In order to prove the lemma we have to look at

$$r(f)\varphi(g) = \int_G \varphi(h)f(g^{-1}h)dh$$

more carefully. Let

$$\varphi_2(g) = \int_{\Gamma\cap N\backslash N} \varphi(ng)dn$$

and define $\varphi_1(g)$ by

$$\varphi(g) = \varphi_1(g) + \varphi_2(g)$$

The expression for $r(f)\varphi(g)$ breaks up then into the sum of two similar expressions, and we want to consider the first

$$\int_G \varphi_1(h)f(g^{-1}h)dh$$

We write it as

$$\int_{\Gamma\cap N\backslash G} \varphi_1(h) \sum_{\delta\in\Gamma\cap N} f(g^{-1}\delta h)dh$$

The qualitative behaviour of the kernel

(8) $$\sum_{\delta\in\Gamma\cap N} f(g^{-1}\delta h)$$

for

$$g = \begin{pmatrix} 1 & x \\ 0 & 1 \end{pmatrix} \begin{pmatrix} \alpha & 0 \\ 0 & \alpha^{-1} \end{pmatrix} k \qquad |x| \leq b, \ \alpha > \varepsilon > 0$$

is easy enough to discover. Let

$$h = \begin{pmatrix} 1 & y \\ 0 & 1 \end{pmatrix} \begin{pmatrix} \beta & 0 \\ 0 & \beta^{-1} \end{pmatrix} k$$

We assume, for it is simply a matter of the proper choice of coordinates in the space defining $SL(2, \mathbb{R})$, that

$$\Gamma \cap N = \left\{ \begin{pmatrix} 1 & k \\ 0 & 1 \end{pmatrix} \middle| \, k \in Z \right\}$$

We may take $b = 1$ and assume that $|y| \leq b$. It is clear that there is a $\delta > 0$ so that each term of the sum (8) is 0 unless

$$(9) \qquad \delta \leq \frac{\alpha}{\beta} \leq \frac{1}{\delta}$$

However when this is so the sum becomes, at least if f is bi-invariant under K,

$$\Sigma f \left(\begin{pmatrix} \alpha^{-1}\beta & 0 \\ 0 & \alpha\beta^{-1} \end{pmatrix} \begin{pmatrix} 1 & \beta^{-2}(k+y-x) \\ 0 & 1 \end{pmatrix} \right)$$

Replacing the sum by an integral, we see that (8) is equal to

$$\int_N f(g^{-1}nh)dn + R(g, h)$$

where $R(g, h)$ is 0 unless (9) is satisfied, and then it goes to zero faster than any power of α as $\alpha \longrightarrow \infty$.

The integral

$$\int_G \varphi_1(h) f(g^{-1}h) dh$$

is equal to

$$\int_{\Gamma\cap N\backslash G} \varphi_1(h)R(g,\ h)dh$$

If

$$|\varphi(h)| \le c\alpha(h)^a$$

for $\beta(h) \ge \varepsilon' > 0$, with ε' sufficiently small, this integral is smaller in absolute value than

$$cd(r)\alpha(g)^{a-r}$$

for any real r. Here $d(r)$ and ε' depend on f, but there is an obvious uniformity.

We return to the proof of the lemma. Choose an f with $\alpha_f(\lambda) \neq 0$. We may as well suppose that $\alpha_f(\lambda_k) \neq 0$ for all k. If

$$f_k = \frac{1}{\alpha_f(\lambda_k)} f$$

then

(10)
$$r(f_k)E_k(g) = E_k(g)$$

The inequality (i) implies a similar inequality

$$|E_k(g)| \le c'c_k\alpha(g)^a$$

for $\alpha(g) \ge \varepsilon'$. Here c' is a constant depending on Γ, ε, ε', and a. Applying the discussion of the previous paragraph to f_k and $\varphi(g) = E_k(g)$, we see that

(11)
$$E_k(g) = A_kF(g,\ \lambda_k) + B_kF(g,\ -\lambda_k) + R_k(g)$$

with

(12)
$$|R_k(g)| \le dc_k\alpha(g)^{a'}$$

Here a' is a real number with

(13) $$a' < -\mathrm{Inf}|\mathrm{Re}\,\lambda_k| \qquad a' < a$$

and d depends on a'.

We choose c_k to be as small as possible and yet still satisfy (i). If the sequence is not bounded we pass to a subsequence and suppose $c_k \uparrow \infty$. Then for some g with $\alpha(g) \geq \varepsilon$

(14) $$|E_k(g)| \geq \frac{c_k}{2}\, \alpha(g)^a$$

It follows from (11), (12), and (13) that there is an R so that for all k any g satisfying (14) also satisfies

(15) $$\alpha(g) \leq R$$

From (10) and Ascoli's lemma we can pass to a subsequence and suppose that $\left\{\frac{1}{c_k} E_k(g)\right\}$ converges uniformly on compact sets to a function $E(g)$. By (15) this function will not be identically zero. On the other hand

$$\int_{\Gamma\cap N\backslash N} E(ng)dn \equiv 0$$

and $E(g)$ is orthogonal to all rapidly decreasing cusp forms. This is a contradiction.

Once we know that the c_k can be taken to be bounded, we can apply (10) and Ascoli's lemma to find convergent subsequences of $\{E_k(g)\}$. If two subsequences converged to different limits then the difference of the limits would again be cusp forms and yet orthogonal to cusp forms. This contradiction yields the second assertion of the lemma.

It also follows from the above proof that

$$E_k(g) - A_k F(g, \lambda_k) - B_k F(g, -\lambda_k)$$

is uniformly rapidly decreasing as $\alpha(g) \longrightarrow \infty$ and in particular is uniformly square integrable. If $E_k(g) = E(g, \lambda_k)$ is an Eisenstein series then for $\alpha(g)$ sufficiently large this difference is just $E''(g, \lambda_k)$. The boundedness of $\|E''(\cdot, \lambda)\|$ in a neighbourhood of a point on the imaginary axis which we asserted above is therefore clear.

We define $E(g, \lambda)$ in the domain $\operatorname{Re} \lambda < 0$, $\lambda \notin [-1, 0)$ by

$$E(g, \lambda) = M(\lambda)E(g, -\lambda)$$

Then

$$\int_{\Gamma \cap N \backslash N} E(ng, \lambda) = F(g, \lambda) + M(\lambda)F(g, -\lambda)$$

and the discussion above allows us to extend by the reflection principle across the imaginary axis.

It remains to treat the interval $[-1, 1]$. Here it is simplest to depart from the methods of the text and to employ instead the Maass-Selberg relations. To verify these it is best to regard a function on G/K as a function of

$$z = x+iy$$

in the upper half-plane. Here

$$gi = z$$

and if

$$g = \begin{pmatrix} 1 & x \\ 0 & 1 \end{pmatrix} \begin{pmatrix} \alpha & 0 \\ 0 & \alpha^{-1} \end{pmatrix} k$$

then

$$gi = x+i\alpha^2$$

If $E(g)$ is a function on $\Gamma \backslash G$ let

$$F(g) = \int_{\Gamma \cap N \backslash N} E(ng)dn$$

If

$$r(f)E = \alpha_f(\lambda)E$$

for compactly supported, bi-invariant f then

$$r(f)F = \alpha_f(\lambda)F$$

for all such f. Moreover if Δ is the operator

$$y^2 \left\{ \frac{\partial^2}{\partial x^2} + \frac{\partial^2}{\partial y^2} \right\}$$

then

$$\Delta E = \frac{\lambda^2 - 1}{4} E$$

and

$$\Delta F = \frac{\lambda^2 - 1}{4} F$$

Thus if $\lambda \neq 0$

$$F(g) = AF(g, \lambda) + BF(g, -\lambda) = Ay^{\frac{\lambda+1}{2}} + By^{\frac{-\lambda+1}{2}}$$

while if $\lambda = 0$

$$F(g) = Ay^{1/2} + By^{1/2} \ln y$$

The proof of the lemma shows that if $E(g)$ does not grow too rapidly as $\alpha(g) \longrightarrow \infty$ then

$$E(g) \sim F(g)$$

Suppose we have two such functions E and E' corresponding to the same λ. Remove from the upper half-plane the region $y > R$, for a sufficiently large R, as well as the transforms under Γ of all such points. Division by Γ then yields a manifold M which may be thought of as a closed manifold with a cylindrical tube protruding from it. The boundary is a circle, the image of $y = R$. If we integrate with respect to the invariant area,

$$0 = \int_M \Delta E \cdot E' - E \cdot \Delta E'$$

Integrating by parts we see easily that the right side is asymptotic as $N \longrightarrow \infty$ to

$$\begin{array}{ll} \lambda(AB' - BA') & \lambda \neq 0 \\ BA' - AB' & \lambda = 0 \end{array}$$

These are the Maass-Selberg relations. We conclude in particular that if E and E' are both orthogonal to cusp forms then they are proportional.

Now choose any point $\lambda_0 \neq 0$ in the interval $[-1, 1]$. Choose a non-singular matrix

$$\begin{pmatrix} a & b \\ c & d \end{pmatrix}$$

so that if $E_0(g)$ is a function as above corresponding to λ_0 and orthogonal to cusp forms then

(16) $$aA_0 + bB_0 = 0$$

If $E(g)$ corresponds to λ and is also orthogonal to cusp forms then for λ close to λ_0

$$cA + dB$$

must dominate

$$aA + dB$$

Otherwise we could choose a sequence $\lambda_k \longrightarrow \lambda_0$ and a sequence $E_k(g)$ with

$$cA_k + dB_k \longrightarrow 0 \qquad aA_k + bB_k \longrightarrow 1$$

Our lemma would then show that $E_k \longrightarrow E_0$, for some E_0 contradicting (16).

To show that $M(\lambda)$ is meromorphic near λ_0 we have only to show that

$$\frac{a+bM(\lambda)}{c+dM(\lambda)}$$

is continuous. We have just observed that it is bounded. If it were not continuous at λ, or rather, since it is only defined in a dense set, if it cannot be extended to be continuous, we could choose two sequences $\{\lambda_k'\}$, $\{\lambda_k''\}$ both approaching λ but with

$$\lim \frac{a+bM(\lambda_k')}{c+dM(\lambda_k')} \neq \lim \frac{a+bM(\lambda_k'')}{c+dM(\lambda_k'')}$$

The lemma would give two functions $E'(g)$ and $E''(g)$ whose difference $E(g)$ would have

$$F(g) = AF(g, \lambda) + BF(g, -\lambda)$$

with

$$aA + bB \neq 0$$

$$cA + dB = 0$$

This is a contradiction.

To show that $M(\lambda)$ is meromorphic at $\lambda_0 = 0$ we use for λ near 0 the representation

$$F(g) = A\alpha(g)\cosh\lambda\alpha(g) + B\alpha(g)\,\frac{\sinh\lambda\alpha(g)}{\lambda}$$

and a simple variant of the basic lemma. Otherwise the argument is the same.

Selective index of terminology

Selective index of notation

399: Functional Analysis and its Applications. Proceedings 3. Edited by H. G. Garnir, K. R. Unni and J. H. Williamson. 34 pages. 1974.

400: A Crash Course on Kleinian Groups. Proceedings 1974. ed by L. Bers and I. Kra. VII, 130 pages. 1974.

401: M. F. Atiyah, Elliptic Operators and Compact Groups. 3 pages. 1974.

402: M. Waldschmidt, Nombres Transcendants. VIII, 277 es. 1974.

403: Combinatorial Mathematics. Proceedings 1972. Edited . A. Holton. VIII, 148 pages. 1974.

404: Théorie du Potentiel et Analyse Harmonique. Edité par raut. V, 245 pages. 1974.

405: K. J. Devlin and H. Johnsbråten, The Souslin Problem. 132 pages. 1974.

406: Graphs and Combinatorics. Proceedings 1973. Edited R. A. Bari and F. Harary. VIII, 355 pages. 1974.

407: P. Berthelot, Cohomologie Cristalline des Schémas de acteristique p > o. II, 604 pages. 1974.

408: J. Wermer, Potential Theory. VIII, 146 pages. 1974.

409: Fonctions de Plusieurs Variables Complexes, Séminaire çois Norguet 1970–1973. XIII, 612 pages. 1974.

410: Séminaire Pierre Lelong (Analyse) Année 1972–1973. 81 pages. 1974.

411: Hypergraph Seminar. Ohio State University, 1972. ed by C. Berge and D. Ray-Chaudhuri. IX, 287 pages. 1974.

412: Classification of Algebraic Varieties and Compact nplex Manifolds. Proceedings 1974. Edited by H. Popp. V, pages. 1974.

413: M. Bruneau, Variation Totale d'une Fonction. XIV, 332 es. 1974.

414: T. Kambayashi, M. Miyanishi and M. Takeuchi, Uniınt Algebraic Groups. VI, 165 pages. 1974.

415: Ordinary and Partial Differential Equations. Proceedings 4. XVII, 447 pages. 1974.

416: M. E. Taylor, Pseudo Differential Operators. IV, 155 es. 1974.

417: H. H. Keller, Differential Calculus in Locally Convex ces. XVI, 131 pages. 1974.

418: Localization in Group Theory and Homotopy Theory Related Topics. Battelle Seattle 1974 Seminar. Edited by P. J. on. VI, 172 pages 1974.

419: Topics in Analysis. Proceedings 1970. Edited by O. E. to, I. S. Louhivaara, and R. H. Nevanlinna. XIII, 392 pages. 1974.

420: Category Seminar. Proceedings 1972/73. Edited by G. M. y. VI, 375 pages. 1974.

421: V. Poénaru, Groupes Discrets. VI, 216 pages. 1974.

422: J.-M. Lemaire, Algèbres Connexes et Homologie des aces de Lacets. XIV, 133 pages. 1974.

423: S. S. Abhyankar and A. M. Sathaye, Geometric Theory Algebraic Space Curves. XIV, 302 pages. 1974.

424: L. Weiss and J. Wolfowitz, Maximum Probability mators and Related Topics. V, 106 pages. 1974.

425: P. R. Chernoff and J. E. Marsden, Properties of Infinite ensional Hamiltonian Systems. IV, 160 pages. 1974.

426: M. L. Silverstein, Symmetric Markov Processes. X, 287 es. 1974.

427: H. Omori, Infinite Dimensional Lie Transformation ups. XII, 149 pages. 1974.

428: Algebraic and Geometrical Methods in Topology, Prodings 1973. Edited by L. F. McAuley. XI, 280 pages. 1974.

Vol. 429: L. Cohn, Analytic Theory of the Harish-Chandra C-Function. III, 154 pages. 1974.

Vol. 430: Constructive and Computational Methods for Differential and Integral Equations. Proceedings 1974. Edited by D. L. Colton and R. P. Gilbert. VII, 476 pages. 1974.

Vol. 431: Séminaire Bourbaki – vol. 1973/74. Exposés 436–452. IV, 347 pages. 1975.

Vol. 432: R. P. Pflug, Holomorphiegebiete, pseudokonvexe Gebiete und das Levi-Problem. VI, 210 Seiten. 1975.

Vol. 433: W. G. Faris, Self-Adjoint Operators. VII, 115 pages. 1975.

Vol. 434: P. Brenner, V. Thomée, and L. B. Wahlbin, Besov Spaces and Applications to Difference Methods for Initial Value Problems. II, 154 pages. 1975.

Vol. 435: C. F. Dunkl and D. E. Ramirez, Representations of Commutative Semitopological Semigroups. VI, 181 pages. 1975.

Vol. 436: L. Auslander and R. Tolimieri, Abelian Harmonic Analysis, Theta Functions and Function Algebras on a Nilmanifold. V, 99 pages. 1975.

Vol. 437: D. W. Masser, Elliptic Functions and Transcendence. XIV, 143 pages. 1975.

Vol. 438: Geometric Topology. Proceedings 1974. Edited by L. C. Glaser and T. B. Rushing. X, 459 pages. 1975.

Vol. 439: K. Ueno, Classification Theory of Algebraic Varieties and Compact Complex Spaces. XIX, 278 pages. 1975

Vol. 440: R. K. Getoor, Markov Processes: Ray Processes and Right Processes. V, 118 pages. 1975.

Vol. 441: N. Jacobson, PI-Algebras. An Introduction. V, 115 pages. 1975.

Vol. 442: C. H. Wilcox, Scattering Theory for the d'Alembert Equation in Exterior Domains. III, 184 pages. 1975.

Vol. 443: M. Lazard, Commutative Formal Groups. II, 236 pages. 1975.

Vol. 444: F. van Oystaeyen, Prime Spectra in Non-Commutative Algebra. V, 128 pages. 1975.

Vol. 445: Model Theory and Topoi. Edited by F. W. Lawvere, C. Maurer, and G. C. Wraith. III, 354 pages. 1975.

Vol. 446: Partial Differential Equations and Related Topics. Proceedings 1974. Edited by J. A. Goldstein. IV, 389 pages. 1975.

Vol. 447: S. Toledo, Tableau Systems for First Order Number Theory and Certain Higher Order Theories. III, 339 pages. 1975.

Vol. 448: Spectral Theory and Differential Equations. Proceedings 1974. Edited by W. N. Everitt. XII, 321 pages. 1975.

Vol. 449: Hyperfunctions and Theoretical Physics. Proceedings 1973. Edited by F. Pham. IV, 218 pages. 1975.

Vol. 450: Algebra and Logic. Proceedings 1974. Edited by J. N. Crossley. VIII, 307 pages. 1975.

Vol. 451: Probabilistic Methods in Differential Equations. Proceedings 1974. Edited by M. A. Pinsky. VII, 162 pages. 1975.

Vol. 452: Combinatorial Mathematics III. Proceedings 1974. Edited by Anne Penfold Street and W. D. Wallis. IX, 233 pages. 1975.

Vol. 453: Logic Colloquium. Symposium on Logic Held at Boston, 1972–73. Edited by R. Parikh. IV, 251 pages. 1975.

Vol. 454: J. Hirschfeld and W. H. Wheeler, Forcing, Arithmetic, Division Rings. VII, 266 pages. 1975.

Vol. 455: H. Kraft, Kommutative algebraische Gruppen und Ringe. III, 163 Seiten. 1975.

Vol. 456: R. M. Fossum, P. A. Griffith, and I. Reiten, Trivial Extensions of Abelian Categories. Homological Algebra of Trivial Extensions of Abelian Categories with Applications to Ring Theory. XI, 122 pages. 1975.

Vol. 457: Fractional Calculus and Its Applications. Proceedings 1974. Edited by B. Ross. VI, 381 pages. 1975.

Vol. 458: P. Walters, Ergodic Theory – Introductory Lectures. VI, 198 pages. 1975.

Vol. 459: Fourier Integral Operators and Partial Differential Equations. Proceedings 1974. Edited by J. Chazarain. VI, 372 pages. 1975.

Vol. 460: O. Loos, Jordan Pairs. XVI, 218 pages. 1975.

Vol. 461: Computational Mechanics. Proceedings 1974. Edited by J. T. Oden. VII, 328 pages. 1975.

Vol. 462: P. Gérardin, Construction de Séries Discrètes p-adiques. »Sur les séries discrètes non ramifiées des groupes réductifs déployés p-adiques«. III, 180 pages. 1975.

Vol. 463: H.-H. Kuo, Gaussian Measures in Banach Spaces. VI, 224 pages. 1975.

Vol. 464: C. Rockland, Hypoellipticity and Eigenvalue Asymptotics. III, 171 pages. 1975.

Vol. 465: Séminaire de Probabilités IX. Proceedings 1973/74. Edité par P. A. Meyer. IV, 589 pages. 1975.

Vol. 466: Non-Commutative Harmonic Analysis. Proceedings 1974. Edited by J. Carmona, J. Dixmier and M. Vergne. VI, 231 pages. 1975.

Vol. 467: M. R. Essén, The Cos $\pi\lambda$ Theorem. With a paper by Christer Borell. VII, 112 pages. 1975.

Vol. 468: Dynamical Systems – Warwick 1974. Proceedings 1973/74. Edited by A. Manning. X, 405 pages. 1975.

Vol. 469: E. Binz, Continuous Convergence on C(X). IX, 140 pages. 1975.

Vol. 470: R. Bowen, Equilibrium States and the Ergodic Theory of Anosov Diffeomorphisms. III, 108 pages. 1975.

Vol. 471: R. S. Hamilton, Harmonic Maps of Manifolds with Boundary. III, 168 pages. 1975.

Vol. 472: Probability-Winter School. Proceedings 1975. Edited by Z. Ciesielski, K. Urbanik, and W. A. Woyczyński. VI, 283 pages. 1975.

Vol. 473: D. Burghelea, R. Lashof, and M. Rothenberg, Groups of Automorphisms of Manifolds. (with an appendix by E. Pedersen) VII, 156 pages. 1975.

Vol. 474: Séminaire Pierre Lelong (Analyse) Année 1973/74. Edité par P. Lelong. VI, 182 pages. 1975.

Vol. 475: Répartition Modulo 1. Actes du Colloque de Marseille-Luminy, 4 au 7 Juin 1974. Edité par G. Rauzy. V, 258 pages. 1975. 1975.

Vol. 476: Modular Functions of One Variable IV. Proceedings 1972. Edited by B. J. Birch and W. Kuyk. V, 151 pages. 1975.

Vol. 477: Optimization and Optimal Control. Proceedings 1974. Edited by R. Bulirsch, W. Oettli, and J. Stoer. VII, 294 pages. 1975.

Vol. 478: G. Schober, Univalent Functions – Selected Topics. V, 200 pages. 1975.

Vol. 479: S. D. Fisher and J. W. Jerome, Minimum Norm Extremals in Function Spaces. With Applications to Classical and Modern Analysis. VIII, 209 pages. 1975.

Vol. 480: X. M. Fernique, J. P. Conze et J. Gani, Ecole d'Eté de Probabilités de Saint-Flour IV-1974. Edité par P.-L. Hennequin. XI, 293 pages. 1975.

Vol. 481: M. de Guzmán, Differentiation of Integrals in R^n. XII, 226 pages. 1975.

Vol. 482: Fonctions de Plusieurs Variables Complexes II. Séminaire François Norguet 1974–1975. IX, 367 pages. 1975.

Vol. 483: R. D. M. Accola, Riemann Surfaces, Theta Functions, and Abelian Automorphisms Groups. III, 105 pages. 1975.

Vol. 484: Differential Topology and Geometry. Proceedings 1974. Edited by G. P. Joubert, R. P. Moussu, and R. H. Roussarie. IX, 287 pages. 1975.

Vol. 485: J. Diestel, Geometry of Banach Spaces – Selected Top XI, 282 pages. 1975.

Vol. 486: S. Stratila and D. Voiculescu, Representations of Algebras and of the Group U (∞). IX, 169 pages. 1975.

Vol. 487: H. M. Reimann und T. Rychener, Funktionen beschrän mittlerer Oszillation. VI, 141 Seiten. 1975.

Vol. 488: Representations of Algebras, Ottawa 1974. Proceedi 1974. Edited by V. Dlab and P. Gabriel. XII, 378 pages. 1975.

Vol. 489: J. Bair and R. Fourneau, Etude Géométrique des Espa Vectoriels. Une Introduction. VII, 185 pages. 1975.

Vol. 490: The Geometry of Metric and Linear Spaces. Proceedi 1974. Edited by L. M. Kelly. X, 244 pages. 1975.

Vol. 491: K. A. Broughan, Invariants for Real-Generated Unif Topological and Algebraic Categories. X, 197 pages. 1975.

Vol. 492: Infinitary Logic: In Memoriam Carol Karp. Edited by D. Kueker. VI, 206 pages. 1975.

Vol. 493: F. W. Kamber and P. Tondeur, Foliated Bundles Characteristic Classes. XIII, 208 pages. 1975.

Vol. 494: A Cornea and G. Licea. Order and Potential Resolv Families of Kernels. IV, 154 pages. 1975.

Vol. 495: A. Kerber, Representations of Permutation Groups I 175 pages. 1975.

Vol. 496: L. H. Hodgkin and V. P. Snaith, Topics in K-Theory. Independent Contributions. III, 294 pages. 1975.

Vol. 497: Analyse Harmonique sur les Groupes de Lie. Proceedi 1973–75. Edité par P. Eymard et al. VI, 710 pages. 1975.

Vol. 498: Model Theory and Algebra. A Memorial Tribute Abraham Robinson. Edited by D. H. Saracino and V. B. Weispfenn X, 463 pages. 1975.

Vol. 499: Logic Conference, Kiel 1974. Proceedings. Edited G. H. Müller, A. Oberschelp, and K. Potthoff. V, 651 pages 19

Vol. 500: Proof Theory Symposion, Kiel 1974. Proceedings. Editec J. Diller and G. H. Müller. VIII, 383 pages. 1975.

Vol. 501: Spline Functions, Karlsruhe 1975. Proceedings. Edited K. Böhmer, G. Meinardus, and W. Schempp. VI, 421 pages. 1976.

Vol. 502: János Galambos, Representations of Real Numbers Infinite Series. VI, 146 pages. 1976.

Vol. 503: Applications of Methods of Functional Analysis to Proble in Mechanics. Proceedings 1975. Edited by P. Germain and Nayroles. XIX, 531 pages. 1976.

Vol. 504: S. Lang and H. F. Trotter, Frobenius Distributions GL_2-Extensions. III, 274 pages. 1976.

Vol. 505: Advances in Complex Function Theory. Proceedin 1973/74. Edited by W. E. Kirwan and L. Zalcman. VIII, 203 pag 1976.

Vol. 506: Numerical Analysis, Dundee 1975. Proceedings. Edit by G. A. Watson. X, 201 pages. 1976.

Vol. 507: M. C. Reed, Abstract Non-Linear Wave Equations. V 128 pages. 1976.

Vol. 508: E. Seneta, Regularly Varying Functions. V, 112 pages. 197

Vol. 509: D. E. Blair, Contact Manifolds in Riemannian Geome VI, 146 pages. 1976.

Vol. 510: V. Poènaru, Singularités C^∞ en Présence de Symétr V, 174 pages. 1976.

Vol. 511: Séminaire de Probabilités X. Proceedings 1974/75. Ed par P. A. Meyer. VI, 593 pages. 1976.

Vol. 512: Spaces of Analytic Functions, Kristiansand, Norway 197 Proceedings. Edited by O. B. Bekken, B. K. Øksendal, and A. Str VIII, 204 pages. 1976.

Vol. 513: R. B. Warfield, Jr. Nilpotent Groups. VIII, 115 pages. 197

Vol. 514: Séminaire Bourbaki vol. 1974/75. Exposés 453 – 470. 276 pages. 1976.

Vol. 515: Bäcklund Transformations. Nashville, Tennessee 19 Proceedings. Edited by R. M. Miura. VIII, 295 pages. 1976.